1979

Elementary Linear Algebra

Elementary Linear Algebra

SECOND EDITION

BERNARD KOLMAN

Department of Mathematics, Drexel University

Macmillan Publishing Co., Inc.
NEW YORK

Collier Macmillan Publishers
LONDON

Macmillan Publishing Co., Inc.
866 Third Avenue, New York, New York 10022

Collier Macmillan Canada, Ltd.

Library of Congress Cataloging in Publication Data

Kolman, Bernard, (date)
  Elementary linear algebra.

  Includes index.
  1. Algebras, Linear. I. Title.
QA184.K668   1977   512'.5   75-45087
ISBN 0-02-365900-9

Printing: 1 2 3 4 5 6 7 8   Year: 7 8 9 0 1 2 3

To Lillie,
Lisa,
and Stephen

Preface

This textbook has been written for students who have completed a calculus course. It thus provides the student with his first experience in postulational or axiomatic mathematics while keeping in close touch with the computational aspects of the subject. Few subjects can claim to have such widespread applications in other areas of mathematics—multivariable calculus, differential equations, and probability theory, for example—as well as in physics, biology, chemistry, economics, psychology, sociology, and all fields of engineering. Engineering, the sciences, and the social sciences today are becoming more analytically oriented; that is, more mathematical in flavor, and the mere ability to manipulate matrices is no longer adequate. Linear algebra affords, at the sophomore level, an excellent opportunity to develop a capability for handling abstract concepts.

The author has learned from experience that at the sophomore level, abstract ideas must be introduced quite gradually and be based on some firm foundations. Thus this book begins with Chapter 0, which introduces the notion of sets as well as some material on functions. This chapter can be skipped by the better prepared student. If necessary, the chapter should be consulted each time an idea discussed in it occurs in the rest of the text.

The book begins the study of linear algebra in Chapter 1 with the treatment of matrices as mere arrays of numbers that arise naturally in the solution of systems of linear equations, a problem already familiar to the student. Methods for solving such systems of equations and matrix properties are considered in this chapter. In Chapter 2, we come to a more abstract notion, that of a vector space. We restrict our attention to finite-dimensional, real vector spaces. Here we tap some of the many geometric ideas that arise naturally. Thus, we prove that an n-dimensional real vector space is isomorphic to R^n, the vector space of all ordered n-tuples of real numbers, or the vector space of all $n \times 1$ matrices with real entries. Since R^n is but a slight generalization of R^2 and R^3, two- and three-dimensional space studied in the calculus, this shows that the notion of a finite-dimensional, real vector space is not as remote as it may have seemed when first introduced. Chapter 3 covers inner product spaces and has a strong geometric orientation. Chapter 4 deals with matrices and linear transformations; in it we consider the dimension theorems and also applications to the solution of systems of linear equations. Chapter 5 introduces the basic properties of determinants and some of their applications. Chapter 6 considers eigenvalues and eigenvectors, and real quadratic forms. In this chapter we completely solve the diagonalization problem for symmetric matrices. Chapter 7 provides an introduction to the important application of linear algebra in the solution of differential equations. It is possible to go from Section 6.1 directly to Chapter 7, providing an immediate application of the material in Section 6.1. Moreover, Section 6.3 provides an application of the material in Section 6.2.

In using the first edition of this book, for a one-quarter linear algebra course meeting four times a week, no difficulty has been encountered in covering the first six chapters (Section 6.3 was omitted and the optional material was not part of the first edition). A suggested pace for covering the basic material, based on seven years of experience with the first edition, follows.

Suggested Pace for Basic Material (Chapters 1 to 6, omitting all optional material and Section 6.3)

Chapter 1	8 lectures
Chapter 2	7 lectures
Chapter 3	3 lectures (For most students Section 3.1 is a review of known material.)
Chapter 4	7 lectures
Chapter 5	5 lectures
Chapter 6	5 lectures
	35 lectures

On most occasions when the author has taught linear algebra, the subject of eigenvalues and eigenvectors had to be covered much too hastily, because of time limitations, to do the material justice. This book has been written so that the topic can be comfortably covered in a one-quarter or one-semester course. The exercises form an integral part of the text. Many of them are numerical in nature, whereas others are of a theoretical type. The answers to selected numerical exercises are provided in the answer section.

This edition differs from the first edition in the following ways:

1. Sections 5.2, 5.3, and 5.4 of the original edition have been grouped as Chapter 3, "Inner Product Spaces," following Chapter 2, "Real Vector Spaces," to provide a somewhat tighter organization of these related topics, as well as to unify the material on eigenvalues and eigenvectors. If an instructor wishes to do so, Chapter 3 may still be covered after Section 6.1 and before Section 6.2, that is, as the material is used in the diagonalization problem.

2. More geometric material, including lines, planes, and cross product, has been added. The material on lines, planes, and cross product is optional (and has been so marked); it can be omitted without any loss of continuity and can be used as individual student projects.

3. Twenty-four additional illustrative examples have been added.

4. Sixty-five additional exercises have been added.

5. Twenty additional figures have been added.

6. A good many minor writing changes have been made through the entire text to achieve greater clarity and an improved pedagogic presentation.

7. A more open format has been used to improve appearance and legibility.

It is a pleasure to express my gratitude to Professors Edward Norman, Florida Technological University, and Charles S. Duris, Drexel University, for their helpful suggestions in connection with the first edition; Professor William F. Trench, Drexel University, for thoroughly reviewing the entire manuscript of the second edition and for making innumerable suggestions and improvements; the students of Mathematics N504 at Drexel University who have been using the first edition; instructors from a diverse number of institutions, who communicated to me their classroom experiences with the first edition; my wife for her continued support and encouragement; and to Mr. Everett W. Smethurst, Senior Editor, Mrs. Elaine W. Wetterau, Production Supervisor, and the entire staff of Macmillan Publishing Co., Inc., for their interest and unfailing cooperation in all phases of this project. To all these goes a sincere expression of thanks.

B. K.

Contents

Elementary Linear Algebra

CHAPTER 0
Preliminaries

The study of any subject requires the setting down of common terminology and notation. This is especially true of a subject as technical as mathematics. In this chapter we present a brief discussion of the basic language of mathematics to be used in this book as well as in all other mathematical work. We discuss sets and functions; many of the ideas are undoubtedly familiar to the reader, but some may be new. This chapter should be read quickly and consulted as the need arises.

0.1. Sets

A **set** is a collection, class, aggregate, or family of objects, which are called **elements** or **members** of the set. A set will be denoted by a capital letter, and an element of a set by a lowercase letter or by a Greek letter. A set S is specified either by describing all the elements of S, or by stating a property that determines, unequivocally, whether an element is or is not an element of S. Thus $S = \{1, 2, 3\}$ is the set of all positive integers < 4. A real number belongs to S if it is a positive integer < 4. Thus S has been described in both ways. Sets A and B are said to be **equal** if each element of A belongs to B and if each element of B belongs to A. We write $A = B$. Thus $\{1, 2, 3\} = \{3, 2, 1\}$ $= \{2, 1, 3\}$, and so on. If A and B are sets such that every element of A belongs to B, then A is said to be a **subset** of B. If A is a subset of B and $A \neq B$, then A is called a **proper subset** of B. The set of all rational numbers is a proper subset of the set of all real numbers; the set $\{1, 2, 3\}$ is a subset of $\{1, 2, 3\}$ and a proper subset of $\{1, 2, 3, 4\}$; the set of all isosceles triangles is a proper subset of the set of all triangles. We can see that every set is a subset of itself. The **empty set** is the set that has no elements in it. The set of all real numbers whose squares equal -1 is empty because the square of a real number is never negative.

0.2. Functions

A **function** f from a set S into a set T is a rule that assigns to each element s of S a unique element t of T. We denote the function f by $f: S \to T$ and write $t = f(s)$. Functions constitute the basic ingredient of the calculus and other branches of mathematics, and the reader has dealt extensively with them. The set S is called the **domain of** f; the subset $f(S)$ of T consisting of all the elements $f(s)$, for s in S, is called the **range of** f or the **image of** S **under** f. As examples of functions we consider the following:

1. Let $S = T =$ the set of all real numbers. Let $f: S \to T$ be defined by the rule $f(s) = s^2$, for s in S.
2. Let $S =$ the set of all real numbers and let $T =$ the set of all nonnegative real numbers. Let $f: S \to T$ be defined by the rule $f(s) = s^2$, for s in S.
3. Let $S =$ three-dimensional space, where each point is described by x-, y-, and z-coordinates (x_1, x_2, x_3). Let $T =$ the (x, y)-plane. Let $f: S \to T$ be defined by the rule $f((x_1, x_2, x_3)) = (x_1, x_2, 0)$. To see what f does, we take a point (x_1, x_2, x_3) in S, draw a line from (x_1, x_2, x_3) perpendicular to T, the (x, y)-plane, and find the point of intersection $(x_1, x_2, 0)$ of this line with the (x, y)-plane. This point is the image of (x_1, x_2, x_3) under f; f is called a **projection function** (Figure 0.1).
4. Let $S = T =$ the set of all real numbers. Let $f: S \to T$ be defined by the rule $f(s) = 2s + 1$, for s in S.
5. Let $S =$ the x-axis in the (x, y)-plane and let $T =$ the (x, y)-plane. Let $f: S \to T$ be defined by the rule $f(s, 0) = (s, 1)$, for s in S.
6. Let $S =$ the set of all real numbers. Let $T =$ the set of all positive real numbers. Let $g: S \to T$ be defined by the rule $g(s) = e^s$, for s in S.

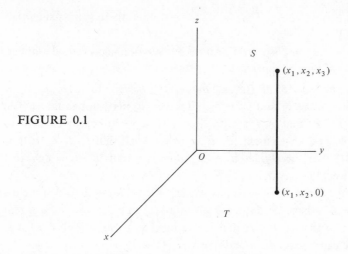

FIGURE 0.1

There are two properties of functions that we need to distinguish. A function $f: S \rightarrow T$ is called **one-one** if $f(s_1) \neq f(s_2)$ whenever s_1 and s_2 are distinct elements of S. An equivalent statement is that if $f(s_1) = f(s_2)$, then we must have $s_1 = s_2$ (see Figure 0.2). A function $f: S \rightarrow T$ is called **onto** if the range of f is all of T, that is, if for any given t in T there is at least one s in S such that $f(s) = t$ (see Figure 0.3).

We now examine the listed functions:

1. f is not one-one for if $f(s_1) = f(s_2)$, it need not follow that $s_1 = s_2$ $[f(2) = f(-2) = 4]$. Since the range of f is the set of nonnegative real

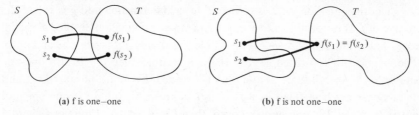

(a) f is one—one   (b) f is not one—one

FIGURE 0.2

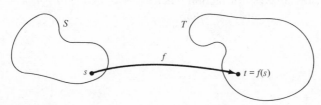

FIGURE 0.3

numbers, f is not onto. That is, if $t = -4$, then there is no s such that $f(s) = -4$.

2. f is not one-one, but is onto. For if t is a given nonnegative real number, then $s = \sqrt{t}$ is in S and $f(s) = t$.

3. f is not one-one for if $f(a_1, a_2, a_3) = f(b_1, b_2, b_3)$, then $(a_1, a_2, 0) = (b_1, b_2, 0)$ or $a_1 = b_1$ and $a_2 = b_2$. However, b_3 need not equal a_3. The range of f is T; that is, f is onto. For let $(x_1, x_2, 0)$ be any element of T. Can we find an element (a_1, a_2, a_3) of S such that $f((a_1, a_2, a_3)) = (x_1, x_2, 0)$? We merely let $a_1 = x_1$, $a_2 = x_2$, and let $a_3 =$ any real number we wish, say $a_3 = 5$.

4. f is one-one for if $f(s_1) = f(s_2)$, then $2s_1 + 1 = 2s_2 + 1$, which means that $s_1 = s_2$. Also, f is onto, for given a real number t we seek a real number s so that $f(s) = t$; that is, we need to solve $2s + 1 = t$ for s, which we can always do, obtaining $s = \frac{1}{2}(t - 1)$.

5. f is one-one but f is not onto because not every element in T has 1 for its y-coordinate.

6. g is one-one and onto because $e^{s_1} \neq e^{s_2}$ if $s_1 \neq s_2$, and for any positive t we can always solve $t = e^s$, obtaining $s = \ln t$.

If $f: S \to T$ and $g: T \to U$ are functions, then we can define a new function, $g \circ f$, by $(g \circ f)(s) = g(f(s))$, for s in S. The function $g \circ f: S \to U$ is called the **composite of f and g**. Thus, if f and g are the functions 4 and 6, then $g \circ f$ is defined by $(g \circ f)(s) = g(f(s)) = e^{2s+1}$, and $f \circ g$ is defined by $(f \circ g)(s) = f(g(s)) = 2e^s + 1$. The function $i: S \to S$ defined by $i(s) = s$, for s in S, is called the **identity function on S**. A function $f: S \to T$ for which there is a function $g: T \to S$ such that $g \circ f = i_S =$ identity function on S and $f \circ g = i_T =$ identity function on T is called an **invertible function** and g is called the **inverse** of f. It is not difficult to show, and we do so in Chapter 4 for a special case, that a function $f: S \to T$ is invertible if and only if it is one-one and onto. The inverse of f is written as f^{-1}. If f is invertible, then f^{-1} is also one-one and onto. We recall that "if and only if" means that both the statement and its converse are true. That is, if $f: S \to T$ is invertible, then f is one-one and onto; if $f: S \to T$ is one-one and onto, then f is invertible. Functions 4 and 6 are invertible; the inverse of function 4 is $g: T \to S$ defined by $g(t) = (t - 1)/2$ for t in T; the inverse of function 6 is $h: T \to S$, defined by $h(t) = \ln t$.

CHAPTER 1

Linear Equations and Matrices

1.1. Systems of Linear Equations

One of the most frequently recurring practical problems in almost all fields of study—such as mathematics, physics, biology, chemistry, economics, all phases of engineering, operations research, the social sciences, and so forth—is that of solving a system of linear equations. The equation

$$y = a_1 x_1 + a_2 x_2 + \cdots + a_n x_n, \tag{1}$$

which expresses y in terms of the variables $x_1, x_2, \ldots, x_n$ and the constants $a_1, a_2, \ldots, a_n$, is called a **linear equation**. In many applications we are given y and must find numbers $x_1, x_2, \ldots, x_n$ satisfying (1).

A **solution** to a linear equation (1) is an ordered collection of n numbers $s_1, s_2, \ldots, s_n$ such that (1) is satisfied when $x_1 = s_1, x_2 = s_2, \ldots, x_n = s_n$ are substituted in (1). Thus $x_1 = 2$, $x_2 = 3$, and $x_3 = -4$ is a solution to the linear equation

$$6x_1 - 3x_2 + 4x_3 = -13,$$

because

$$6(2) - 3(3) + 4(-4) = -13.$$

More generally, a **system of** m **linear equations in** n **unknowns**, or a **linear system**, is a set of m linear equations each in n unknowns. A linear system can be conveniently written as

$$
\begin{aligned}
a_{11}x_1 + a_{12}x_2 + \cdots + a_{1n}x_n &= b_1 \\
a_{21}x_1 + a_{22}x_2 + \cdots + a_{2n}x_n &= b_2 \\
\vdots \qquad\qquad \vdots \qquad\qquad \vdots \qquad \vdots \\
a_{m1}x_1 + a_{m2}x_2 + \cdots + a_{mn}x_n &= b_m.
\end{aligned}
\tag{2}
$$

Thus the ith equation is

$$a_{i1}x_1 + a_{i2}x_2 + \cdots + a_{in}x_n = b_i.$$

In (2) the a_{ij} are known constants. Given values of $b_1, b_2, \ldots, b_n$, we want to find values of $x_1, x_2, \ldots, x_n$ that will satisfy each equation in (2).

A **solution** to a linear system (2) is an ordered collection of n numbers $s_1, s_2, \ldots, s_n$ which have the property that each equation in (2) is satisfied when $x_1 = s_1, x_2 = s_2, \ldots, x_n = s_n$ are substituted.

If the linear system (2) has no solution, it is said to be **inconsistent**; if it has a solution, it is called **consistent**. If $b_1 = b_2 = \cdots = b_m = 0$, then (2) is called a **homogeneous system**. The solution $x_1 = x_2 = \cdots = x_n = 0$ to a homogeneous system is called the **trivial solution**. A solution to a homogeneous system in which not all of $x_1, x_2, \ldots, x_n$ are zero is called a **nontrivial solution**.

Consider another system of r linear equations in n unknowns:

$$
\begin{aligned}
c_{11}x_1 + c_{12}x_2 + \cdots + c_{1n}x_n &= d_1 \\
c_{21}x_1 + c_{22}x_2 + \cdots + c_{2n}x_n &= d_2 \\
\vdots \qquad\qquad \vdots \qquad\qquad \vdots \qquad \vdots \\
c_{r1}x_1 + c_{r2}x_2 + \cdots + c_{rn}x_n &= d_r.
\end{aligned}
\tag{3}
$$

We say that (2) and (3) are **equivalent** if they both have exactly the same solutions.

EXAMPLE 1. The linear system

$$
\begin{aligned}
x_1 - 3x_2 &= -7 \\
2x_1 + x_2 &= 7
\end{aligned}
\tag{4}
$$

has only the solution $x_1 = 2$ and $x_2 = 3$. The linear system

$$\begin{aligned} 8x_1 - 3x_2 &= 7 \\ 3x_1 - 2x_2 &= 0 \\ 10x_1 - 2x_2 &= 14 \end{aligned} \tag{5}$$

also has only the solution $x_1 = 2$ and $x_2 = 3$. Thus (4) and (5) are equivalent.

We know from experience that one method for solving linear systems is elimination; that is, we eliminate some variables by adding a multiple of one equation to another equation. Elimination merely amounts to the development of a new linear system which is equivalent to the original system but is much simpler to solve. Readers have probably confined their earlier work in this area to linear systems in which $m = n$, that is, linear systems having as many equations as unknowns. In this course we shall broaden our outlook by dealing with systems in which we have $m = n$, $m < n$, and $m > n$. Indeed, there are numerous applications in which $m \neq n$.

EXAMPLE 2. Consider the linear system

$$\begin{aligned} x_1 - 3x_2 &= -3 \\ 2x_1 + x_2 &= 8. \end{aligned} \tag{6}$$

To eliminate x_1, we subtract twice the first equation from the second, obtaining

$$7x_2 = 14,$$

an equation having no x_1 term. Thus we have eliminated the unknown x_1. Then solving for x_2, we have

$$x_2 = 2,$$

and substituting into the first equation of (6) we obtain

$$x_1 = 3.$$

Then $x_1 = 3$, $x_2 = 2$ is the only solution to the given linear system.

EXAMPLE 3. Consider the linear system

$$\begin{aligned} x_1 - 3x_2 &= -7 \\ 2x_1 - 6x_2 &= 7. \end{aligned} \tag{7}$$

Again, we decide to eliminate x_1. We subtract twice the first equation from the second one, obtaining

$$0 = 21,$$

which makes no sense. This means that (7) has no solution; it is inconsistent. We could have come to the same conclusion from observing that in (7) the left side of the second equation is twice the left side of the first equation, but the right side of the second equation is not twice the right side of the first equation.

EXAMPLE 4. Consider the linear system

$$\begin{aligned} x_1 + 2x_2 + 3x_3 &= 6 \\ 2x_1 - 3x_2 + 2x_3 &= 14 \\ 3x_1 + x_2 - x_3 &= -2. \end{aligned} \tag{8}$$

To eliminate x_1, we subtract twice the first equation from the second and three times the first equation from the third, obtaining

$$\begin{aligned} -7x_2 - 4x_3 &= 2 \\ -5x_2 - 10x_3 &= -20. \end{aligned} \tag{9}$$

This is a system of two equations in the unknowns x_2 and x_3. We divide the second equation of (9) by -5, obtaining

$$\begin{aligned} -7x_2 - 4x_3 &= 2 \\ x_2 + 2x_3 &= 4, \end{aligned}$$

which we write, by interchanging equations, as

$$\begin{aligned} x_2 + 2x_3 &= 4 \\ -7x_2 - 4x_3 &= 2. \end{aligned} \tag{10}$$

We now eliminate x_2 in (10) by adding 7 times the first equation to the second one, to obtain

$$10x_3 = 30,$$

or

$$x_3 = 3. \tag{11}$$

Substituting this value of x_3 into the first equation of (10), we find that $x_2 = -2$. Substituting these values of x_2 and x_3 into the first equation of (8), we find that $x_1 = 1$. We might observe further that our elimination procedure has actually produced the linear system

$$
\begin{aligned}
x_1 + 2x_2 + 3x_3 &= 6 \\
x_2 + 2x_3 &= 4 \\
x_3 &= 3,
\end{aligned} \tag{12}
$$

obtained by using the first equations of (8) and (10) as well as (11). The importance of the procedure is that although the linear systems (8) and (12) are equivalent, (12) is easier to solve.

If we examine the method of elimination more closely, we find that it involves three manipulations that can be performed on a linear system to convert it into an equivalent system. These manipulations are as follows:

1. Interchange the ith and jth equations.
2. Multiply an equation by a nonzero constant.
3. Replace the ith equation by c times the jth equation plus the ith equation, $i \neq j$. That is, replace

$$
a_{i1}x_1 + a_{i2}x_2 + \cdots + a_{in}x_n = b_i
$$

by

$$
(a_{i1} + ca_{j1})x_1 + (a_{i2} + ca_{j2})x_2 + \cdots + (a_{in} + ca_{jn})x_n = b_i + cb_j.
$$

It is not difficult to prove that performing these manipulations on a linear system leads to an equivalent system. Before continuing the study of solving linear systems, we now introduce the notion of a matrix, which will greatly simplify our notational problems and develop tools to solve many important applied problems.

1.1. Exercises

1. Show that the linear system obtained by interchanging two equations in (2) is equivalent to (2).
2. Show that the linear system obtained by multiplying an equation in (2) by a nonzero constant is equivalent to (2).
3. Show that the linear system obtained by adding a multiple of an equation in (2) to another equation is equivalent to (2).

1.2. Matrices; Matrix Operations

In this section we define matrices and study some of their properties. Matrices enable us to write linear systems in a compact form and make it easy to automate the elimination method, on a computer, to obtain a fast and efficient procedure for solution. Their use, however, is not merely that of a convenient notation. We now develop operations on matrices and will work with matrices according to the rules they obey; this will enable us to solve linear systems and do other computational problems in a fast and efficient manner. Of course, as any good definition should do, the notion of a matrix provides not only a new way of looking at old problems but also gives rise to many new questions, some of which we study in this book.

Definition 1.1. A **matrix** is a rectangular array of numbers denoted by

$$A = \begin{bmatrix} a_{11} & a_{12} & \cdots & a_{1n} \\ a_{21} & a_{22} & \cdots & a_{2n} \\ \vdots & \vdots & & \vdots \\ a_{m1} & a_{m2} & \cdots & a_{mn} \end{bmatrix}.$$

Unless stated otherwise, we shall assume that all our matrices are composed entirely of real numbers. The *i*th **row** of A is

$$[a_{i1} \quad a_{i2} \quad \cdots \quad a_{in}] \qquad (1 \le i \le m)$$

while the *j*th **column** of A is

$$\begin{bmatrix} a_{1j} \\ a_{2j} \\ \vdots \\ a_{mj} \end{bmatrix} \qquad (1 \le j \le n).$$

If a matrix A has m rows and n columns, we shall say that A is an m **by** n $(m \times n)$ **matrix**. If $m = n$, we say that A is a **square matrix of order** n and that the elements $a_{11}, a_{22}, \ldots, a_{nn}$ are on the **main diagonal** of A. We refer to a_{ij} as the (i, j) **entry** (entry in *i*th row and *j*th column) or *i, j*th **element** and we often write

$$A = [a_{ij}].$$

We shall also write $_mA_n$ to indicate that A has m rows and n columns. If A is $n \times n$, we merely write A_n.

EXAMPLE 1. The following are matrices:

$$A = \begin{bmatrix} 1 & 2 & 3 \\ 2 & -1 & 4 \\ 0 & -3 & 2 \end{bmatrix}, \qquad B = [1 \quad 3 \quad -7],$$

$$C = \begin{bmatrix} 2 \\ -1 \\ 3 \\ 4 \end{bmatrix}, \qquad \text{and} \qquad D = \begin{bmatrix} 0 & 3 \\ -1 & -2 \end{bmatrix}.$$

In A, $a_{32} = -3$; in C, $c_{41} = 4$. Here A is 3×3, B is 1×3, C is 4×1, and D is 2×2.

Whenever a new object is introduced in mathematics, one must determine when two such objects are equal. For example, in the set of all rational numbers the numbers $\frac{2}{3}$ and $\frac{4}{6}$ are called equal, although they are not represented in the same manner. What we have in mind is the definition that a/b equals c/d when $ad = bc$. Accordingly, we now have the following definition.

Definition 1.2. Two $m \times n$ matrices $A = [a_{ij}]$ and $B = [b_{ij}]$ are **equal** if they agree entry by entry, that is, if $a_{ij} = b_{ij}$ for $i = 1, 2, \ldots, m$ and $j = 1, 2, \ldots, n$.

EXAMPLE 2. The matrices

$$A = \begin{bmatrix} 1 & 2 & -1 \\ 2 & -3 & 4 \\ 0 & -4 & 5 \end{bmatrix} \quad \text{and} \quad B = \begin{bmatrix} 1 & 2 & w \\ 2 & x & 4 \\ y & -4 & z \end{bmatrix}$$

are equal if and only if $w = -1$, $x = -3$, $y = 0$, and $z = 5$.

We next define a number of operations that will produce new matrices out of given matrices; this will enable us to compute with the matrices and not deal with the equations from which they arise. These operations are also useful in the applications of matrices.

Definition 1.3. If $A = [a_{ij}]$ and $B = [b_{ij}]$ are both $m \times n$ matrices, then the **sum** $A + B$ is an $m \times n$ matrix $C = [c_{ij}]$ defined by $c_{ij} = a_{ij} + b_{ij}$, $i = 1, 2, \ldots, m; j = 1, 2, \ldots, n$. Thus, to obtain the sum of A and B, we merely add corresponding entries.

EXAMPLE 3. Let

$$A = \begin{bmatrix} 1 & -2 & 3 \\ 2 & -1 & 4 \end{bmatrix} \quad \text{and} \quad B = \begin{bmatrix} 0 & 2 & 1 \\ 1 & 3 & -4 \end{bmatrix}.$$

Then

$$A + B = \begin{bmatrix} 1+0 & -2+2 & 3+1 \\ 2+1 & -1+3 & 4-4 \end{bmatrix} = \begin{bmatrix} 1 & 0 & 4 \\ 3 & 2 & 0 \end{bmatrix}.$$

It should be noted that the sum of the matrices A and B is defined only when A and B have the same number of rows and the same number of columns, that is, only when A and B are of the same size. Thus we shall now establish the convention that when $A + B$ is formed, both A and B are of the same size.

Definition 1.4. If $A = [a_{ij}]$ is an $m \times n$ matrix and c is a real number, then the **scalar multiple** of A by c, cA, is the $m \times n$ matrix $C = [c_{ij}]$, where $c_{ij} = ca_{ij}, i = 1, 2, \ldots, m, j = 1, 2, \ldots, n$; that is, the matrix C is obtained by multiplying each entry of A by c.

EXAMPLE 4. We have

$$2 \begin{bmatrix} 4 & -2 & -3 \\ 7 & -3 & 2 \end{bmatrix} = \begin{bmatrix} 8 & -4 & -6 \\ 14 & -6 & 4 \end{bmatrix}.$$

If A and B are $m \times n$ matrices, we write $A + (-1)B$ as $A - B$ and call this the **difference between A and B**.

We shall often use the **summation notation**, and we now review this useful and compact notation.

By $\sum_{i=1}^{n} r_i a_i$ we mean $r_1 a_1 + r_2 a_2 + \cdots + r_n a_n$. The letter i is called the **index of summation**; it is a dummy variable that can be replaced by another letter so that we can then write

$$\sum_{i=1}^{n} r_i a_i = \sum_{j=1}^{n} r_j a_j = \sum_{k=1}^{n} r_k a_k.$$

Thus

$$\sum_{i=1}^{4} r_i a_i = r_1 a_1 + r_2 a_2 + r_3 a_3 + r_4 a_4.$$

The summation notation satisfies the following properties:

1. $\displaystyle\sum_{i=1}^{n} (r_i + s_i)a_i = \sum_{i=1}^{n} r_i a_i + \sum_{i=1}^{n} s_i a_i.$

2. $\displaystyle\sum_{i=1}^{n} c(r_i a_i) = c \sum_{i=1}^{n} r_i a_i.$

3. $\displaystyle\sum_{j=1}^{m} \sum_{i=1}^{n} a_{ij} = \sum_{i=1}^{n} \sum_{j=1}^{m} a_{ij}.$

Definition 1.5. If $A = [a_{ij}]$ is an $m \times n$ matrix and $B = [b_{ij}]$ is an $n \times p$ matrix, then the **product** of A and B, $AB = C = [c_{ij}]$, is an $m \times p$ matrix defined by

$$c_{ij} = \sum_{k=1}^{n} a_{ik} b_{kj} = a_{i1} b_{1j} + a_{i2} b_{2j} + \cdots + a_{in} b_{nj} \qquad i = 1, 2, \ldots, m$$

$$j = 1, 2, \ldots, p.$$

Note that AB can be defined only when the number of columns of A is the same as the number of rows of B. We also observe that the (i, j) entry in C is obtained by using the ith row of A and the jth column of B. Thus

$$\begin{bmatrix} a_{11} & a_{12} & \cdots & a_{1n} \\ a_{21} & a_{22} & \cdots & a_{2n} \\ \vdots & \vdots & & \vdots \\ a_{i1} & a_{i2} & \cdots & a_{in} \\ \vdots & \vdots & & \vdots \\ a_{m1} & a_{m2} & \cdots & a_{mn} \end{bmatrix} \begin{bmatrix} b_{11} & b_{12} & \cdots & b_{1j} & \cdots & b_{1p} \\ b_{21} & b_{22} & \cdots & b_{2j} & \cdots & b_{2p} \\ \vdots & \vdots & & \vdots & & \vdots \\ b_{n1} & b_{n2} & \cdots & b_{nj} & \cdots & b_{np} \end{bmatrix}$$

$$= \begin{bmatrix} c_{11} & c_{12} & \cdots & c_{1p} \\ c_{21} & c_{22} & \cdots & c_{2p} \\ \vdots & \vdots & c_{ij} & \vdots \\ c_{m1} & c_{m2} & \cdots & c_{mp} \end{bmatrix}.$$

EXAMPLE 5. Let

$$A = \begin{bmatrix} 1 & 2 & -1 \\ 3 & 1 & 4 \end{bmatrix} \quad \text{and} \quad B = \begin{bmatrix} -2 & 5 \\ 4 & -3 \\ 2 & 1 \end{bmatrix}.$$

Then

$$AB = \begin{bmatrix} (1)(-2) + (2)(4) + (-1)(2) & (1)(5) + (2)(-3) + (-1)(1) \\ (3)(-2) + (1)(4) + (4)(2) & (3)(5) + (1)(-3) + (4)(1) \end{bmatrix}$$

$$= \begin{bmatrix} 4 & -2 \\ 6 & 16 \end{bmatrix}.$$

If A is an $m \times n$ matrix and B is an $n \times p$ matrix, then AB is defined. In considering BA, several situations may occur:

1. BA may not be defined. This will take place if $p \neq m$.
2. If BA is defined, BA will be $n \times n$ and AB will be $m \times m$, and if $m \neq n$, AB and BA are of different sizes.
3. If BA and AB are of the same size, they may be unequal.

As in the case of addition, we establish the convention that when AB is written, it is defined.

EXAMPLE 6. Let A be a 2×3 matrix and let B be a 3×4 matrix. Then AB is 2×4 and BA is not defined.

EXAMPLE 7. Let A be 2×3 and let B be 3×2. Then AB is 2×2 and BA is 3×3.

EXAMPLE 8. Let

$$A = \begin{bmatrix} 1 & 2 \\ -1 & 3 \end{bmatrix} \quad \text{and} \quad B = \begin{bmatrix} 2 & 1 \\ 0 & 1 \end{bmatrix}.$$

Then

$$AB = \begin{bmatrix} 2 & 3 \\ -2 & 2 \end{bmatrix} \quad \text{while} \quad BA = \begin{bmatrix} 1 & 7 \\ -1 & 3 \end{bmatrix}.$$

We now return to the linear system (2) in Section 1.1 and define the following matrices:

$$A = \begin{bmatrix} a_{11} & a_{12} & \cdots & a_{1n} \\ a_{21} & a_{22} & \cdots & a_{2n} \\ \vdots & \vdots & & \vdots \\ a_{m1} & a_{m2} & \cdots & a_{mn} \end{bmatrix}, \qquad X = \begin{bmatrix} x_1 \\ x_2 \\ \vdots \\ x_n \end{bmatrix}, \qquad B = \begin{bmatrix} b_1 \\ b_2 \\ \vdots \\ b_m \end{bmatrix}.$$

We can then write the linear system (2) as $AX = B$. The matrix A is called the **coefficient matrix** of the system and the matrix

$$\begin{bmatrix} a_{11} & a_{12} & \cdots & a_{1n} & \vdots & b_1 \\ a_{21} & a_{22} & \cdots & a_{2n} & \vdots & b_2 \\ \vdots & \vdots & & \vdots & \vdots & \vdots \\ a_{m1} & a_{m2} & \cdots & a_{mn} & \vdots & b_m \end{bmatrix}.$$

is called the **augmented matrix** of the system. The coefficient and augmented matrices of a linear system will play key roles in our methods of solving linear systems.

EXAMPLE 9. Consider the following linear system

$$\begin{aligned} 2x_1 + 3x_2 - 4x_3 + x_4 &= 5 \\ -2x_1 \quad\quad\quad + x_3 \quad\quad &= 7 \\ 3x_1 + 2x_2 \quad\quad\quad - 4x_4 &= 3. \end{aligned}$$

We can write this in matrix form as

$$\begin{bmatrix} 2 & 3 & -4 & 1 \\ -2 & 0 & 1 & 0 \\ 3 & 2 & 0 & -4 \end{bmatrix} \begin{bmatrix} x_1 \\ x_2 \\ x_3 \\ x_4 \end{bmatrix} = \begin{bmatrix} 5 \\ 7 \\ 3 \end{bmatrix}.$$

The coefficient matrix of this system is

$$\begin{bmatrix} 2 & 3 & -4 & 1 \\ -2 & 0 & 1 & 0 \\ 3 & 2 & 0 & -4 \end{bmatrix},$$

and the augmented matrix is

$$\begin{bmatrix} 2 & 3 & -4 & 1 & \vdots & 5 \\ -2 & 0 & 1 & 0 & \vdots & 7 \\ 3 & 2 & 0 & -4 & \vdots & 3 \end{bmatrix}.$$

Definition 1.6. If $A = [a_{ij}]$ is an $m \times n$ matrix, then the **transpose** of A, $A' = [a_{ij}']$, is an $n \times m$ matrix defined by $a_{ij}' = a_{ji}$. Thus the transpose of A is obtained from A by interchanging the rows and columns of A.

EXAMPLE 10. If

$$A = \begin{bmatrix} 1 & 2 & -1 \\ -3 & 2 & 7 \end{bmatrix}, \quad \text{then} \quad A' = \begin{bmatrix} 1 & -3 \\ 2 & 2 \\ -1 & 7 \end{bmatrix}.$$

1.2. Exercises

Consider the following matrices for Exercises 1, 2, and 3:

$$A = \begin{bmatrix} 1 & 2 & 3 \\ 2 & 1 & 4 \end{bmatrix}, \quad B = \begin{bmatrix} 1 & 0 \\ 2 & 1 \\ 3 & 2 \end{bmatrix}, \quad C = \begin{bmatrix} 3 & -1 & 3 \\ 4 & 1 & 5 \\ 2 & 1 & 3 \end{bmatrix},$$

$$D = \begin{bmatrix} 3 & -2 \\ 2 & 5 \end{bmatrix}, \quad \text{and} \quad E = \begin{bmatrix} 2 & -4 & 5 \\ 0 & 1 & 4 \\ 3 & 2 & 1 \end{bmatrix}.$$

1. If possible compute:
   (a) $C + E$.   (b) AB and BA.   (c) $2C - 3E$.
   (d) $CB + D$.   (e) $AB + D^2$, where $D^2 = DD$.   (f) $(3)(2A)$ and $6A$.
2. If possible, compute:
   (a) $A(BD)$.   (b) $(AB)D$.   (c) $A(C + E)$.
   (d) $AC + AE$.   (e) $3A + 2A$ and $5A$.
3. If possible, compute:
   (a) A'.   (b) $(A')'$.   (c) $(AB)'$.
   (d) $B'A'$.   (e) $(C + E)'$ and $C' + E'$.   (f) $A(2B)$ and $2(AB)$.
4. (a) Let A be an $m \times n$ matrix with a row consisting entirely of zeros. Prove that if B is an $n \times p$ matrix, then AB has a row of zeros.
   (b) Let A be an $m \times n$ matrix with a column consisting entirely of zeros and let B be $p \times m$. Prove that BA has a column of zeros.
5. Let $A = \begin{bmatrix} 1 & 2 \\ 3 & 2 \end{bmatrix}$ and $B = \begin{bmatrix} 2 & -1 \\ -3 & 4 \end{bmatrix}$. Show that $AB \neq BA$.

6. Consider the following linear system:

$$2x_1 + 3x_2 - 3x_3 + x_4 + x_5 = 7$$
$$3x_1 + 2x_3 + 3x_5 = -2$$
$$2x_1 + 3x_2 - 4x_4 = 3$$
$$x_3 + x_4 + x_5 = 5.$$

(a) Find the coefficient matrix.

(b) Write the linear system in matrix form.

(c) Find the augmented matrix.

7. Write the linear system whose augmented matrix is

$$\begin{bmatrix} -2 & -1 & 0 & 4 & \vdots & 5 \\ -3 & 2 & 7 & 8 & \vdots & 3 \\ 1 & 0 & 0 & 2 & \vdots & 4 \\ 3 & 0 & 1 & 3 & \vdots & 6 \end{bmatrix}.$$

8. If $\begin{bmatrix} a+b & c+d \\ c-d & a-b \end{bmatrix} = \begin{bmatrix} 4 & 6 \\ 10 & 2 \end{bmatrix}$, find a, b, c, and d.

9. Write the following linear system in matrix form.

$$2x_1 + 3x_2 = 0$$
$$3x_2 + x_3 = 0$$
$$2x_1 - x_3 = 0.$$

10. Write the linear system whose augmented matrix is

(a) $\begin{bmatrix} 2 & 1 & 3 & 4 & \vdots & 0 \\ 3 & -1 & 2 & 0 & \vdots & 3 \\ -2 & 1 & -4 & 3 & \vdots & 2 \end{bmatrix}.$ (b) $\begin{bmatrix} 2 & 1 & 3 & 4 & \vdots & 0 \\ 3 & -1 & 2 & 0 & \vdots & 3 \\ -2 & 1 & -4 & 3 & \vdots & 2 \\ 0 & 0 & 0 & 0 & \vdots & 0 \end{bmatrix}.$

11. How are the linear systems obtained in Exercise 10 related?

12. If $A = [a_{ij}]$ is an $n \times n$ matrix, then the **trace** of A, Tr (A), is defined as the sum of all elements on the main diagonal of A, Tr $(A) = \sum_{i=1}^{n} a_{ii}$. Prove:

(a) Tr $(cA) = c$ Tr (A), where c is a real number.

(b) Tr $(A + B) = $ Tr $(A) + $ Tr (B).

(c) Tr $(AB) = $ Tr (BA).

13. Compute the trace of each of the following matrices.

(a) $\begin{bmatrix} 1 & 0 \\ 2 & 3 \end{bmatrix}.$ (b) $\begin{bmatrix} 2 & 2 & 3 \\ 2 & 4 & 4 \\ 3 & -2 & -5 \end{bmatrix}.$ (c) $\begin{bmatrix} 1 & 0 & 0 \\ 0 & 1 & 0 \\ 0 & 0 & 1 \end{bmatrix}.$

14. Show that there are no 2×2 matrices A and B such that $AB - BA = \begin{bmatrix} 1 & 0 \\ 0 & 1 \end{bmatrix}.$

15. Show that the jth column of the matrix product AB is equal to the matrix product AB_j, where B_j is the jth column of B.
16. Show that if $AX = B$ has more than one solution, then it has infinitely many solutions. (*Hint*: If X_1 and X_2 are solutions, consider $X_3 = rX_1 + sX_2$, where $r + s = 1$.)

1.3. Algebraic Properties of Matrix Operations

In this section we consider the algebraic properties of the matrix operations just defined. Many of these properties are similar to the familiar properties holding for the real numbers. However, there will be striking differences between the set of real numbers and the set of matrices in their algebraic behavior under certain operations, for example, under multiplication (as seen in Section 1.2). Most of the properties will be stated as theorems, whose proofs will be left as exercises.

Theorem 1.1. *If A and B are $m \times n$ matrices, then $A + B = B + A$.*

Proof: Let $A = [a_{ij}]$, $B = [b_{ij}]$, $A + B = C = [c_{ij}]$, and $B + A = D = [d_{ij}]$. We must show that $c_{ij} = d_{ij}$ for all i, j. Now $c_{ij} = a_{ij} + b_{ij}$ and $d_{ij} = b_{ij} + a_{ij}$ for all i, j. Since a_{ij} and b_{ij} are real numbers, we have $a_{ij} + b_{ij} = b_{ij} + a_{ij}$, which implies that $c_{ij} = d_{ij}$ for all i, j.

EXAMPLE 1. We have

$$\begin{bmatrix} 1 & 2 & 3 \\ -2 & 1 & 4 \end{bmatrix} + \begin{bmatrix} 3 & 2 & -1 \\ 3 & -1 & 2 \end{bmatrix} = \begin{bmatrix} 4 & 4 & 2 \\ 1 & 0 & 6 \end{bmatrix}$$

$$= \begin{bmatrix} 3 & 2 & -1 \\ 3 & -1 & 2 \end{bmatrix} + \begin{bmatrix} 1 & 2 & 3 \\ -2 & 1 & 4 \end{bmatrix}.$$

Theorem 1.2. *If A, B, and C are $m \times n$ matrices, then $A + (B + C) = (A + B) + C$.*

Proof: Exercise.

EXAMPLE 2. We have

$$\begin{bmatrix} 1 & 2 \\ 3 & 4 \end{bmatrix} + \left(\begin{bmatrix} 2 & 1 \\ 3 & -2 \end{bmatrix} + \begin{bmatrix} 3 & 1 \\ -2 & 1 \end{bmatrix} \right) = \begin{bmatrix} 6 & 4 \\ 4 & 3 \end{bmatrix}$$

$$= \left(\begin{bmatrix} 1 & 2 \\ 3 & 4 \end{bmatrix} + \begin{bmatrix} 2 & 1 \\ 3 & -2 \end{bmatrix} \right) + \begin{bmatrix} 3 & 1 \\ -2 & 1 \end{bmatrix}.$$

Theorem 1.3. *There exists a unique $m \times n$ matrix $_mO_n$ such that*

$$A + {}_mO_n = {}_mO_n + A = A \qquad \text{for any } m \times n \text{ matrix } A.$$

Proof: Let $U = [u_{ij}]$. Then $A + U = A$ if and only if $a_{ij} + u_{ij} = a_{ij}$, which holds if and only if $u_{ij} = 0$. Thus U is the $m \times n$ matrix all of whose elements are zero; U is denoted by $_mO_n$.

We call $_mO_n$ the $m \times n$ **zero matrix**. When $m = n$, we write O_n. When m and n are understood, we shall write $_mO_n$ merely as O.

Theorem 1.4. *Given any $m \times n$ matrix A, there exists an $m \times n$ matrix B such that $A + B = {}_mO_n$.*

Proof: Exercise.

We can now show that B is unique and that it is $(-1)A$, which we have already agreed to write as $-A$ and call it the **negative** of A.

EXAMPLE 3. If $A = \begin{bmatrix} 1 & 3 & -2 \\ -2 & 4 & 3 \end{bmatrix}$, then $-A = \begin{bmatrix} -1 & -3 & 2 \\ 2 & -4 & -3 \end{bmatrix}$.

Theorem 1.5. *If A is an $m \times n$ matrix, B is an $n \times p$ matrix, and C is a $p \times q$ matrix, then $A(BC) = (AB)C$.*

Proof: We shall prove the result for $m = 2$, $n = 3$, $p = 4$, and $q = 3$. The general proof is completely analogous.

Let $A = [a_{ij}]$, $B = [b_{ij}]$, $C = [c_{ij}]$, $AB = D = [d_{ij}]$, $BC = E = [e_{ij}]$, $(AB)C = F = [f_{ij}]$, and $A(BC) = G = [g_{ij}]$. We must show that $f_{ij} = g_{ij}$ for all i, j. Now

$$f_{ij} = \sum_{k=1}^{4} d_{ik}c_{kj} = \sum_{k=1}^{4} \left(\sum_{r=1}^{3} a_{ir}b_{rk} \right) c_{kj}$$

and

$$g_{ij} = \sum_{r=1}^{3} a_{ir}e_{rj} = \sum_{r=1}^{3} a_{ir} \left(\sum_{k=1}^{4} b_{rk}c_{kj} \right).$$

Then

$$f_{ij} = \sum_{k=1}^{4} (a_{i1}b_{1k} + a_{i2}b_{2k} + a_{i3}b_{3k})c_{kj}$$

$$= a_{i1} \sum_{k=1}^{4} b_{1k}c_{kj} + a_{i2} \sum_{k=1}^{4} b_{2k}c_{kj} + a_{i3} \sum_{k=1}^{4} b_{3k}c_{kj}$$

$$= \sum_{r=1}^{3} a_{ir} \left(\sum_{k=1}^{4} b_{rk}c_{kj} \right) = g_{ij}.$$

EXAMPLE 4. Let

$$A = \begin{bmatrix} 5 & 2 & 3 \\ 2 & -3 & 4 \end{bmatrix}, \qquad B = \begin{bmatrix} 2 & -1 & 1 & 0 \\ 0 & 2 & 2 & 2 \\ 3 & 0 & -1 & 3 \end{bmatrix},$$

and

$$C = \begin{bmatrix} 1 & 0 & 2 \\ 2 & -3 & 0 \\ 0 & 0 & 3 \\ 2 & 1 & 0 \end{bmatrix}.$$

Then

$$A(BC) = \begin{bmatrix} 5 & 2 & 3 \\ 2 & -3 & 4 \end{bmatrix} \begin{bmatrix} 0 & 3 & 7 \\ 8 & -4 & 6 \\ 9 & 3 & 3 \end{bmatrix} = \begin{bmatrix} 43 & 16 & 56 \\ 12 & 30 & 8 \end{bmatrix}$$

and

$$(AB)C = \begin{bmatrix} 19 & -1 & 6 & 13 \\ 16 & -8 & -8 & 6 \end{bmatrix} \begin{bmatrix} 1 & 0 & 2 \\ 2 & -3 & 0 \\ 0 & 0 & 3 \\ 2 & 1 & 0 \end{bmatrix} = \begin{bmatrix} 43 & 16 & 56 \\ 12 & 30 & 8 \end{bmatrix}.$$

Theorem 1.6. *If r and s are real numbers, A is an m × n matrix, and B is an n × p matrix, then*
(a) $r(sA) = (rs)A = s(rA)$.
(b) $A(rB) = r(AB)$.

Proof: Exercise.

EXAMPLE 5. Let

$$A = \begin{bmatrix} 4 & 2 & 3 \\ 2 & -3 & 4 \end{bmatrix} \quad \text{and} \quad B = \begin{bmatrix} 3 & -2 & 1 \\ 2 & 0 & -1 \\ 0 & 1 & 2 \end{bmatrix}.$$

Then

$$2(3A) = 2\begin{bmatrix} 12 & 6 & 9 \\ 6 & -9 & 12 \end{bmatrix} = \begin{bmatrix} 24 & 12 & 18 \\ 12 & -18 & 24 \end{bmatrix} = 6A.$$

We also have

$$A(2B) = \begin{bmatrix} 4 & 2 & 3 \\ 2 & -3 & 4 \end{bmatrix}\begin{bmatrix} 6 & -4 & 2 \\ 4 & 0 & -2 \\ 0 & 2 & 4 \end{bmatrix} = \begin{bmatrix} 32 & -10 & 16 \\ 0 & 0 & 26 \end{bmatrix} = 2(AB).$$

Theorem 1.7

(a) *If A and B are m × n matrices and C is an n × p matrix, then*

$$(A + B)C = AC + BC.$$

(b) *If C is an m × n matrix and A and B are n × p matrices, then*

$$C(A + B) = CA + CB.$$

Proof: Exercise.

EXAMPLE 6. Let

$$A = \begin{bmatrix} 2 & 2 & 3 \\ 3 & -1 & 2 \end{bmatrix}, \quad B = \begin{bmatrix} 0 & 0 & 1 \\ 2 & 3 & -1 \end{bmatrix}, \quad \text{and} \quad C = \begin{bmatrix} 1 & 0 \\ 2 & 2 \\ 3 & -1 \end{bmatrix}.$$

Then

$$(A + B)C = \begin{bmatrix} 2 & 2 & 4 \\ 5 & 2 & 1 \end{bmatrix}\begin{bmatrix} 1 & 0 \\ 2 & 2 \\ 3 & -1 \end{bmatrix} = \begin{bmatrix} 18 & 0 \\ 12 & 3 \end{bmatrix}$$

and

$$AC + BC = \begin{bmatrix} 15 & 1 \\ 7 & -4 \end{bmatrix} + \begin{bmatrix} 3 & -1 \\ 5 & 7 \end{bmatrix} = \begin{bmatrix} 18 & 0 \\ 12 & 3 \end{bmatrix}.$$

Theorem 1.8. *If a and b are real numbers and A is an m × n matrix, then* $(a + b)A = aA + bA$.

Proof: Exercise.

Theorem 1.9. *If A and B are m × n matrices and a is a real number, then* $a(A + B) = aA + aB$.

Proof: Exercise.

So far we have seen that multiplication and addition of matrices have much in common with multiplication and addition of real numbers. We now look at some properties of the transpose.

Theorem 1.10. *If A is an m × n matrix, then* $(A')' = A$.

Proof: Exercise.

Theorem 1.11
   (a) *If A is an m × n matrix and c is a real number, then* $(cA)' = cA'$.
   (b) *If A and B are m × n matrices, then* $(A + B)' = A' + B'$.

Proof: Exercise.

EXAMPLE 7. Let

$$A = \begin{bmatrix} 1 & 2 & 3 \\ -2 & 0 & 1 \end{bmatrix} \quad \text{and} \quad B = \begin{bmatrix} 3 & -1 & 2 \\ 3 & 2 & -1 \end{bmatrix}.$$

Then

$$A' = \begin{bmatrix} 1 & -2 \\ 2 & 0 \\ 3 & 1 \end{bmatrix} \quad \text{and} \quad B' = \begin{bmatrix} 3 & 3 \\ -1 & 2 \\ 2 & -1 \end{bmatrix}.$$

Also,

$$A + B = \begin{bmatrix} 4 & 1 & 5 \\ 1 & 2 & 0 \end{bmatrix} \quad \text{and} \quad (A + B)' = \begin{bmatrix} 4 & 1 \\ 1 & 2 \\ 5 & 0 \end{bmatrix}.$$

Now

$$A' + B' = \begin{bmatrix} 4 & 1 \\ 1 & 2 \\ 5 & 0 \end{bmatrix} = (A + B)'.$$

Theorem 1.12. *If A is an $m \times n$ matrix and B is an $n \times p$ matrix, then $(AB)' = B'A'$.*

Proof: Let $A = [a_{ij}]$ and $B = [b_{ij}]$; let $AB = C = [c_{ij}]$. We must prove that c_{ij}' is the (i, j) entry in $B'A'$. Now

$$c_{ij}' = c_{ji} = \sum_{k=1}^{n} a_{jk}b_{ki} = \sum_{k=1}^{n} a_{kj}'b_{ik}' = \sum_{k=1}^{n} b_{ik}'a_{kj}' = \text{the } (i, j) \text{ entry in } B'A'.$$

EXAMPLE 8. Let

$$A = \begin{bmatrix} 1 & 3 & 2 \\ 2 & -1 & 3 \end{bmatrix} \quad \text{and} \quad B = \begin{bmatrix} 0 & 1 \\ 2 & 2 \\ 3 & -1 \end{bmatrix}.$$

Then

$$AB = \begin{bmatrix} 12 & 5 \\ 7 & -3 \end{bmatrix} \quad \text{and} \quad (AB)' = \begin{bmatrix} 12 & 7 \\ 5 & -3 \end{bmatrix}.$$

On the other hand,

$$A' = \begin{bmatrix} 1 & 2 \\ 3 & -1 \\ 2 & 3 \end{bmatrix} \quad \text{and} \quad B' = \begin{bmatrix} 0 & 2 & 3 \\ 1 & 2 & -1 \end{bmatrix},$$

and then

$$B'A' = \begin{bmatrix} 12 & 7 \\ 5 & -3 \end{bmatrix} = (AB)'.$$

We also note another peculiarity of matrix multiplication. If a and b are real numbers, then $ab = 0$ can hold only if a or b is zero. However, this is not true for matrices.

EXAMPLE 9. If $A = \begin{bmatrix} 1 & 2 \\ 2 & 4 \end{bmatrix}$ and $B = \begin{bmatrix} 4 & -6 \\ -2 & 3 \end{bmatrix}$, then neither A nor B is the zero matrix, but $AB = \begin{bmatrix} 0 & 0 \\ 0 & 0 \end{bmatrix}$.

1.3. Exercises

1. Prove Theorem 1.2.
2. Prove Theorem 1.4.
3. Verify Theorem 1.5 for the following matrices:

$$A = \begin{bmatrix} 1 & 3 \\ 2 & -1 \end{bmatrix}, \quad B = \begin{bmatrix} -1 & 3 & 2 \\ 1 & -3 & 4 \end{bmatrix}, \quad \text{and} \quad C = \begin{bmatrix} 1 & 0 \\ 3 & -1 \\ 1 & 2 \end{bmatrix}.$$

4. Prove Theorem 1.6.
5. Verify Theorem 1.6(b) for the following matrices:

$$A = \begin{bmatrix} 1 & 3 \\ 2 & -1 \end{bmatrix}, \quad B = \begin{bmatrix} -1 & 3 & 2 \\ 1 & -3 & 4 \end{bmatrix}, \quad \text{and} \quad r = -3.$$

6. Prove Theorem 1.7.
7. Verify Theorem 1.7(b) for the following matrices:

$$A = \begin{bmatrix} 2 & -3 & 2 \\ 3 & -1 & -2 \end{bmatrix}, \quad B = \begin{bmatrix} 0 & 1 & 2 \\ 1 & 3 & -2 \end{bmatrix}, \quad \text{and} \quad C = \begin{bmatrix} 1 & -3 \\ -3 & 4 \end{bmatrix}.$$

8. Find a pair of unequal 2×2 matrices A and B, other than those given in Example 9, such that $AB = O_2$.
9. Find two different 2×2 matrices A such that $A^2 = O_2$.
10. Prove Theorem 1.8.

11. Verify Theorem 1.8 for $a = 4$, $b = -2$, and $A = \begin{bmatrix} 2 & -3 \\ 4 & 2 \end{bmatrix}$.

12. Find two different 2×2 matrices A such that $A^2 = \begin{bmatrix} 1 & 0 \\ 0 & 1 \end{bmatrix}$.

13. Prove Theorem 1.9.
14. Verify Theorem 1.9 for $a = -3$ and

$$A = \begin{bmatrix} 4 & 2 \\ 1 & -3 \\ 3 & 2 \end{bmatrix} \quad \text{and} \quad B = \begin{bmatrix} 0 & 2 \\ 4 & 3 \\ -2 & 1 \end{bmatrix}.$$

15. Find two unequal 2×2 matrices A and B such that $AB = \begin{bmatrix} 1 & 0 \\ 0 & 1 \end{bmatrix}$.

16. Prove Theorem 1.10.

17. Find three 2×2 matrices, A, B, and C, such that $AB = AC$ with $B \neq C$ and $A \neq O$.

18. Prove Theorem 1.11.

19. Verify Theorem 1.11 for $A = \begin{bmatrix} 1 & 3 & 2 \\ 2 & 1 & -3 \end{bmatrix}$, $B = \begin{bmatrix} 4 & 2 & -1 \\ -2 & 1 & 5 \end{bmatrix}$, and $c = -4$.

20. Verify Theorem 1.12 for $A = \begin{bmatrix} 1 & 3 & 2 \\ 2 & 1 & -3 \end{bmatrix}$ and $B = \begin{bmatrix} 3 & -1 \\ 2 & 4 \\ 1 & 2 \end{bmatrix}$.

21. Let A be an $m \times n$ matrix and c a real number. Show that if $cA = O$, then $c = 0$ or $A = O$.

1.4. Special Types of Matrices and Partitioned Matrices

We have already introduced one special type of matrix, $_mO_n$, the matrix all of whose entries are zero. We now consider several other types of matrices whose structure is rather specialized and for which it will be convenient to have special names.

An $n \times n$ matrix $A = [a_{ij}]$ is called a **diagonal matrix** if $a_{ij} = 0$ for $i \neq j$. Thus, for a diagonal matrix, the terms *off* the main diagonal are all zero. A **scalar matrix** is a diagonal matrix whose diagonal elements are equal. The scalar matrix $I_n = [a_{ij}]$, where $a_{ii} = 1$ and $a_{ij} = 0$ for $i \neq j$, is called the **identity matrix**.

EXAMPLE 1. Let

$$A = \begin{bmatrix} 1 & 0 & 0 \\ 0 & 2 & 0 \\ 0 & 0 & 3 \end{bmatrix}, \qquad B = \begin{bmatrix} 2 & 0 & 0 \\ 0 & 2 & 0 \\ 0 & 0 & 2 \end{bmatrix}, \qquad \text{and} \qquad I_3 = \begin{bmatrix} 1 & 0 & 0 \\ 0 & 1 & 0 \\ 0 & 0 & 1 \end{bmatrix}.$$

Then A, B, and I_3 are diagonal matrices; B and I_3 are scalar matrices; and I_3 is the 3×3 identity matrix.

It is easy to show that if A is any $m \times n$ matrix, then $AI_n = A$ and $I_mA = A$. Also, if A is a scalar matrix, then $A = rI_n$ for some scalar r.

An $n \times n$ matrix $A = [a_{ij}]$ is called **upper triangular** if $a_{ij} = 0$ for $i > j$. It is called **lower triangular** if $a_{ij} = 0$ for $i < j$.

EXAMPLE 2. The matrix

$$A = \begin{bmatrix} 1 & 3 & 3 \\ 0 & 3 & 5 \\ 0 & 0 & 2 \end{bmatrix}$$

is upper triangular and

$$B = \begin{bmatrix} 1 & 0 & 0 \\ 2 & 3 & 0 \\ 3 & 5 & 2 \end{bmatrix}$$

is lower triangular.

Definition 1.7. A matrix A is called **symmetric** if $A' = A$.

Definition 1.8. A matrix A is called **skew symmetric** if $A' = -A$.

EXAMPLE 3. $A = \begin{bmatrix} 1 & 2 & 3 \\ 2 & 4 & 5 \\ 3 & 5 & 6 \end{bmatrix}$ is a symmetric matrix.

EXAMPLE 4. $B = \begin{bmatrix} 0 & 2 & 3 \\ -2 & 0 & -4 \\ -3 & 4 & 0 \end{bmatrix}$ is a skew-symmetric matrix.

We can make a few observations about symmetric and skew-symmetric matrices; the proofs of most of these statements will be left as exercises.

It follows from the above definitions that if A is symmetric or skew symmetric, then A is a square matrix. If A is a symmetric matrix, then the entries of A are symmetric with respect to the main diagonal of A. Also, A is symmetric if and only if $a_{ij} = a_{ji}$ and A is skew symmetric if and only if $a_{ij} = -a_{ji}$. Moreover, if A is skew symmetric, then the entries on the main diagonal of A are all zero. An important property of symmetric and skew-symmetric matrices is the following.

Theorem 1.13. *If A is an $n \times n$ matrix, then $A = S + K$, where S is symmetric and K is skew symmetric. Moreover, this decomposition is unique.*

Proof: Assume that there is such a decomposition $A = S + K$, where S is symmetric and K is skew symmetric. We shall determine S and K. Now $A' = S' + K' = S - K$. Thus we have the expressions

$$A = S + K$$
$$A' = S - K.$$

Adding these two expressions, we obtain $A + A' = 2S$, so

$$S = \tfrac{1}{2}(A + A').$$

Subtracting instead of adding leads to

$$K = \tfrac{1}{2}(A - A').$$

It is easy to verify that $A = S + K$, that S is symmetric, and that K is skew symmetric. Thus we have shown that such a representation is possible and that the expressions for S and K are unique.

EXAMPLE 5. Let $A = \begin{bmatrix} 1 & 3 & -2 \\ 4 & 6 & 2 \\ 5 & 1 & 3 \end{bmatrix}$. Then

$$S = \tfrac{1}{2}(A + A') = \begin{bmatrix} 1 & \tfrac{7}{2} & \tfrac{3}{2} \\ \tfrac{7}{2} & 6 & \tfrac{3}{2} \\ \tfrac{3}{2} & \tfrac{3}{2} & 3 \end{bmatrix},$$

$$K = \tfrac{1}{2}(A - A') = \begin{bmatrix} 0 & -\tfrac{1}{2} & -\tfrac{7}{2} \\ \tfrac{1}{2} & 0 & \tfrac{1}{2} \\ \tfrac{7}{2} & -\tfrac{1}{2} & 0 \end{bmatrix},$$

and $A = S + K$.

If we start out with an $m \times n$ matrix $A = [a_{ij}]$ and cross out some, but not all, of its rows or columns, we obtain a **submatrix** of A.

EXAMPLE 6. Let $A = \begin{bmatrix} 1 & 2 & 3 & 4 \\ -2 & 4 & -3 & 5 \\ 3 & 0 & 5 & -3 \end{bmatrix}$. If we cross out the second row and third column, we obtain the submatrix $\begin{bmatrix} 1 & 2 & 4 \\ 3 & 0 & -3 \end{bmatrix}$.

We may now consider a matrix A as partitioned into submatrices. Of course, the partitioning may be done in many different ways.

EXAMPLE 7. The matrix

$$
A = \begin{bmatrix}
a_{11} & a_{12} & a_{13} & a_{14} & a_{15} \\
a_{21} & a_{22} & a_{23} & a_{24} & a_{25} \\
a_{31} & a_{32} & a_{33} & a_{34} & a_{35} \\
a_{41} & a_{42} & a_{43} & a_{44} & a_{45}
\end{bmatrix}
$$

is partitioned as

$$
A = \begin{bmatrix}
A_{11} & A_{12} \\
A_{21} & A_{22}
\end{bmatrix}.
$$

We could also write

$$
A = \begin{bmatrix}
a_{11} & a_{12} & a_{13} & a_{14} & a_{15} \\
a_{21} & a_{22} & a_{23} & a_{24} & a_{25} \\
a_{31} & a_{32} & a_{33} & a_{34} & a_{35} \\
a_{41} & a_{42} & a_{43} & a_{44} & a_{45}
\end{bmatrix} = \begin{bmatrix}
A_{11} & A_{12} & A_{13} \\
A_{21} & A_{22} & A_{23}
\end{bmatrix},
$$

which gives another partitioning of A. We will thus speak of **partitioned matrices**.

EXAMPLE 8. The augmented matrix (defined in Section 1.2) of a linear system is a partitioned matrix. Thus, if $AX = B$, we can write the augmented matrix of this system as $[A \mid B]$.

If A is partitioned as shown last and

$$
B = \begin{bmatrix}
b_{11} & b_{12} & b_{13} & b_{14} \\
b_{21} & b_{22} & b_{23} & b_{24} \\
b_{31} & b_{32} & b_{33} & b_{34} \\
b_{41} & b_{42} & b_{43} & b_{44} \\
b_{51} & b_{52} & b_{53} & b_{54}
\end{bmatrix} = \begin{bmatrix}
B_{11} & B_{12} \\
B_{21} & B_{22} \\
B_{31} & B_{32}
\end{bmatrix},
$$

then by straightforward computation we can show that

$$AB = \begin{bmatrix} (A_{11}B_{11} + A_{12}B_{21} + A_{13}B_{31}) & (A_{11}B_{12} + A_{12}B_{22} + A_{13}B_{32}) \\ \hline (A_{21}B_{11} + A_{22}B_{21} + A_{23}B_{31}) & (A_{21}B_{12} + A_{22}B_{22} + A_{23}B_{32}) \end{bmatrix}.$$

EXAMPLE 9. Let

$$A = \begin{bmatrix} 1 & 0 & 1 & 0 \\ 0 & 2 & 3 & -1 \\ \hline 2 & 0 & -4 & 0 \\ 0 & 1 & 0 & 3 \end{bmatrix} = \begin{bmatrix} A_{11} & A_{12} \\ A_{21} & A_{22} \end{bmatrix};$$

let

$$B = \begin{bmatrix} 2 & 0 & 0 & 1 & 1 & -1 \\ 0 & 1 & 1 & -1 & 2 & 2 \\ \hline 1 & 3 & 0 & 0 & 1 & 0 \\ -3 & -1 & 2 & 1 & 0 & -1 \end{bmatrix} = \begin{bmatrix} B_{11} & B_{12} \\ B_{21} & B_{22} \end{bmatrix}.$$

Then

$$AB = C = \begin{bmatrix} 3 & 3 & 0 & 1 & 2 & -1 \\ 6 & 12 & 0 & -3 & 7 & 5 \\ \hline 0 & -12 & 0 & 2 & -2 & -2 \\ -9 & -2 & 7 & 2 & 2 & -1 \end{bmatrix} = \begin{bmatrix} C_{11} & C_{12} \\ C_{21} & C_{22} \end{bmatrix},$$

where C_{11} should be $A_{11}B_{11} + A_{12}B_{21}$. Verifying that C_{11} is this expression, we have

$$A_{11}B_{11} + A_{12}B_{21} = \begin{bmatrix} 1 & 0 \\ 0 & 2 \end{bmatrix}\begin{bmatrix} 2 & 0 & 0 \\ 0 & 1 & 1 \end{bmatrix} + \begin{bmatrix} 1 & 0 \\ 3 & -1 \end{bmatrix}\begin{bmatrix} 1 & 3 & 0 \\ -3 & -1 & 2 \end{bmatrix}$$

$$= \begin{bmatrix} 2 & 0 & 0 \\ 0 & 2 & 2 \end{bmatrix} + \begin{bmatrix} 1 & 3 & 0 \\ 6 & 10 & -2 \end{bmatrix}$$

$$= \begin{bmatrix} 3 & 3 & 0 \\ 6 & 12 & 0 \end{bmatrix} = C_{11}.$$

Partitioned matrices can be used to great advantage in dealing with matrices that exceed the memory capacity of a computer. Thus, in multiplying two partitioned matrices, one can keep the matrices on disk and only bring into memory the submatrices required to form the submatrix products. The latter, of course, can be put out on disk as they are formed. The partitioning must be done so that the products of corresponding submatrices are defined. The addition of partitioned matrices is performed in the obvious manner.

Nonsingular Matrices

We now come to a special type of square matrix that will play a very important role throughout most of this book.

Definition 1.9. An $n \times n$ matrix A is called **nonsingular**, or **invertible**, if there exists an $n \times n$ matrix B such that $AB = BA = I_n$. Otherwise, A is called **singular**, or **noninvertible**; B is called an **inverse** of A.

EXAMPLE 10. Let $A = \begin{bmatrix} 2 & 3 \\ 2 & 2 \end{bmatrix}$ and let $B = \begin{bmatrix} -1 & \frac{3}{2} \\ 1 & -1 \end{bmatrix}$. Since $AB =$

$BA = I_2$, we conclude that B is an inverse of A.

Theorem 1.14. *The inverse of a matrix, if it exists, is unique.*

Proof: Let B and C be inverses of A. Then $AB = BA = I_n$ and $AC = CA = I_n$. We now have $B = BI_n = B(AC) = (BA)C = I_nC = C$, which proves that the inverse of a matrix, if it exists, is unique.

We now write the inverse of a nonsingular matrix A, as A^{-1}. Thus $AA^{-1} = A^{-1}A = I_n$. The next example shows that a matrix may fail to have an inverse.

EXAMPLE 11. Let $A = \begin{bmatrix} 1 & 2 \\ 2 & 4 \end{bmatrix}$. To find A^{-1}, we need $B = \begin{bmatrix} a & b \\ c & d \end{bmatrix}$ such that $AB = BA = \begin{bmatrix} 1 & 0 \\ 0 & 1 \end{bmatrix}$; that is, such that $\begin{bmatrix} 1 & 2 \\ 2 & 4 \end{bmatrix}\begin{bmatrix} a & b \\ c & d \end{bmatrix} = \begin{bmatrix} 1 & 0 \\ 0 & 1 \end{bmatrix}$. Then

$$
\begin{aligned}
a + 2c &= 1 \qquad \text{and} \qquad b + 2d = 0 \\
2a + 4c &= 0 \qquad\qquad\qquad\quad 2b + 4d = 1.
\end{aligned}
$$

These linear systems have no solutions, and so A has no inverse.

EXAMPLE 12. Let $A = \begin{bmatrix} 1 & 2 \\ 3 & 4 \end{bmatrix}$ and $A^{-1} = \begin{bmatrix} a & b \\ c & d \end{bmatrix}$; then we must have

$$\begin{bmatrix} 1 & 2 \\ 3 & 4 \end{bmatrix} \begin{bmatrix} a & b \\ c & d \end{bmatrix} = \begin{bmatrix} a & b \\ c & d \end{bmatrix} \begin{bmatrix} 1 & 2 \\ 3 & 4 \end{bmatrix} = \begin{bmatrix} 1 & 0 \\ 0 & 1 \end{bmatrix},$$

so that

$$\begin{array}{ccc} a + 2c = 1 & & b + 2d = 0 \\ 3a + 4c = 0 & \text{and} & 3b + 4d = 1. \end{array}$$

Solving these two linear systems we find that $A^{-1} = \begin{bmatrix} -2 & 1 \\ \frac{3}{2} & -\frac{1}{2} \end{bmatrix}$.

We next establish several properties of inverses of matrices.

Theorem 1.15. *If A and B are both nonsingular $n \times n$ matrices, then AB is nonsingular and $(AB)^{-1} = B^{-1}A^{-1}$.*

Proof: We have $(AB)(B^{-1}A^{-1}) = A(BB^{-1})A^{-1} = (AI_n)A^{-1} = AA^{-1} = I_n$. Similarly, $(B^{-1}A^{-1})(AB) = I_n$. Therefore, AB is nonsingular. Since the inverse of a matrix is unique, we conclude that $(AB)^{-1} = B^{-1}A^{-1}$.

Corollary 1.1. *If $A_1, A_2, \ldots, A_r$ are $n \times n$ nonsingular matrices, then $A_1A_2 \cdots A_r$ is nonsingular and $(A_1A_2 \cdots A_r)^{-1} = A_r^{-1}A_{r-1}^{-1} \cdots A_1^{-1}$.*

Proof: Exercise.

Theorem 1.16. *If A is a nonsingular matrix, then A^{-1} is nonsingular and $(A^{-1})^{-1} = A$.*

Proof: Exercise.

Theorem 1.17. *If A is a nonsingular matrix, then A' is nonsingular and $(A')^{-1} = (A^{-1})'$.*

Proof: We have $AA^{-1} = I_n$. Taking transposes of both sides, we obtain $(A^{-1})'A' = I_n' = I_n$. Taking transposes of both sides of the equation $A^{-1}A = I_n$, we find, similarly, that $(A')(A^{-1})' = I_n$. These equations imply that $(A^{-1})' = (A')^{-1}$.

It follows from Theorem 1.17 that if A is a symmetric nonsingular matrix, then A^{-1} is symmetric (see Exercise 14).

We have now seen that for some matrices A, we can find a matrix A^{-1}, called the inverse of A, such that $AA^{-1} = A^{-1}A = I_n$. The inverse of a matrix corresponds to the notion of reciprocal of a number. We shall soon develop efficient methods for finding A^{-1}, when it exists. At present we can find inverses by the brute-force method used in Example 12. In the meantime, note that if $AX = B$ is a system of n linear equations in n unknowns (A is $n \times n$), and if A has an inverse, then $A^{-1}(AX) = (A^{-1}A)X = A^{-1}B$ or $I_nX = X = A^{-1}B$. Moreover, $X = A^{-1}B$ is clearly a solution. Thus, if we have A^{-1}, we get a unique solution X. This observation is useful in industrial problems. Many physical models are described by linear systems. This means that if n values are used as inputs (which can be arranged as the $n \times 1$ matrix X), then m values are obtained as outputs (which can be arranged as the $m \times 1$ matrix B) by the rule $AX = B$. The matrix A is inherently tied to the process. Thus suppose that a chemical process has a certain matrix A associated with it. Any change in the process may result in a new matrix. The problem frequently encountered in systems analysis is that of determining the input to be used to obtain a desired output. That is, we want to solve the linear system $AX = B$ for X as we vary B. If A is a nonsingular square matrix, an efficient way of handling this is as follows. Compute A^{-1} once; then whenever we change B we find the corresponding solution X by forming $A^{-1}B$.

1.4. Exercises

1. (a) Show that if A is any $m \times n$ matrix, then $I_mA = A$ and $AI_n = A$.
   (b) Show that if A is an $n \times n$ scalar matrix, then $A = rI_n$ for some real number r.
2. Prove that the sum, product, and scalar multiple of diagonal, scalar, and upper (lower) triangular matrices is diagonal, scalar, upper (lower) triangular, respectively.
3. (a) Show that A is symmetric if and only if $a_{ij} = a_{ji}$ for all i, j.
   (b) Show that A is skew symmetric if and only if $a_{ij} = -a_{ji}$ for all i, j.
   (c) Show that if A is skew symmetric, then the elements on the main diagonal of A are all zero.
4. Show that if A is a symmetric matrix, then A' is symmetric.
5. Show that if A is any $n \times n$ matrix, then
   (a) AA' and $A'A$ are symmetric.
   (b) $A + A'$ is symmetric.
   (c) $A - A'$ is skew symmetric.

6. Let A and B be symmetric matrices.
   (a) Show that $A + B$ is symmetric.
   (b) Show that AB is symmetric if and only if $AB = BA$.

7. Write the matrix $A = \begin{bmatrix} 3 & -2 & 1 \\ 5 & 2 & 3 \\ -1 & 6 & 2 \end{bmatrix}$ as a sum of a symmetric and a skew-symmetric matrix.

8. Show that if $AB = AC$ and A is nonsingular, then $B = C$.

9. Find the inverse of $A = \begin{bmatrix} 1 & 3 \\ 5 & 2 \end{bmatrix}$; that is, find a 2×2 matrix B such that $AB = BA = I_2$.

10. Formulate the method for adding partitioned matrices and verify your method by partitioning the matrices

$$A = \begin{bmatrix} 1 & 3 & -1 \\ 2 & 1 & 0 \\ 2 & -3 & 1 \end{bmatrix} \quad \text{and} \quad B = \begin{bmatrix} 3 & 2 & 1 \\ -2 & 3 & 1 \\ 4 & 1 & 5 \end{bmatrix}$$

in two different ways and finding their sum.

11. Show that if A is nonsingular and $AB = O_n$ for an $n \times n$ matrix B, then $B = O_n$.

12. Let $A = \begin{bmatrix} a & b \\ c & d \end{bmatrix}$. Show that A is nonsingular if and only if $ad - bc \neq 0$.

13. Consider the linear system $AX = B$, where A is the matrix defined in Exercise 9.
    (a) Find a solution if $B = \begin{bmatrix} 3 \\ 4 \end{bmatrix}$.
    (b) Find a solution if $B = \begin{bmatrix} 5 \\ 6 \end{bmatrix}$.

14. Prove that if A is symmetric and nonsingular, then A^{-1} is symmetric.

15. Consider the homogeneous system $AX = O$, where A is $n \times n$. If A is nonsingular, show that the only solution is the trivial one, $X = O$.

16. Prove that if one row (column) of the $n \times n$ matrix A consists entirely of zeros, then A is singular. (*Hint:* Assume that A is nonsingular; that is, there exists an $n \times n$ matrix B such that $AB = BA = I_n$. Establish a contradiction.)

17. Let A and B be the following matrices:

$$A = \begin{bmatrix} 2 & 1 & 3 & 4 & 2 \\ 1 & 2 & 3 & -1 & 4 \\ 2 & 3 & 2 & 1 & 4 \\ 5 & -1 & 3 & 2 & 6 \\ 3 & 1 & 2 & 4 & 6 \\ 2 & -1 & 3 & 5 & 7 \end{bmatrix} \quad \text{and} \quad B = \begin{bmatrix} 1 & 2 & 3 & 4 & 1 \\ 2 & 1 & 3 & 2 & -1 \\ 1 & 5 & 4 & 2 & 3 \\ 2 & 1 & 3 & 5 & 7 \\ 3 & 2 & 4 & 6 & 1 \end{bmatrix}.$$

Find AB by partitioning A and B in two different ways.

18. Prove Corollary 1.1.

19. Prove Theorem 1.16.

20. Show that the matrix $A = \begin{bmatrix} 2 & 3 \\ 4 & 6 \end{bmatrix}$ is singular.

21. Let

$$A = \begin{bmatrix} 3 & 2 & -1 \\ 0 & -4 & 3 \\ 0 & 0 & 0 \end{bmatrix} \quad \text{and} \quad B = \begin{bmatrix} 6 & -3 & 2 \\ 0 & 2 & 4 \\ 0 & 0 & 3 \end{bmatrix}.$$

Verify that $A + B$ and AB are upper triangular.

22. Find two 2×2 singular matrices whose sum is nonsingular.

23. Find two 2×2 nonsingular matrices whose sum is singular.

24. If A is a nonsingular matrix whose inverse is $\begin{bmatrix} 2 & 1 \\ 4 & 1 \end{bmatrix}$, find A.

1.5. Echelon Form of a Matrix

In this section we take the elimination method for solving linear systems, learned in high school, and systematize it by introducing the language of matrices. This will result in two methods for solving a system of m linear equations in n unknowns. These methods take the augmented matrix of the linear system, perform certain operations on it, and obtain a new matrix that represents an equivalent linear system (that is, has the same solutions as the original linear system). The important point here is that the latter linear system can be solved very easily. We begin with the following definition.

Definition 1.10. An $m \times n$ matrix A is said to be in **row echelon form** if:
  (a) There exists an integer k, $1 \le k \le m$, such that the elements in rows $k + 1, k + 2, \ldots, m$ are all zero and none of the first k rows consist entirely of zeros.
  (b) If we count from left to right, the first nonzero element in row i is 1, $i = 1, 2, \ldots, k$.
  (c) If the 1 in row i occurs in column c_i, then $c_1 < c_2 < \cdots < c_k$.
  (d) All elements of column c_i appearing in rows $i + 1, i + 2, \ldots, m$, are zero.

If in addition to the above requirements, all the elements of column c_i in any row other than row i are zero, then A is said to be in **reduced row echelon form**. A similar definition can be formulated in the obvious manner for **column echelon form** and **reduced column echelon form**. We might also note that (d) follows from (a), (b), and (c).

In Definition 1.10, if $k = m$, there are no rows all of whose elements are zero.

EXAMPLE 1. The following are matrices in row echelon form:

$$A = \begin{bmatrix} 1 & 5 & 0 & 2 & -2 & 4 \\ 0 & 1 & 0 & 3 & 4 & 8 \\ 0 & 0 & 0 & 1 & 7 & -2 \\ 0 & 0 & 0 & 0 & 0 & 0 \\ 0 & 0 & 0 & 0 & 0 & 0 \end{bmatrix}, \quad B = \begin{bmatrix} 1 & 0 & 0 & 0 \\ 0 & 1 & 0 & 0 \\ 0 & 0 & 1 & 0 \\ 0 & 0 & 0 & 1 \end{bmatrix},$$

and

$$C = \begin{bmatrix} 0 & 0 & 1 & 3 & 5 & 7 & 9 \\ 0 & 0 & 0 & 0 & 1 & -2 & 3 \\ 0 & 0 & 0 & 0 & 0 & 1 & 2 \\ 0 & 0 & 0 & 0 & 0 & 0 & 1 \\ 0 & 0 & 0 & 0 & 0 & 0 & 0 \end{bmatrix}.$$

For A, B, and C the integer k is 3, 4, and 4, respectively; for C, $c_1 = 3$, $c_2 = 5$, $c_3 = 6$, and $c_4 = 7$.

EXAMPLE 2. The following are matrices in reduced row echelon form

$$B = \begin{bmatrix} 1 & 0 & 0 & 0 \\ 0 & 1 & 0 & 0 \\ 0 & 0 & 1 & 0 \\ 0 & 0 & 0 & 1 \end{bmatrix}, \quad D = \begin{bmatrix} 1 & 0 & 0 & 0 & -2 & 4 \\ 0 & 1 & 0 & 0 & 4 & 8 \\ 0 & 0 & 0 & 1 & 7 & -2 \\ 0 & 0 & 0 & 0 & 0 & 0 \\ 0 & 0 & 0 & 0 & 0 & 0 \end{bmatrix},$$

and

$$E = \begin{bmatrix} 1 & 2 & 0 & 0 & 1 \\ 0 & 0 & 1 & 2 & 3 \\ 0 & 0 & 0 & 0 & 0 \end{bmatrix}.$$

We shall now show that every matrix can be put into row (column) echelon form, or into reduced row (column) echelon form, by means of certain row or column operations.

Definition 1.11. An **elementary row (column) operation** on a matrix A is any one of the following operations:

(a) Interchange rows (columns) i and j of A. This is called a **type I operation**.

(b) Multiply row (column) i by $c \neq 0$. This is called a **type II operation**.

(c) Add c times row (column) i to row (column) j, $i \neq j$. This is called a **type III operation**.

Definition 1.12. An $m \times n$ matrix A is said to be **row (column) equivalent** to an $m \times n$ matrix B if B results from A by a finite sequence of elementary row (column) operations.

EXAMPLE 3. The matrix

$$A = \begin{bmatrix} 1 & 2 & 4 & 3 \\ 2 & 1 & 3 & 2 \\ 1 & -1 & 2 & 3 \end{bmatrix}$$

is row equivalent to

$$D = \begin{bmatrix} 2 & 4 & 8 & 6 \\ 1 & -1 & 2 & 3 \\ 4 & -1 & 7 & 8 \end{bmatrix},$$

because if we add twice row 3 of A to its second row, we obtain

$$B = \begin{bmatrix} 1 & 2 & 4 & 3 \\ 4 & -1 & 7 & 8 \\ 1 & -1 & 2 & 3 \end{bmatrix}.$$

Interchanging rows 2 and 3 of B, we obtain

$$C = \begin{bmatrix} 1 & 2 & 4 & 3 \\ 1 & -1 & 2 & 3 \\ 4 & -1 & 7 & 8 \end{bmatrix}.$$

Multiplying row 1 of C by 2, we obtain D.

We can easily show (see Exercise 1) that (a) every matrix is row equivalent to itself; (b) if A is row equivalent to B, then B is row equivalent to A; and

(c) if A is row equivalent to B and B is row equivalent to C, then A is row equivalent to C. In view of (b), both statements "A is row equivalent to B" and "B is row equivalent to A" can be replaced by "A and B are row equivalent." A similar statement holds for column equivalence.

Theorem 1.18. *Every nonzero $m \times n$ matrix $A = [a_{ij}]$ is row (column) equivalent to a matrix in row (column) echelon form.*

Proof: We shall prove that A is row equivalent to a matrix in row echelon form, that is, that by using only elementary row operations we can transform A into a matrix in row echelon form. A completely analogous proof using elementary column operations establishes the result for column equivalence. We look in matrix A for the first column with a nonzero entry; say this is column j and say that this first nonzero entry in column j occurs in row i. Now interchange, if necessary, rows 1 and i, getting matrix $B = [b_{ij}]$. Thus $b_{1j} \neq 0$. Divide all entries in row 1 of B by b_{1j}, obtaining $C = [c_{ij}]$. Note that $c_{1j} = 1$. Now if c_{hj}, $2 \leq h \leq m$, is not zero, then to row h of C we add $-c_{hj}$ times row 1; all elements in column j, rows 2, 3, ..., m are zero. Denote the resulting matrix by D. Note that we have used only elementary row operations. Next, consider the $(m - 1) \times n$ submatrix A_1 of D obtained by deleting the first row of D. We now repeat the above procedure with matrix A_1 instead of matrix A. Continuing this way, we obtain a matrix in row echelon form which is row equivalent to A.

EXAMPLE 4. Let

$$A = \begin{bmatrix} 0 & 2 & 3 & -4 & 1 \\ 0 & 0 & 2 & 3 & 4 \\ 2 & 2 & -5 & 2 & 4 \\ 2 & 0 & -6 & 9 & 7 \end{bmatrix}.$$

Column 1 is the first (counting from left to right) column in A with a nonzero entry. The first (counting from top to bottom) nonzero entry in the first column occurs in the third row. We interchange the first and third rows of A, obtaining

$$B = \begin{bmatrix} 2 & 2 & -5 & 2 & 4 \\ 0 & 0 & 2 & 3 & 4 \\ 0 & 2 & 3 & -4 & 1 \\ 2 & 0 & -6 & 9 & 7 \end{bmatrix}.$$

Divide the first row of B by $b_{11} = 2$, to obtain

$$C = \begin{bmatrix} 1 & 1 & -\frac{5}{2} & 1 & 2 \\ 0 & 0 & 2 & 3 & 4 \\ 0 & 2 & 3 & -4 & 1 \\ 2 & 0 & -6 & 9 & 7 \end{bmatrix}.$$

Add -2 times the first row of C to the fourth row of C to produce a matrix D in which the only nonzero entry in the first column is $d_{11} = 1$:

$$D = \begin{bmatrix} 1 & 1 & -\frac{5}{2} & 1 & 2 \\ 0 & 0 & 2 & 3 & 4 \\ 0 & 2 & 3 & -4 & 1 \\ 0 & -2 & -1 & 7 & 3 \end{bmatrix}.$$

Identify A_1 as the submatrix of D obtained by deleting the first row of D; do not erase the first row of D. Repeat the above steps with A_1 instead of A.

$$A_1 = \begin{matrix} 1 & 1 & -\frac{5}{2} & 1 & 2 \\ \begin{bmatrix} 0 & 0 & 2 & 3 & 4 \\ 0 & 2 & 3 & -4 & 1 \\ 0 & -2 & -1 & 7 & 3 \end{bmatrix} \end{matrix}$$

Interchange the first and second rows of A_1 to obtain

$$B_1 = \begin{matrix} 1 & 1 & -\frac{5}{2} & 1 & 2 \\ \begin{bmatrix} 0 & 2 & 3 & -4 & 1 \\ 0 & 0 & 2 & 3 & 4 \\ 0 & -2 & -1 & 7 & 3 \end{bmatrix} \end{matrix}$$

Divide the first row of B_1 by 2 to obtain

$$C_1 = \begin{matrix} 1 & 1 & -\frac{5}{2} & 1 & 2 \\ \begin{bmatrix} 0 & 1 & \frac{3}{2} & -2 & \frac{1}{2} \\ 0 & 0 & 2 & 3 & 4 \\ 0 & -2 & -1 & 7 & 3 \end{bmatrix} \end{matrix}$$

Add two times the first row of C_1 to its third row to obtain

$$D_1 = \begin{matrix} 1 & 1 & -\frac{5}{2} & 1 & 2 \\ \begin{bmatrix} 0 & 1 & \frac{3}{2} & -2 & \frac{1}{2} \\ 0 & 0 & 2 & 3 & 4 \\ 0 & 0 & 2 & 3 & 4 \end{bmatrix} \end{matrix}$$

Deleting the first row of D_1 yields the matrix A_2. We repeat the above procedure with A_2 instead of A. No rows of A_2 have to be interchanged.

$$A_2 = \begin{bmatrix} 1 & 1 & -\frac{5}{2} & 1 & 2 \\ 0 & 1 & \frac{3}{2} & -2 & \frac{1}{2} \\ 0 & 0 & 2 & 3 & 4 \\ 0 & 0 & 2 & 3 & 4 \end{bmatrix} = B_2$$

Divide the first row of B_2 by 2 to obtain

$$C_2 = \begin{bmatrix} 1 & 1 & -\frac{5}{2} & 1 & 2 \\ 0 & 1 & \frac{3}{2} & -2 & \frac{1}{2} \\ 0 & 0 & 1 & \frac{3}{2} & 2 \\ 0 & 0 & 2 & 3 & 4 \end{bmatrix}$$

Finally, add -2 times the first row of C_2 to its second row to obtain

$$D_2 = \begin{bmatrix} 1 & 1 & -\frac{5}{2} & 1 & 2 \\ 0 & 1 & \frac{3}{2} & -2 & \frac{1}{2} \\ 0 & 0 & 1 & \frac{3}{2} & 2 \\ 0 & 0 & 0 & 0 & 0 \end{bmatrix}$$

The matrix

$$H = \begin{bmatrix} 1 & 1 & -\frac{5}{2} & 1 & 2 \\ 0 & 1 & \frac{3}{2} & -2 & \frac{1}{2} \\ 0 & 0 & 1 & \frac{3}{2} & 2 \\ 0 & 0 & 0 & 0 & 0 \end{bmatrix}$$

is in row echelon form and is row equivalent to A.

Theorem 1.19. *Every nonzero $m \times n$ matrix $A = [a_{ij}]$ is row (column) equivalent to a matrix in reduced row (column) echelon form.*

Proof: We proceed as in Theorem 1.18, obtaining a matrix H in row echelon form which is row equivalent to A. In H, if row i contains a nonzero element, then its first (counting from left to right) nonzero element is 1. Suppose that it occurs in column c_i. Then $c_1 < c_2 < \cdots < c_k$, where k ($1 \leq k \leq m$) is the integer discussed in Definition 1.10 of row echelon form of a matrix. Now subtract suitable multiples of row i of H to make all elements in column c_i and rows $i - 1, i - 2, \ldots, 1$ of H equal to zero. The result is a matrix K in reduced row echelon form which has been obtained from H by elementary row operations, and is thus row equivalent to H. Since A is row equivalent to H and H is row equivalent to K, then A is row equivalent to K. An analogous proof can be given to show that A is column equivalent to a matrix in reduced column echelon form.

It can be shown, with some difficulty, that there is only one matrix in reduced row echelon form that is row equivalent to a given matrix. We omit the proof.

EXAMPLE 5. Suppose that we wish to find a matrix that is row equivalent to the matrix A of Example 4. Starting with the matrix H obtained there, we add -1 times the second row to the first row, obtaining

$$\begin{bmatrix} 1 & 0 & -4 & 3 & \frac{3}{2} \\ 0 & 1 & \frac{3}{2} & -2 & \frac{1}{2} \\ 0 & 0 & 1 & \frac{3}{2} & 2 \\ 0 & 0 & 0 & 0 & 0 \end{bmatrix}.$$

In this matrix we add $-\frac{3}{2}$ times the third row to its second row and 4 times the third row to its first row. This yields

$$K = \begin{bmatrix} 1 & 0 & 0 & 9 & \frac{19}{2} \\ 0 & 1 & 0 & -\frac{17}{4} & -\frac{5}{2} \\ 0 & 0 & 1 & \frac{3}{2} & 2 \\ 0 & 0 & 0 & 0 & 0 \end{bmatrix},$$

which is in reduced row echelon form and is row equivalent to A.

We now apply these results to the solution of linear systems.

Theorem 1.20. *Let* $AX = B$ *and* $CX = D$ *be two linear systems each of* m *equations in* n *unknowns. If the augmented matrices* $[A \mid B]$ *and* $[C \mid D]$ *are row equivalent, then the linear systems are equivalent; that is, they have exactly the same solutions.*

Proof: This follows from the definition of row equivalence and from the fact that the three elementary row operations on the augmented matrix are the three manipulations on linear systems, discussed in Section 1.1, which yield equivalent linear systems. We also note that if one system has no solution, then the other system has no solution.

Corollary 1.2. *If A and B are row equivalent m × n matrices, then the homogeneous systems AX = O and BX = O are equivalent.*

Proof: Exercise.

We now pause to observe that we have developed the essential features of two very straightforward methods for solving linear systems. The idea consists of starting with the linear system $AX = B$, then obtaining a partitioned matrix $[C \mid D]$ in either row echelon form or reduced row echelon form that is row equivalent to the augmented matrix $[A \mid B]$. Now $[C \mid D]$ represents the linear system $CX = D$, which is quite simple to solve because of the structure of $[C \mid D]$, and the set of solutions to this system gives precisely the set of solutions to $AX = B$. The method where $[C \mid D]$ is in row echelon form is called **Gaussian elimination;** the method where $[C \mid D]$ is in reduced row echelon form is called **Gauss–Jordan reduction**. These methods are used often and computer codes of their implementations are widely available.

We thus consider $CX = D$, where C is $m \times n$, and $[C \mid D]$ is in row echelon form. Then, for example, $[C \mid D]$ is of the following form:

$$\left[\begin{array}{ccccccc|c}
1 & c_{12} & c_{13} & \cdots & & c_{1n} & & d_1 \\
0 & 0 & 1 & c_{24} & \cdots & c_{2n} & & d_2 \\
\vdots & & & & & \vdots & & \vdots \\
0 & 0 & \cdots & & 0 & 1 & c_{k-1\,n} & d_{k-1} \\
0 & \cdots & & & & 0 & 1 & d_k \\
0 & \cdots & & & & & 0 & d_{k+1} \\
\vdots & & & & & & \vdots & \vdots \\
0 & \cdots & & & & & 0 & d_m
\end{array}\right].$$

This augmented matrix represents the linear system

$$x_1 + c_{12}x_2 + c_{13}x_3 + \cdots \qquad\qquad + c_{1n}\ x_n = d_1$$
$$x_3 + c_{24}x_4 + \cdots \qquad + c_{2n}\ x_n = d_2$$
$$\vdots$$
$$x_{n-1} + c_{k-1\,n}x_n = d_{k-1}$$
$$x_n = d_k$$
$$0x_1 + \cdots \qquad\qquad + \quad 0\,x_n = d_{k+1}$$
$$\vdots$$
$$0x_1 + \cdots \qquad\qquad + \quad 0\,x_n = d_m.$$

First, if $d_{k+1} = 1$, then $CX = D$ has no solution, for at least one equation is not satisfied. If $d_{k+1} = 0$, which implies that $d_{k+2} = \cdots = d_m = 0$, we then obtain $x_n = d_k$, $x_{n-1} = d_{k-1} - c_{k-1}\,_n x_n = d_{k-1} - c_{k-1}\,_n d_k$, and continue using backward substitution to find the remaining unknowns. Of course, in the solution some of the unknowns may be expressed in terms of others that can take on any values whatever. This merely indicates that $CX = D$ has infinitely many solutions. On the other hand, every unknown may have a determined value, indicating that the solution is unique.

EXAMPLE 6. Let

$$[C \mid D] = \begin{bmatrix} 1 & 2 & 3 & 4 & 5 & \vdots & 6 \\ 0 & 1 & 2 & 3 & -1 & \vdots & 7 \\ 0 & 0 & 1 & 2 & 3 & \vdots & 7 \\ 0 & 0 & 0 & 1 & 2 & \vdots & 9 \end{bmatrix}.$$

Then

$$\begin{aligned} x_4 &= 9 - 2x_5 \\ x_3 &= 7 - 2x_4 - 3x_5 = 7 - 2(9 - 3x_5) - 3x_5 = -11 + x_5 \\ x_2 &= 7 - 2x_3 - 3x_4 + x_5 = 2 + 5x_5 \\ x_1 &= 6 - 2x_2 - 3x_3 - 4x_4 - 5x_5 = -1 - 10x_5 \\ x_5 &= \text{any real number.} \end{aligned}$$

Thus all solutions are of the form

$$\begin{aligned} x_1 &= -1 - 10r \\ x_2 &= 2 + 5r \\ x_3 &= -11 + r \\ x_4 &= 9 - 2r \\ x_5 &= r, \text{ any real number.} \end{aligned}$$

EXAMPLE 7. If

$$[C \mid D] = \begin{bmatrix} 1 & 2 & 3 & 4 & \vdots & 5 \\ 0 & 1 & 2 & 3 & \vdots & 6 \\ 0 & 0 & 0 & 0 & \vdots & 1 \end{bmatrix},$$

then $CX = D$ has no solution, for the last equation is $0x_1 + 0x_2 + 0x_3 + 0x_4 = 1$, which can never be satisfied.

EXAMPLE 8. If

$$[C \mid D] = \begin{bmatrix} 1 & 2 & 3 & 4 & \vdots & 5 \\ 0 & 1 & 2 & 3 & \vdots & 6 \\ 0 & 0 & 1 & 2 & \vdots & 7 \\ 0 & 0 & 0 & 1 & \vdots & 8 \end{bmatrix},$$

then

$$x_1 = 0$$
$$x_2 = 0$$
$$x_3 = -9$$
$$x_4 = 8.$$

The solution to $CX = D$ is unique.

If $[C \mid D]$ is in reduced row echelon form, then we can solve $CX = D$ without backward substitution, but, of course, it takes more effort to put a matrix in reduced row echelon form than in row echelon form.

EXAMPLE 9. If

$$[C \mid D] = \begin{bmatrix} 1 & 0 & 0 & 0 & \vdots & 5 \\ 0 & 1 & 0 & 0 & \vdots & 6 \\ 0 & 0 & 1 & 0 & \vdots & 7 \\ 0 & 0 & 0 & 1 & \vdots & 8 \end{bmatrix},$$

then

$$x_1 = 5$$
$$x_2 = 6$$
$$x_3 = 7$$
$$x_4 = 8.$$

EXAMPLE 10. If

$$[C \mid D] = \begin{bmatrix} 1 & 1 & 2 & 0 & -\frac{5}{2} & \vdots & \frac{2}{3} \\ 0 & 0 & 0 & 1 & \frac{1}{2} & \vdots & \frac{1}{2} \\ 0 & 0 & 0 & 0 & 0 & \vdots & 0 \end{bmatrix},$$

then

$$x_4 = \tfrac{1}{2} - \tfrac{1}{2}x_5$$
$$x_1 = \tfrac{2}{3} - x_2 - 2x_3 + \tfrac{5}{2}x_5,$$

where x_2, x_3, and x_5 can take on any real numbers. Thus a solution is of the form

$$
\begin{aligned}
x_1 &= \tfrac{2}{3} - r - 2s + \tfrac{5}{2}t \\
x_2 &= r \\
x_3 &= s \\
x_4 &= \tfrac{1}{2} - \tfrac{1}{2}t \\
x_5 &= t,
\end{aligned}
$$

where r, s, and t are any real numbers.

We now solve a linear system both by Gaussian elimination and by Gauss–Jordan reduction.

EXAMPLE 11. Consider the linear system

$$
\begin{aligned}
x_1 + 2x_2 + 3x_3 &= 6 \\
2x_1 - 3x_2 + 2x_3 &= 14 \\
3x_1 + x_2 - x_3 &= -2.
\end{aligned}
$$

We form the augmented matrix

$$
\left[\begin{array}{ccc:c}
1 & 2 & 3 & 6 \\
2 & -3 & 2 & 14 \\
3 & 1 & -1 & -2
\end{array}\right].
$$

Subtract twice the first row from the second row:

$$
\left[\begin{array}{ccc:c}
1 & 2 & 3 & 6 \\
0 & -7 & -4 & 2 \\
3 & 1 & -1 & -2
\end{array}\right].
$$

Subtract three times the first row from the third row:

$$
\left[\begin{array}{ccc:c}
1 & 2 & 3 & 6 \\
0 & -7 & -4 & 2 \\
0 & -5 & -10 & -20
\end{array}\right].
$$

Divide the third row by -5 and interchange the second and third rows:

$$
\left[\begin{array}{ccc:c}
1 & 2 & 3 & 6 \\
0 & 1 & 2 & 4 \\
0 & -7 & -4 & 2
\end{array}\right].
$$

Add seven times the second row to the third row:

$$\begin{bmatrix} 1 & 2 & 3 & \vdots & 6 \\ 0 & 1 & 2 & \vdots & 4 \\ 0 & 0 & 10 & \vdots & 30 \end{bmatrix}.$$

Divide the third row by 10:

$$\begin{bmatrix} 1 & 2 & 3 & \vdots & 6 \\ 0 & 1 & 2 & \vdots & 4 \\ 0 & 0 & 1 & \vdots & 3 \end{bmatrix}.$$

This matrix is in row echelon form. This means that $x_3 = 3$ and $x_2 + 2x_3 = 4$, so

$$x_2 = 4 - 2(3) = -2$$

$x_1 + 2x_2 + 3x_3 = 6$ implies that

$$x_1 = 6 - 2x_2 - 3x_3 = 6 - 2(-2) - 3(3) = 1.$$

Thus $x_1 = 1$, $x_2 = -2$, and $x_3 = 3$ is the solution. This gives the solution by Gaussian elimination.

If, instead, we wish to use Gauss–Jordan reduction, we would transform the last matrix into reduced row echelon form by the following steps:

Subtract twice the second row from the first row:

$$\begin{bmatrix} 1 & 0 & -1 & \vdots & -2 \\ 0 & 1 & 2 & \vdots & 4 \\ 0 & 0 & 1 & \vdots & 3 \end{bmatrix}.$$

Subtract twice the third row from the second row:

$$\begin{bmatrix} 1 & 0 & -1 & \vdots & -2 \\ 0 & 1 & 0 & \vdots & -2 \\ 0 & 0 & 1 & \vdots & 3 \end{bmatrix}.$$

Add the third row to the first row:

$$\begin{bmatrix} 1 & 0 & 0 & \vdots & 1 \\ 0 & 1 & 0 & \vdots & -2 \\ 0 & 0 & 1 & \vdots & 3 \end{bmatrix}.$$

The solution is $x_1 = 1$, $x_2 = -2$, and $x_3 = 3$, as before.

Now we consider a homogeneous system $AX = O$ of m linear equations in n unknowns.

EXAMPLE 12. Consider the homogeneous system whose augmented matrix is

$$\left[\begin{array}{ccccc:c} 1 & 0 & 0 & 0 & 2 & 0 \\ 0 & 0 & 1 & 0 & 3 & 0 \\ 0 & 0 & 0 & 1 & 4 & 0 \\ 0 & 0 & 0 & 0 & 0 & 0 \end{array}\right].$$

Since the augmented matrix is in reduced row echelon form, the solution is easily seen to be

$$x_1 = -2r$$
$$x_2 = s$$
$$x_3 = -3r$$
$$x_4 = -4r$$
$$x_5 = r,$$

where r and s are any real numbers.

In Example 12 we solved a homogeneous system of $m(=4)$ linear equations in $n(=5)$ unknowns, where $m < n$ and the augmented matrix A was in reduced row echelon form. We can ignore any row of the augmented matrix that consists entirely of zeros. Thus let rows $1, 2, \ldots, r$ of A be the nonzero rows, and let the 1 in row i occur in column c_i. We are then solving a homogeneous system of r equations in n unknowns, $r < n$, and in this special case (A is in reduced row echelon form) we can solve for $x_{c_1}, x_{c_2}, \ldots, x_{c_r}$ in terms of the remaining $n - r$ unknowns. Since the latter can take on any real values, there are infinitely many solutions to the system $AX = O$; in particular, there is a nontrivial solution. We now show that this situation holds whenever we have $m < n$; A does not have to be in reduced row echelon form.

Theorem 1.21. *If A is an $m \times n$ matrix with $m < n$, then the homogeneous system $AX = O$ has a nontrivial solution.*

Proof: Let B be a matrix in reduced row echelon form which is row equivalent to A. Then the homogeneous systems $AX = O$ and $BX = O$ are equivalent. If we let r be the number of nonzero rows of B, then $r \le m$. If

$m < n$, we conclude that $r < n$. We are then solving r equations in n unknowns and can solve for r unknowns in terms of the remaining $n - r$ unknowns, the latter being free to take on any values we please. Thus $BX = O$, and hence $AX = O$ has a nontrivial solution.

We shall soon use this result in the following equivalent form: If A is $m \times n$ and $AX = O$ has only the trivial solution, then $m \geq n$.

EXAMPLE 13. Consider the homogeneous system

$$\begin{aligned} x_1 + \ \ x_2 + x_3 + x_4 &= 0 \\ x_1 \qquad\qquad\ \ + x_4 &= 0 \\ x_1 + 2x_2 + x_3 \qquad &= 0. \end{aligned}$$

The augmented matrix

$$A = \begin{bmatrix} 1 & 1 & 1 & 1 & \vdots & 0 \\ 1 & 0 & 0 & 1 & \vdots & 0 \\ 1 & 2 & 1 & 0 & \vdots & 0 \end{bmatrix}$$

is row equivalent to

$$\begin{bmatrix} 1 & 0 & 0 & 1 & \vdots & 0 \\ 0 & 1 & 0 & -1 & \vdots & 0 \\ 0 & 0 & 1 & 1 & \vdots & 0 \end{bmatrix}.$$

Hence the solution is

$$\begin{aligned} x_1 &= -r \\ x_2 &= r \\ x_3 &= -r \\ x_4 &= r, \text{ any real number.} \end{aligned}$$

A useful property of matrices in reduced row echelon form (see Exercise 3) is that if A is an $n \times n$ matrix in reduced row echelon form $\neq I_n$, then A has a row consisting entirely of zeros.

1.5. Exercises

1. Prove the following:
   (a) Every matrix is row equivalent to itself.

(b) If A is row equivalent to B, then B is row equivalent to A.

(c) If A is row equivalent to B and B is row equivalent to C, then A is row equivalent to C.

2. Let

$$A = \begin{bmatrix} 0 & 0 & -1 & 2 & 3 \\ 0 & 2 & 3 & 4 & 5 \\ 0 & 1 & 3 & -1 & 2 \\ 0 & 3 & 2 & 4 & 1 \end{bmatrix}.$$

(a) Find a matrix B in row echelon form that is row equivalent to A.

(b) Find a matrix C in reduced row echelon form that is row equivalent to A.

3. Let A be an $n \times n$ matrix in reduced row echelon form. Prove that if $A \neq I_n$, then A has a row consisting entirely of zeros.

4. Let

$$A = \begin{bmatrix} 1 & -2 & 0 & 2 \\ 2 & -3 & -1 & 5 \\ 1 & 3 & 2 & 5 \\ 1 & 1 & 0 & 2 \end{bmatrix}.$$

(a) Find a matrix B in row echelon form that is row equivalent to A.

(b) Find a matrix C in reduced row echelon form that is row equivalent to A.

5. Consider the linear system

$$\begin{aligned} x_1 + x_2 + 2x_3 &= -1 \\ x_1 - 2x_2 + x_3 &= -5 \\ 3x_1 + x_2 + x_3 &= 3. \end{aligned}$$

(a) Find all solutions, if any exist, by using the Gaussian elimination method.

(b) Find all solutions, if any exist, by using the Gauss–Jordan reduction method.

6. Repeat Exercise 5 for each of the following linear systems.

(a) $\begin{aligned} x_1 + x_2 + 2x_3 + 3x_4 &= 13 \\ x_1 - 2x_2 + x_3 + x_4 &= 8 \\ 3x_1 + x_2 + x_3 - x_4 &= 1. \end{aligned}$ (b) $\begin{aligned} x_1 + x_2 + x_3 &= 1 \\ x_1 + x_2 - 2x_3 &= 3 \\ 2x_1 + x_2 + x_3 &= 2. \end{aligned}$

(c) $\begin{aligned} 2x_1 + x_2 + x_3 - 2x_4 &= 1 \\ 3x_1 - 2x_2 + x_3 - 6x_4 &= -2 \\ x_1 + x_2 - x_3 - x_4 &= -1 \\ 6x_1 \quad\quad + x_3 - 9x_4 &= -2 \\ 5x_1 - x_2 + 2x_3 - 8x_4 &= 3. \end{aligned}$

In Exercises 7, 8, and 9 solve the linear system, if it is consistent, with given *augmented* matrix.

7. (a) $\begin{bmatrix} 1 & 1 & 1 & 0 \\ 1 & 1 & 0 & 3 \\ 0 & 1 & 1 & 1 \end{bmatrix}.$ (b) $\begin{bmatrix} 1 & 2 & 3 & 0 \\ 1 & 1 & 1 & 0 \\ 1 & 1 & 2 & 0 \end{bmatrix}.$

(c) $\begin{bmatrix} 1 & 2 & 3 & 0 \\ 1 & 1 & 1 & 0 \\ 5 & 7 & 9 & 0 \end{bmatrix}$.   (d) $\begin{bmatrix} 1 & 2 & 3 & 0 \\ 1 & 2 & 1 & 0 \end{bmatrix}$.

8. (a) $\begin{bmatrix} 1 & 2 & 3 & 1 & 8 \\ 1 & 3 & 0 & 1 & 7 \\ 1 & 0 & 2 & 1 & 3 \end{bmatrix}$.   (b) $\begin{bmatrix} 1 & 1 & 3 & -3 & 0 \\ 0 & 2 & 1 & -3 & 3 \\ 1 & 0 & 2 & -1 & -1 \end{bmatrix}$.

9. (a) $\begin{bmatrix} 1 & 2 & 1 & 7 \\ 2 & 0 & 1 & 4 \\ 1 & 0 & 2 & 5 \\ 1 & 2 & 3 & 11 \\ 2 & 1 & 4 & 12 \end{bmatrix}$.   (b) $\begin{bmatrix} 1 & 2 & 1 & 0 \\ 2 & 3 & 0 & 0 \\ 0 & 1 & 2 & 0 \\ 2 & 1 & 4 & 0 \end{bmatrix}$.

10. Let $A = \begin{bmatrix} a & b \\ c & d \end{bmatrix}$ and $X = \begin{bmatrix} x_1 \\ x_2 \end{bmatrix}$. Show that the linear system $AX = O$ has only the trivial solution if and only if $ad - bc \neq 0$.

11. Show that $A = \begin{bmatrix} a & b \\ c & d \end{bmatrix}$ is row equivalent to I_2 if and only if $ad - bc \neq 0$.

12. In the following linear system determine all values of a for which the resulting linear system has:
    (1) No solution.
    (2) A unique solution.
    (3) Infinitely many solutions.

$$\begin{aligned} x_1 + x_2 - x_3 &= 2 \\ x_1 + 2x_2 + x_3 &= 3 \\ x_1 + x_2 + (a^2 - 5)x_3 &= a. \end{aligned}$$

13. Repeat Exercise 12 for the linear system

$$\begin{aligned} x_1 + x_2 + x_3 &= 2 \\ 2x_1 + 3x_2 + 2x_3 &= 5 \\ 2x_1 + 3x_2 + (a^2 - 1)x_3 &= a + 1. \end{aligned}$$

14. (a) Formulate the definitions of column echelon form and reduced column echelon form of a matrix.
    (b) Prove that every $m \times n$ matrix is column equivalent to a matrix in column echelon form.

15. Prove that every $m \times n$ matrix is column equivalent to a matrix in reduced column echelon form.

16. Repeat Exercise 12 for the linear system

$$\begin{aligned} x_1 + x_2 &= 3 \\ x_1 + (a^2 - 8)x_2 &= a. \end{aligned}$$

17. Let A be the matrix in Exercise 2.
    (a) Find a matrix in column echelon form that is column equivalent to A.
    (b) Find a matrix in reduced column echelon form that is column equivalent to A.

18. Repeat Exercise 12 for the linear system

$$\begin{aligned} x_1 + x_2 + \quad\quad x_3 &= 2 \\ x_1 + 2x_2 + \quad\quad x_3 &= 3 \\ x_1 + x_2 + (a^2 - 5)x_3 &= a. \end{aligned}$$

19. Repeat Exercise 17 for the matrix

$$\begin{bmatrix} 1 & 2 & 3 & 4 & 5 \\ 2 & 1 & 3 & -1 & 2 \\ 3 & 1 & 2 & 4 & 1 \end{bmatrix}.$$

20. Show that if the homogeneous system

$$\begin{aligned} (a - r)x_1 + \quad\quad dx_2 &= 0 \\ cx_1 + (b - r)x_2 &= 0 \end{aligned}$$

has a nontrivial solution, then r satisfies the equation $(a - r)(b - r) - cd = 0$.

1.6. Elementary Matrices

In this section we develop a method for finding the inverse of a matrix if it exists. This method is such that we do not have to first find out whether A^{-1} exists. We start to find A^{-1}; if in the course of the computation we hit a certain situation, then we know that A^{-1} does not exist. Otherwise, we proceed to the end and obtain A^{-1}. This method requires that elementary row operations of types I, II, and III be performed on A. We clarify these notions by starting with the following definition.

Definition 1.13. An $n \times n$ **elementary matrix of type I, type II, or type III** is a matrix obtained from the identity matrix I_n by performing an elementary row or elementary column operation of type I, type II, or type III.

EXAMPLE 1. The following are elementary matrices:

$$E_1 = \begin{bmatrix} 0 & 0 & 1 \\ 0 & 1 & 0 \\ 1 & 0 & 0 \end{bmatrix}, \quad E_2 = \begin{bmatrix} 1 & 0 & 0 \\ 0 & -2 & 0 \\ 0 & 0 & 1 \end{bmatrix},$$

$$E_3 = \begin{bmatrix} 1 & 2 & 0 \\ 0 & 1 & 0 \\ 0 & 0 & 1 \end{bmatrix}, \quad \text{and} \quad E_4 = \begin{bmatrix} 1 & 0 & 3 \\ 0 & 1 & 0 \\ 0 & 0 & 1 \end{bmatrix}.$$

Matrix E_1 is of type I—we interchanged the first and third rows of I_3; E_2 is of type II—we multiplied the second row of I_3 by -2; E_3 is of type III—we added twice the second row of I_3 to the first row of I_3; and E_4 is of type III—we added three times the first column of I_3 to the third column of I_3.

Theorem 1.22. *Let A be an $m \times n$ matrix, and let an elementary row (column) operation of type I, type II, or type III be performed on A to yield matrix B. Let E be the elementary matrix obtained from I_m (I_n) by performing the same elementary row (column) operation as was performed on A. Then $B = EA(B = AE)$.*

Proof: Exercise.

Theorem 1.22 says that an elementary row operation on A can be achieved by premultiplying A (multiplying A on the left) by the corresponding elementary matrix E; an elementary column operation on A can be obtained by postmultiplying A (multiplying A on the right) by the corresponding elementary matrix.

EXAMPLE 2. Let $A = \begin{bmatrix} 1 & 3 & 2 & 1 \\ -1 & 2 & 3 & 4 \\ 3 & 0 & 1 & 2 \end{bmatrix}$ and let B result from A by adding -2 times the third row of A to the first row of A. Thus $B = \begin{bmatrix} -5 & 3 & 0 & -3 \\ -1 & 2 & 3 & 4 \\ 3 & 0 & 1 & 2 \end{bmatrix}$. Now let E be the matrix that is obtained from I_3 by adding -2 times the third row of I_3 to the first row of I_3. Thus $E = \begin{bmatrix} 1 & 0 & -2 \\ 0 & 1 & 0 \\ 0 & 0 & 1 \end{bmatrix}$. It is easy to verify that $B = EA$.

Theorem 1.23. *If A and B are $m \times n$ matrices, then A is row (column) equivalent to B if and only if $B = E_k E_{k-1} \cdots E_2 E_1 A$ ($B = AE_1 E_2 \cdots E_{k-1} E_k$), where $E_1, E_2, \ldots, E_{k-1}, E_k$ are elementary matrices.*

Proof: We prove only the theorem for row equivalence. If A is row equivalent to B, then B results from A by a sequence of elementary row operations. This implies that there exist elementary matrices $E_1, E_2, \ldots, E_k$ such that $B = E_k E_{k-1} \cdots E_2 E_1 A$.

Conversely, if $B = E_k E_{k-1} \cdots E_2 E_1 A$, where the E_i are elementary matrices, then B results from A by a sequence of elementary row operations, which implies that A is row equivalent to B.

Theorem 1.24. *An elementary matrix E is nonsingular and its inverse is an elementary matrix of the same type.*

Proof: Exercise.

Thus an elementary row operation can be "undone" by another elementary row operation of the same type.

We now obtain an algorithm for finding A^{-1} if it exists; first, we prove the following lemma.

Lemma 1.1.* *Let A be an $n \times n$ matrix and let the homogeneous system $AX = O$ have only the trivial solution $X = O$. Then A is row equivalent to I_n.*

Proof: Let B be a matrix in reduced row echelon form which is row equivalent to A. Then the homogeneous systems $AX = O$ and $BX = O$ are equivalent, and thus $BX = 0$ also has only the trivial solution. It is clear that if r is the number of nonzero rows of B, then the homogeneous system $BX = O$ is equivalent to the homogeneous system whose coefficient matrix consists of the nonzero rows of B and is therefore $r \times n$. Since this last homogeneous system only has the trivial solution, we conclude from Theorem 1.21 that $r \geq n$. Since B is $n \times n$, $r \leq n$. Hence $r = n$, which means that B has no zero rows. Thus $B = I_n$.

Theorem 1.25. *A is nonsingular if and only if A is a product of elementary matrices.*

Proof: If A is a product of elementary matrices $E_1, E_2, \ldots, E_k$, then $A = E_1 E_2 \cdots E_k$. Now each elementary matrix is nonsingular, and the product of nonsingular matrices is nonsingular; therefore, A is nonsingular.

Conversely, if A is nonsingular, then $AX = O$ implies that $A^{-1}(AX) = A^{-1}O = O$, so $I_n X = O$ or $X = O$. Thus $AX = O$ has only the trivial solution. Lemma 1.1 then implies that A is row equivalent to I_n. This means that there exist elementary matrices $E_1, E_2, \ldots, E_k$ such that $I_n = E_k E_{k-1} \cdots E_2 E_1 A$. It then follows that $A = E_1^{-1} E_2^{-1} \cdots E_{k-1}^{-1} E_k^{-1}$. Since the inverse of an elementary matrix is an elementary matrix, we have established the result.

---

* A lemma is a theorem that is established for the purpose of proving another theorem.

Corollary 1.3. *A is nonsingular if and only if A is row equivalent to I_n.*

Proof: If A is row equivalent to I_n, then $I_n = E_k E_{k-1} \cdots E_2 E_1 A$, where $E_1, E_2, \ldots, E_k$ are elementary matrices. It then follows that $A = E_1^{-1} E_2^{-1} \cdots E_k^{-1}$. Now the inverse of an elementary matrix is an elementary matrix, and so by Theorem 1.25 A is nonsingular.

Conversely, if A is nonsingular, then A is a product of elementary matrices, $A = E_k E_{k-1} \cdots E_2 E_1$. Now $A = A I_n = E_k E_{k-1} \cdots E_2 E_1 I_n$, which implies that A is row equivalent to I_n.

We can see that Lemma 1.1 and Corollary 1.3 imply that if the homogeneous system $AX = O$, where A is $n \times n$, has only the trivial solution $X = O$, then A is nonsingular. Conversely, consider $AX = O$, where A is $n \times n$, and let A be nonsingular. Then A^{-1} exists and we form $A^{-1}(AX) = A^{-1}O = O$. Thus $X = O$, which means that the homogeneous system has only the trivial solution. We have thus proved the following important theorem.

Theorem 1.26. *The homogeneous system of n linear equations in n unknowns $AX = O$ has a nontrivial solution if and only if A is singular.*

EXAMPLE 3. Let $A = \begin{bmatrix} 1 & 2 \\ 2 & 4 \end{bmatrix}$ be the matrix defined in Example 11 of Section 1.4, which is singular. Consider the homogeneous system $AX = O$; that is, $\begin{bmatrix} 1 & 2 \\ 2 & 4 \end{bmatrix}\begin{bmatrix} x_1 \\ x_2 \end{bmatrix} = \begin{bmatrix} 0 \\ 0 \end{bmatrix}$. The reduced row echelon form of the augmented matrix is $\begin{bmatrix} 1 & 2 & \vdots & 0 \\ 0 & 0 & \vdots & 0 \end{bmatrix}$, and so a solution is

$$x_1 = -2r$$
$$x_2 = r,$$

where r is any real number. Thus the homogeneous system has a nontrivial solution.

At the end of the proof of Theorem 1.25, we had $A = E_1^{-1} E_2^{-1} \cdots E_{k-1}^{-1} E_k^{-1}$, from which it follows that $A^{-1} = (E_1^{-1} E_2^{-1} E_{k-1}^{-1} E_k^{-1})^{-1} = E_k E_{k-1} \cdots E_2 E_1$. This now provides an algorithm for finding A^{-1}. Thus we perform elementary row operations on A until we get I_n; the product of the elementary matrices $E_k E_{k-1} \cdots E_2 E_1$ then gives A^{-1}. A convenient way of organizing the computing process is to write down the partitioned matrix $[A \mid I_n]$. Then $(E_k E_{k-1} \cdots E_2 E_1)[A \mid I_n] = [E_k E_{k-1} \cdots E_2 E_1 A \mid E_k E_{k-1} \cdots E_2 E_1] = [I_n \mid A^{-1}]$.

EXAMPLE 4. Let $A = \begin{bmatrix} 1 & 1 & 1 \\ 0 & 2 & 3 \\ 5 & 5 & 1 \end{bmatrix}$. Then

$$[A \mid I_3] = \begin{bmatrix} 1 & 1 & 1 & \vdots & 1 & 0 & 0 \\ 0 & 2 & 3 & \vdots & 0 & 1 & 0 \\ 5 & 5 & 1 & \vdots & 0 & 0 & 1 \end{bmatrix}.$$

We now perform elementary row operations that transform $[A \mid I_3]$ to $[I_3 \mid A^{-1}]$; we consider $[A \mid I_3]$ as a 3×6 matrix, and whatever we do to a row of A we also do to the corresponding row of I_3. Thus we arrange our computations as follows:

	A			I_3		
1	1	1	1	0	0	Subtract five times first row from third row to obtain
0	2	3	0	1	0	
5	5	1	0	0	1	
1	1	1	1	0	0	
0	2	3	0	1	0	Divide second row by 2 to obtain
0	0	-4	-5	0	1	
1	1	1	1	0	0	
0	1	$\frac{3}{2}$	0	$\frac{1}{2}$	0	Subtract second row from first row to obtain
0	0	-4	-5	0	1	
1	0	$-\frac{1}{2}$	1	$-\frac{1}{2}$	0	
0	1	$\frac{3}{2}$	0	$\frac{1}{2}$	0	Divide third row by -4 to obtain
0	0	-4	-5	0	1	
1	0	$-\frac{1}{2}$	1	$-\frac{1}{2}$	0	
0	1	$\frac{3}{2}$	0	$\frac{1}{2}$	0	Add $-\frac{3}{2}$ times third row to second row to obtain
0	0	1	$\frac{5}{4}$	0	$-\frac{1}{4}$	
1	0	$-\frac{1}{2}$	1	$-\frac{1}{2}$	0	
0	1	0	$-\frac{15}{8}$	$\frac{1}{2}$	$\frac{3}{8}$	Add $\frac{1}{2}$ times third row to first row to obtain
0	0	1	$\frac{5}{4}$	0	$-\frac{1}{4}$	
1	0	0	$\frac{13}{8}$	$-\frac{1}{2}$	$-\frac{1}{8}$	
0	1	0	$-\frac{15}{8}$	$\frac{1}{2}$	$\frac{3}{8}$	
0	0	1	$\frac{5}{4}$	0	$-\frac{1}{4}$	

Hence

$$A^{-1} = \begin{bmatrix} \frac{13}{8} & -\frac{1}{2} & -\frac{1}{8} \\ -\frac{15}{8} & \frac{1}{2} & \frac{3}{8} \\ \frac{5}{4} & 0 & -\frac{1}{4} \end{bmatrix}.$$

It is easy to verify that $AA^{-1} = A^{-1}A = I_3$.

The question that arises at this point is how to tell when A is singular; that is, when does the above algorithm fail? The answer is that A is singular if and only if A is row equivalent to a matrix B having at least one row that consists entirely of zeros. We now prove this result.

Theorem 1.27. *An $n \times n$ matrix A is singular if and only if A is row equivalent to a matrix B that has a row of zeros.*

Proof: First, let A be row equivalent to a matrix B that has a row consisting entirely of zeros. From Exercise 16 of Section 1.4 it follows that B is singular. Now $B = E_k E_{k-1} \cdots E_1 A$, where $E_1, E_2, \ldots, E_k$ are elementary matrices. If A is nonsingular, then B is nonsingular, a contradiction. Thus A is singular.

Conversely, if A is singular, then A is not row equivalent to I_n, by Corollary 1.3. Thus A is row equivalent to a matrix $B \neq I_n$, which is in reduced row echelon form. From Exercise 3 of Section 1.5 it follows that B must have a row of zeros.

This means that in order to find A^{-1}, we do not have to determine, in advance, whether or not it exists. We merely start to calculate A^{-1}; if at any point in the computation we find a matrix B that is row equivalent to A and has a row of zeros, then A^{-1} does not exist.

EXAMPLE 5. Let $A = \begin{bmatrix} 1 & 2 & -3 \\ 1 & -2 & 1 \\ 5 & -2 & -3 \end{bmatrix}$. To find A^{-1}, we proceed as

follows:

	A			I_3		
1	2	−3	1	0	0	Subtract first row from second row
1	−2	1	0	1	0	to obtain
5	−2	−3	0	0	1	

$$\begin{array}{ccc|ccc}
1 & 2 & -3 & 1 & 0 & 0 \\
0 & -4 & 4 & -1 & 1 & 0 \\
5 & -2 & -3 & 0 & 0 & 1
\end{array}$$

Subtract 5 times first row from third row to obtain

$$\begin{array}{ccc|ccc}
1 & 2 & -3 & 1 & 0 & 0 \\
0 & -4 & 4 & -1 & 1 & 0 \\
0 & -12 & 12 & -5 & 0 & 1
\end{array}$$

Subtract 3 times second row from third row to obtain

$$\begin{array}{ccc|ccc}
1 & 2 & -3 & 1 & 0 & 0 \\
0 & -4 & 4 & -1 & 1 & 0 \\
0 & 0 & 0 & -2 & -3 & 1
\end{array}$$

At this point A is row equivalent to

$$B = \begin{bmatrix} 1 & 2 & -3 \\ 0 & -4 & 4 \\ 0 & 0 & 0 \end{bmatrix},$$

the last matrix under A. Since B has a row of zeros, we stop and conclude that A is a singular matrix.

In Section 1.4 we defined an $n \times n$ matrix B to be the inverse of the $n \times n$ matrix A if $AB = I_n$ and $BA = I_n$. We now show that one of these equations follows from the other.

Theorem 1.28. *If A and B are $n \times n$ matrices such that $AB = I_n$, then $BA = I_n$. Thus $B = A^{-1}$.*

Proof: We first show that if $AB = I_n$, then A is nonsingular. Suppose that A is singular. Then A is row equivalent to a matrix C with a row of zeros. Now $C = E_k E_{k-1} \cdots E_1 A$, where $E_1, E_2, \ldots, E_k$ are elementary matrices. Then $CB = E_k E_{k-1} \cdots E_1 AB$, so AB is row equivalent to CB. Since CB has a row of zeros, we conclude from Theorem 1.27 that AB is singular. Then $AB = I_n$ is impossible, because I_n is nonsingular. This contradiction shows that A is nonsingular, and so A^{-1} exists. Multiplying both sides of the equation $AB = I_n$ by A^{-1} on the left, we then obtain (verify) $B = A^{-1}$.

1.6. Exercises

1. Prove Theorem 1.22.
2. Let A be a 4×3 matrix. Find the elementary matrix E, which as a premultiplier of A, that is, as EA, performs the following elementary row operations on A:

(a) Multiplies the second row of A by -2.

(b) Adds 3 times the third row of A to the fourth row of A.

(c) Interchanges the first and third rows of A.

3. Let A be a 3×4 matrix. Find the elementary matrix F, which as a post-multiplier of A, that is, as AF, performs the following elementary column operations on A:

(a) Adds -4 times the first column of A to the second column of A.

(b) Interchanges the second and third columns of A.

(c) Multiplies the third column of A by 4.

4. Prove Theorem 1.24. (*Hint:* Find the inverse of the elementary matrix of type I, type II, or type III.)

5. Find the inverse, if it exists, of

(a) $\begin{bmatrix} 1 & 1 & 1 \\ 1 & 2 & 3 \\ 0 & 1 & 1 \end{bmatrix}$.

(b) $\begin{bmatrix} 1 & 1 & 1 & 1 \\ 1 & 2 & -1 & 2 \\ 1 & -1 & 2 & 1 \\ 1 & 3 & 3 & 2 \end{bmatrix}$.

(c) $\begin{bmatrix} 1 & 1 & 1 & 1 \\ 1 & 3 & 1 & 2 \\ 1 & 2 & -1 & 1 \\ 5 & 9 & 1 & 6 \end{bmatrix}$.

(d) $\begin{bmatrix} 1 & 2 & 1 \\ 1 & 3 & 2 \\ 1 & 0 & 1 \end{bmatrix}$.

(e) $\begin{bmatrix} 1 & 2 & 2 \\ 1 & 3 & 1 \\ 1 & 1 & 3 \end{bmatrix}$.

6. Prove that $A = \begin{bmatrix} a & b \\ c & d \end{bmatrix}$ is nonsingular if and only if $ad - bc \neq 0$. If this condition holds, show that

$$A^{-1} = \begin{bmatrix} \dfrac{d}{ad - bc} & \dfrac{-b}{ad - bc} \\ \dfrac{-c}{ad - bc} & \dfrac{a}{ad - bc} \end{bmatrix}.$$

7. Let $A = \begin{bmatrix} 2 & 3 & -1 \\ 1 & 0 & 3 \\ 0 & 2 & -3 \\ -2 & 1 & 3 \end{bmatrix}$. Find the elementary matrix that as a post-multiplier of A performs the following elementary column operations on A:

(a) Multiplies the third column of A by -3.

(b) Interchanges the second and third columns of A.

(c) Adds -5 times the first column of A to the third column of A.

8. Prove that $A = \begin{bmatrix} 1 & 2 & 3 \\ 0 & 1 & 2 \\ 1 & 0 & 3 \end{bmatrix}$ is nonsingular and write it as a product of

elementary matrices. (*Hint:* First, write the inverse as a product of elementary matrices, then use Theorem 1.24.)

9. Which of the following homogeneous systems have a nontrivial solution?

(a) $x_1 + 2x_2 + 3x_3 = 0$ (b) $2x_1 + x_2 - x_3 = 0$

        $2x_2 + 2x_3 = 0$ $x_1 - 2x_2 - 3x_3 = 0$

   $x_1 + 2x_2 + 3x_3 = 0.$ $-3x_1 - x_2 + 2x_3 = 0.$

(c) $3x_1 + x_2 + 3x_3 = 0$

  $-2x_1 + 2x_2 - 4x_3 = 0$

    $2x_1 - 3x_2 + 5x_3 = 0.$

10. Find out which of the following matrices are singular. For the nonsingular ones find the inverse.

(a) $\begin{bmatrix} 1 & 3 \\ 2 & 6 \end{bmatrix}$.   (b) $\begin{bmatrix} 1 & 3 \\ -2 & 6 \end{bmatrix}$.   (c) $\begin{bmatrix} 1 & 2 & 3 \\ 1 & 1 & 2 \\ 0 & 1 & 2 \end{bmatrix}$.   (d) $\begin{bmatrix} 1 & 2 & 3 \\ 1 & 1 & 2 \\ 0 & 1 & 1 \end{bmatrix}$.

11. Find the inverse of $A = \begin{bmatrix} 1 & 3 \\ 2 & 4 \end{bmatrix}$.

12. Find the inverse of $A = \begin{bmatrix} 1 & 2 & 3 \\ 0 & 2 & 3 \\ 1 & 2 & 4 \end{bmatrix}$.

13. Show that $\begin{bmatrix} 1 & 2 \\ 3 & 4 \end{bmatrix}$ is nonsingular and write it as a product of elementary matrices. (See the hint in Exercise 8.)

14. Invert the following matrices, if possible.

(a) $\begin{bmatrix} 1 & 2 & -3 & 1 \\ -1 & 3 & -3 & -2 \\ 2 & 0 & 1 & 5 \\ 3 & 1 & -2 & 5 \end{bmatrix}$.   (b) $\begin{bmatrix} 3 & 1 & 2 \\ 2 & 1 & 2 \\ 1 & 2 & 2 \end{bmatrix}$.

(c) $\begin{bmatrix} 1 & 2 & 3 \\ 1 & 1 & 2 \\ 1 & 1 & 0 \end{bmatrix}$.   (d) $\begin{bmatrix} 2 & 1 & 3 \\ 0 & 1 & 2 \\ 1 & 0 & 3 \end{bmatrix}$.

15. Find the inverse of $A = \begin{bmatrix} 1 & 1 & 2 & 1 \\ 0 & -2 & 0 & 0 \\ 1 & 2 & 1 & -2 \\ 0 & 3 & 2 & 1 \end{bmatrix}$.

16. If A is a nonsingular matrix whose inverse is $\begin{bmatrix} 4 & 2 \\ 1 & 1 \end{bmatrix}$, find A.

17. Prove that two $m \times n$ matrices A and B are row equivalent if and only if there exists a nonsingular matrix P such that $B = PA$. (*Hint:* Use Theorems 1.23 and 1.25.)

18. Let A and B be row equivalent $n \times n$ matrices. Prove that A is nonsingular if and only if B is nonsingular.

19. Let A and B be $n \times n$ matrices. Show that if AB is nonsingular, then A and B must be nonsingular. (*Hint:* Use Theorem 1.26.)

20. Let A be an $m \times n$ matrix. Show that A is row equivalent to $_mO_n$ if and only if $A = {}_mO_n$.

21. Let A and B be $m \times n$ matrices. Show that A is row equivalent to B if and only if A' is column equivalent to B'.

1.7. Equivalent Matrices

We have so far considered A to be row (column) equivalent to B if B results from A by a finite sequence of elementary row (column) operations. A natural extension of this idea is that of considering B to arise from A by a finite sequence of elementary row *or* elementary column operations. This leads to the notion of equivalence of matrices. The material discussed in this section will also be used in Section 2.4.

Definition 1.14. If A and B are two $m \times n$ matrices, then A is **equivalent** to B if we can obtain B from A by a finite sequence of elementary row and elementary column operations.

As we have seen in the case of row equivalence, we can easily show (see Exercise 1) that (a) every matrix is equivalent to itself; (b) if A is equivalent to B, then B is equivalent to A; (c) if A is equivalent to B and B is equivalent to C, then A is equivalent to C. In view of (b) both statements "A is equivalent to B" and "B is equivalent to A" can be replaced by "A and B are equivalent." It is also easy to show that if two matrices are row equivalent, then they are equivalent (see Exercise 3).

Theorem 1.29. *If A is any nonzero $m \times n$ matrix, then A is equivalent to a partitioned matrix of the form*

$$\begin{bmatrix} I_k & {}_kO_{n-k} \\ {}_{m-k}O_k & {}_{m-k}O_{n-k} \end{bmatrix}.$$

Proof: By Theorem 1.19, A is row equivalent to a matrix B that is in reduced row echelon form. Using elementary column operations of type I, we get B to be equivalent to a matrix C of the form

$$\begin{bmatrix} I_k & {}_kU_{n-k} \\ {}_{m-k}O_k & {}_{m-k}O_{n-k} \end{bmatrix},$$

where k is the number of nonzero rows in B. By elementary column operations of type III, C is equivalent to a matrix D of the form

$$\begin{bmatrix} I_k & {}_kO_{n-k} \\ {}_{m-k}O_k & {}_{m-k}O_{n-k} \end{bmatrix}.$$

From Exercise 1 it then follows that A is equivalent to D.

Of course, in Theorem 1.29, k may equal m, in which case there will not be any zero rows at the bottom of the matrix. (What happens if $k = n$? If $k = m = n$?)

EXAMPLE 1. Let $A = \begin{bmatrix} 1 & 1 & 2 & -1 \\ 1 & 2 & 1 & 0 \\ -1 & -4 & 1 & -2 \\ 1 & -2 & 5 & -4 \end{bmatrix}$. To find a matrix of the

form described in Theorem 1.29, which is equivalent to A, we proceed as follows.

Subtract the first row of A from the second row of A to obtain:

$$\begin{bmatrix} 1 & 1 & 2 & -1 \\ 0 & 1 & -1 & 1 \\ -1 & -4 & 1 & -2 \\ 1 & -2 & 5 & -4 \end{bmatrix}$$ Add the first row to the third to obtain

$$\begin{bmatrix} 1 & 1 & 2 & -1 \\ 0 & 1 & -1 & 1 \\ 0 & -3 & 3 & -3 \\ 1 & -2 & 5 & -4 \end{bmatrix}$$ Subtract the first row from the fourth to obtain

$$\begin{bmatrix} 1 & 1 & 2 & -1 \\ 0 & 1 & -1 & 1 \\ 0 & -3 & 3 & -3 \\ 0 & -3 & 3 & -3 \end{bmatrix}$$ Subtract the third row from the fourth to obtain

$$\begin{bmatrix} 1 & 1 & 2 & -1 \\ 0 & 1 & -1 & 1 \\ 0 & -3 & 3 & -3 \\ 0 & 0 & 0 & 0 \end{bmatrix}$$ Multiply the third row by $-\frac{1}{3}$ to obtain

$$\begin{bmatrix} 1 & 1 & 2 & -1 \\ 0 & 1 & -1 & 1 \\ 0 & 1 & -1 & 1 \\ 0 & 0 & 0 & 0 \end{bmatrix}$$

Subtract the second row from the third to obtain

$$\begin{bmatrix} 1 & 1 & 2 & -1 \\ 0 & 1 & -1 & 1 \\ 0 & 0 & 0 & 0 \\ 0 & 0 & 0 & 0 \end{bmatrix}$$

Subtract the second row from the first to obtain

$$\begin{bmatrix} 1 & 0 & 3 & -2 \\ 0 & 1 & -1 & 1 \\ 0 & 0 & 0 & 0 \\ 0 & 0 & 0 & 0 \end{bmatrix}$$

Subtract 3 times the first column from the third to obtain

$$\begin{bmatrix} 1 & 0 & 0 & -2 \\ 0 & 1 & -1 & 1 \\ 0 & 0 & 0 & 0 \\ 0 & 0 & 0 & 0 \end{bmatrix}$$

Add 2 times the first column to the fourth to obtain

$$\begin{bmatrix} 1 & 0 & 0 & 0 \\ 0 & 1 & -1 & 1 \\ 0 & 0 & 0 & 0 \\ 0 & 0 & 0 & 0 \end{bmatrix}$$

Add the second column to the third to obtain

$$\begin{bmatrix} 1 & 0 & 0 & 0 \\ 0 & 1 & 0 & 1 \\ 0 & 0 & 0 & 0 \\ 0 & 0 & 0 & 0 \end{bmatrix}$$

Subtract the second column from the fourth to obtain

$$\begin{bmatrix} 1 & 0 & 0 & 0 \\ 0 & 1 & 0 & 0 \\ 0 & 0 & 0 & 0 \\ 0 & 0 & 0 & 0 \end{bmatrix}$$

This is the desired matrix.

The following theorem gives another useful way to look at the equivalence of matrices.

Theorem 1.30. *Two* $m \times n$ *matrices* A *and* B *are equivalent if and only if* $B = PAQ$ *for some nonsingular matrices* P *and* Q.

Proof: Exercise.

We next prove a theorem that is analogous to Corollary 1.3.

Theorem 1.31. *An* $n \times n$ *matrix* A *is nonsingular if and only if* A *is equivalent to* I_n.

Proof: If A is equivalent to I_n, then I_n arises from A by a sequence of elementary row or elementary column operations. Thus there exist elementary matrices $E_1, E_2, \ldots, E_r, F_1, F_2, \ldots, F_s$ such that $I_n = E_r E_{r-1} \cdots E_2 E_1 A F_1 F_2 \cdots F_s$. Let $E_r, E_{r-1} \cdots E_2 E_1 = P$, and $F_1 F_2 \cdots F_s = Q$. Then $I_n = PAQ$, where P and Q are nonsingular. It then follows that $A = P^{-1} Q^{-1}$, and since P^{-1} and Q^{-1} are nonsingular, A is nonsingular.

Conversely, if A is nonsingular, then from Corollary 1.3 it follows that A is row equivalent to I_n. Hence A is equivalent to I_n.

1.7. Exercises

1. (a) Prove that every matrix A is equivalent to itself.
   (b) Prove that if A is equivalent to B, then B is equivalent to A.
   (c) Prove that if A is equivalent to B and B is equivalent to C, then A is equivalent to C.

2. For each of the following matrices, find a matrix of the form described in Theorem 1.29, equivalent to the given matrix.

(a) $\begin{bmatrix} 1 & 2 & -1 & 4 \\ 5 & 1 & 2 & -3 \\ 2 & 1 & 4 & 3 \\ 2 & 0 & 1 & 2 \\ 5 & 1 & 2 & 3 \end{bmatrix}$. (b) $\begin{bmatrix} 1 & 2 & 1 \\ 2 & 3 & 1 \\ 2 & 1 & 3 \end{bmatrix}$.

(c) $\begin{bmatrix} 1 & -2 & 1 \\ 2 & 3 & 2 \\ 3 & 1 & 3 \end{bmatrix}$. (d) $\begin{bmatrix} 1 & 3 & -1 & 2 \\ 2 & -4 & -2 & 1 \\ 3 & 1 & 2 & -3 \\ 7 & 3 & -2 & 5 \end{bmatrix}$.

3. Show that if A and B are row equivalent, then they are equivalent.

4. Prove Theorem 1.30.

5. Let $A = \begin{bmatrix} 1 & 2 & 3 \\ 1 & 1 & 2 \\ 0 & 1 & 1 \end{bmatrix}$. Find a matrix B of the form described in Theorem 1.29

which is equivalent to A. Also, find nonsingular matrices P and Q such that $B = PAQ$.

6. Let A be an $m \times n$ matrix. Show that A is equivalent to $_mO_n$ if and only if $A = {_mO_n}$.

7. Let A and B be $m \times n$ matrices. Show that A is equivalent to B if and only if A' is equivalent to B'.

8. For each of the following matrices A, find a matrix $B \neq A$ that is equivalent to A:

(a) $A = \begin{bmatrix} 1 & -2 & 3 & 1 \\ 0 & -1 & 4 & 3 \\ 1 & 0 & -2 & -1 \end{bmatrix}$.   (b) $A = \begin{bmatrix} 1 & 3 \\ 2 & 6 \end{bmatrix}$.

(c) $A = \begin{bmatrix} 1 & 2 & 3 & 4 & 3 \\ 0 & 1 & -2 & 0 & 2 \\ -1 & 3 & 2 & 0 & 1 \end{bmatrix}$.

9. Let A and B be equivalent matrices. Prove that A is nonsingular if and only if B is nonsingular.

CHAPTER **2**

Real Vector Spaces

Vectors in the Plane—A Review

In many applications we deal with measurable quantities, such as pressure, mass, and speed, which can be completely described by giving their magnitude. They are called **scalars** and will be denoted by lowercase Latin letters. There are many other measurable quantities, such as velocity, force, and acceleration, which require for their description not only magnitude, but also a sense of direction. These are called **vectors** and their study comprises this chapter. Vectors will be denoted by lowercase Greek letters, such as α, β, γ, and δ. The reader has already encountered vectors in elementary physics and in the calculus.

We quickly review the definition of a vector in the plane. We draw two perpendicular lines, which are labeled the x- and y-axes and choose a unit of length on the x-axis and a unit of length on the y-axis. With each point P in the plane we associate an ordered pair (x, y) of real numbers, its **coordinates**. Conversely, we can associate a point in the plane with each ordered pair of real numbers. The point P with coordinates (x, y) is denoted as $P(x, y)$, or simply as (x, y). The set of all points in the plane is denoted by R^2; it is called **2-space**.

Consider the 2 × 1 matrix

$$X = \begin{bmatrix} x \\ y \end{bmatrix},$$

where x and y are real numbers. With X we associate the directed line segment with the initial point at the origin and the terminal point at $P(x, y)$. The directed line segment from O to P is denoted by $\overrightarrow{OP}$; O is called its **tail** and P its **head**. We distinguish the tail and the head by placing an arrow at the head (Figure 2.1). A directed line segment has a **direction**, indicated by the arrow at its head. The **magnitude** of a directed line segment is its length. Thus a directed line segment can be used to describe force, velocity, or acceleration. Conversely, with a directed line segment $\overrightarrow{OP}$ with tail $O(0, 0)$ and head $P(x, y)$ we can associate the matrix

$$\begin{bmatrix} x \\ y \end{bmatrix}.$$

A vector in the plane is a 2 × 1 matrix $\alpha = \begin{bmatrix} x \\ y \end{bmatrix}$, where x and y are real numbers, which are called the **components** of α. We refer to a vector in the plane merely as a **vector**.

Thus with every vector we can associate a directed line segment and,

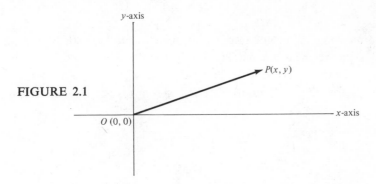

FIGURE 2.1

conversely, with every directed line segment we can associate a vector. Frequently, the notions of directed line segments and vectors are used interchangeably, and a directed line segment is called a **vector**. Since a vector is a matrix, the vectors

$$\alpha_1 = \begin{bmatrix} x_1 \\ y_1 \end{bmatrix} \quad \text{and} \quad \alpha_2 = \begin{bmatrix} x_2 \\ y_2 \end{bmatrix}$$

are said to be **equal** if $x_1 = x_2$ and $y_1 = y_2$. That is, two vectors are equal if their respective components are equal.

With each vector $\alpha = \begin{bmatrix} x \\ y \end{bmatrix}$ we can also associate the unique point $P(x, y)$; conversely, with each point $P(x, y)$ we associate the unique vector $\begin{bmatrix} x \\ y \end{bmatrix}$. Thus we also write the vector α as (x, y). Of course, this association is carried out by means of the directed line segment $\overrightarrow{OP}$, where O is the origin and P is the point with coordinates (x, y) (Figure 2.1).

Thus the plane may be viewed both as the set of all points or as the set of all vectors. For this reason, and depending upon the context, we sometimes take R^2 as the set of all ordered pairs (x, y) and sometimes as the set of all 2×1 matrices $\begin{bmatrix} x \\ y \end{bmatrix}$ (or directed line segments).

Let

$$\alpha = \begin{bmatrix} x_1 \\ y_1 \end{bmatrix} \quad \text{and} \quad \beta = \begin{bmatrix} x_2 \\ y_2 \end{bmatrix}$$

be two vectors in the plane. The **sum** of the vectors α and β is the vector

$$\alpha + \beta = \begin{bmatrix} x_1 + x_2 \\ y_1 + y_2 \end{bmatrix}.$$

We can interpret vector addition geometrically, as follows. In Figure 2.2 the directed line segment γ is parallel to β, it has the same length as β, and its tail is the head (x_1, y_1) of α; so its head is $(x_1 + x_2, y_1 + y_2)$. Thus the vector with tail at O and head at $(x_1 + x_2, y_1 + y_2)$ is $\alpha + \beta$. We can also

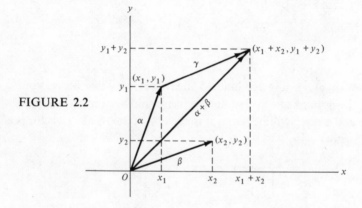

FIGURE 2.2

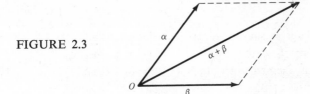

FIGURE 2.3

describe $\alpha + \beta$ as the diagonal of the parallelogram defined by α and β, as shown in Figure 2.3.

If $\alpha = \begin{bmatrix} x \\ y \end{bmatrix}$ is a vector and c is a scalar (a real number), then the **scalar multiple** $c\alpha$ of α by c is the vector $\begin{bmatrix} cx \\ cy \end{bmatrix}$. Thus the scalar multiple $c\alpha$ is obtained by multiplying each component of α by c.

The vector $\alpha + (-1)\beta$ is written as $\alpha - \beta$ and is called the **difference between** α and β. It is shown in Figure 2.4.

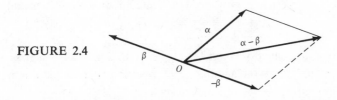

FIGURE 2.4

In mathematics we consider a vector as an ordered pair of real numbers or as a 2×1 matrix; whereas in physics we often treat a vector as a directed line segment. There are three very different interpretations, and one can then ask why all three can be legitimately designated "vector."

To answer this question, we first observe that, mathematically speaking, the only thing that concerns us is the behavior of the object we call "vector." It turns out that all three objects behave, from an algebraic point of view, in exactly the same manner. Moreover, many other objects that arise naturally in applied problems behave, algebraically speaking, as do the above-mentioned objects. To a mathematician this is a perfect situation. For we can now abstract those features that all such objects have in common (that is, those properties that make them all behave alike) and define a new structure. The great advantage of doing this is that we can now talk about properties of all such objects at the same time without having to refer to any one object in particular. This, of course, is much more efficient than studying the properties of each object separately. We shall discuss these notions in the following section.

2.1. Vector Spaces and Subspaces

Definition 2.1. A **real vector space** is a set V of elements such that we have two operations $\oplus$ and $\odot$ defined with the following properties:

(a) If α and β are any elements in V, then $\alpha \oplus \beta$ is in V (that is, V is closed under the operation $\oplus$).

   (1) $\alpha \oplus \beta = \beta \oplus \alpha$ for all α, β in V.

   (2) $\alpha \oplus (\beta \oplus \gamma) = (\alpha \oplus \beta) \oplus \gamma$ for all α, β, γ in V.

   (3) There exists a unique element θ in V such that $\alpha \oplus \theta = \theta \oplus \alpha = \alpha$ for any α in V.

   (4) For each α in V there exists a unique β in V such that $\alpha \oplus \beta = \beta \oplus \alpha = \theta$. We denote β by $-\alpha$ and call it the **negative** of α.

(b) If α is any element in V and c is any real number, then $c \odot \alpha$ is in V.

   (5) $c \odot (\alpha \oplus \beta) = c \odot \alpha \oplus c \odot \beta$ for any α, β in V and any real number c.

   (6) $(c + d) \odot \alpha = c \odot \alpha \oplus d \odot \alpha$ for any α in V and any real numbers c and d.

   (7) $c \odot (d \odot \alpha) = (cd) \odot \alpha$ for any α in V and any real numbers c and d.

   (8) $1 \odot \alpha = \alpha$ for any α in V.

The elements of V are called **vectors**; the elements of R are called **scalars**. The operation $\oplus$ is called **vector addition**; the operation $\odot$ is called **scalar multiplication**. The vector θ is called the **zero vector**.

In order to specify a vector space, we must be given a set V and two operations $\oplus$ and $\odot$ satisfying all the properties of the definition. We shall often refer to a real vector space merely as a **vector space**. Thus a "vector" is now an element of a vector space and no longer needs to be interpreted as a directed line segment. In our examples we shall see, however, how this name came about in a natural manner. We now consider some examples of vector spaces, leaving it to the reader to verify that all the properties of Definition 2.1 hold.

EXAMPLE 1. Consider R^n, the set of all $n \times 1$ matrices $\begin{bmatrix} a_1 \\ a_2 \\ \vdots \\ a_n \end{bmatrix}$ with real entries. Let the operation $\oplus$ be matrix addition and let the operation $\odot$ be multiplication of a matrix by a real number.

By the use of the properties of matrices established in Section 1.3, it is not difficult to show that R^n is a vector space by verifying that the properties of

Definition 2.1 hold. Thus the matrix $\begin{bmatrix} a_1 \\ a_2 \\ \vdots \\ a_n \end{bmatrix}$, as an element of R^n, is now

called a *vector*. We have already discussed R^2 from a geometric point of view. A completely analogous geometric discussion can be carried out for R^3 in 3-space (the world we live in) with the aid of the x-, y-, and z-axes. Although we shall see later that many geometric notions such as length and the angle between vectors can be defined in R^n for $n > 3$, we cannot draw pictures in these cases.

For R^2 and R^3 we can also verify geometrically that all the properties of Definition 2.1 hold. For example, in Figure 2.5 we verify Property 1. The vector spaces R^2 and R^3 are the vector spaces that motivate the abstract definition of a vector space that we have just presented. The elements in R^2 and R^3 are vectors in the traditional sense; R^2 and R^3 do "occupy" space.

FIGURE 2.5

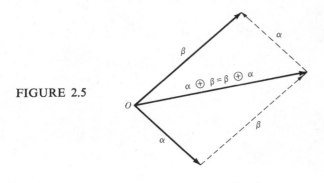

EXAMPLE 2. The set of all $m \times n$ matrices with matrix addition as $\oplus$ and multiplication of a matrix by a real number as $\odot$ is a vector space (verify). We denote this vector by $_mR_n$.

EXAMPLE 3. The set of all real numbers with $\oplus$ as the usual addition of real numbers and $\odot$ as the usual multiplication of real numbers is a vector space (verify). In this case the real numbers play the dual roles of both vectors and scalars.

EXAMPLE 4. Let R_n be the set of all $1 \times n$ matrices $[a_1 \quad a_2 \quad \cdots \quad a_n]$, where we define $\oplus$ by $[a_1 \quad a_2 \quad \cdots \quad a_n] \oplus [b_1 \quad b_2 \quad \cdots \quad b_n] = [a_1 + b_1 \quad a_2 + b_2 \cdots \quad a_n + b_n]$ and $\odot$ by $c \odot [a_1 \quad a_2 \quad \cdots \quad a_n] = [ca_1 \quad ca_2 \quad \cdots \quad ca_n]$. Then R_n is a vector space (verify). This is just a special case of Example 2.

EXAMPLE 5. Another source of examples will be sets of polynomials; therefore, we recall some well-known facts about such functions. A **polynomial** (in t) is an expression of the form $p(t) = a_0 t^n + a_1 t^{n-1} + \cdots + a_{n-1} t + a_n$, where $a_0, a_1, \ldots, a_n$ are real numbers. If $a_0 \neq 0$, then $p(t)$ is said to have **degree n**. Thus the degree of a polynomial is the highest power having a nonzero coefficient; $2t + 1$ has degree 1 and the constant polynomial 3 has degree 0. The zero polynomial has no degree. We now let P be the set of all polynomials. If

$$p(t) = a_0 t^n + a_1 t^{n-1} + \cdots + a_{n-1} t + a_n$$

and

$$q(t) = b_0 t^m + b_1 t^{m-1} + \cdots + b_{m-1} t + b_m,$$

we define $p(t) \oplus q(t)$ as

$$p(t) \oplus q(t) = (a_0 t^n + a_1 t^{n-1} + \cdots + a_{n-1} t + a_n)$$
$$+ (b_0 t^m + b_1 t^{m-1} + \cdots + b_{m-1} t + b_m).$$

If c is a scalar, we also define $c \odot p(t)$ as

$$c \odot p(t) = (ca_0) t^n + (ca_1) t^{n-1} + \cdots + (ca_{n-1}) t + (ca_n).$$

Then P is a vector space (verify).

EXAMPLE 6. Let V be the set of all real-valued continuous functions defined on R^1. If f and g are in V, we define $f \oplus g$ by $(f \oplus g)(t) = f(t) + g(t)$. If f is in V and c is a scalar, we define $c \odot f$ by $(c \odot f)(t) = cf(t)$. Then V is a vector space (verify).

EXAMPLE 7. Let V be the set of all ordered triples of real numbers (x, y, z) with the operations $(x, y, z) \oplus (x', y', z') = (x', y + y', z + z')$ and $c \odot (x, y, z) = (cx, cy, cz)$. It is easy to verify that properties 1, 3, 4, and 6 of Definition 2.1 fail to hold. Thus V is not a vector space.

To verify that a given set V with two operations $\oplus$ and $\odot$ is a real vector space, we must show that it satisfies all the properties of Definition 2.1. The first thing to check is whether (a) and (b) hold, for, if either of these fails, we do not have a vector space.

EXAMPLE 8. Let V be the set of all integers; define $\oplus$ as ordinary addition and $\odot$ as ordinary multiplication. Here V is not a vector space because if α is any vector in V and $c = \sqrt{3}$, then $c \odot \alpha$ is not in V.

The following theorem presents some useful properties common to all vector spaces.

Theorem 2.1. *If V is a vector space, then*
(a) $0 \odot \alpha = \theta$ *for any vector α in V.*
(b) $c \odot \theta = \theta$ *for any scalar c.*
(c) *If $c \odot \alpha = \theta$, then either $c = 0$ or $\alpha = \theta$.*
(d) $(-1) \odot \alpha = -\alpha$ *for any α in V.*

Proof: We prove (d) and leave (a) to (c) as exercises. Now $(-1) \odot \alpha \oplus \alpha = (-1) \odot \alpha \oplus 1 \odot \alpha = (-1 + 1) \odot \alpha = 0 \odot \alpha = \theta$, so $(-1) \odot \alpha = -\alpha$.

We now begin to analyze the structure of a vector space. First, it is convenient to have a name for a subset of a given vector space which is itself a vector space with respect to the same operations as those in V. Thus we have a definition.

Subspaces

Definition 2.2. Let V be a vector space and W a subset of V. If W is a vector space with respect to the operations in V, then W is called a **subspace** of V.

Examples of subspaces of a given vector space occur very frequently. We shall list several of these, leaving the verifications that they are subspaces to the reader. More examples will be found in the exercises.

EXAMPLE 9. Every vector space has at least two subspaces, itself and the subspace $\{\theta\}$ consisting only of the zero vector (recall that $\theta \oplus \theta = \theta$ and $c \odot \theta = \theta$ in any vector space). These subspaces are called **trivial subspaces**.

EXAMPLE 10. Let P_2 be the set consisting of all polynomials of degree ≤ 2 and the zero polynomial; P_2 is a subset of P, the vector space of all polynomials (see Example 5). It is easy to verify that P_2 is a *subspace* of P. In general, the set P_n of all polynomials of degree $\leq n$ and the zero polynomial is a subspace of P.

EXAMPLE 11. Let V be the set of all polynomials of degree exactly $=2$; V is a *subset* of P, the vector space of all polynomials, but not a *subspace* of P, for the sum of the polynomials $2t^2 + 3t + 1$ and $-2t^2 + t + 2$ is not in V, since it is a polynomial of degree 1.

We now pause in our listing of subspaces to develop a labor-saving result. We just noted that to verify that a subset W of a vector space V is a subspace, one must check that (a), (b), and 1 to 8 of Definition 2.1 hold. However, the following theorem says that it is enough to merely check that (a) and (b) hold.

Theorem 2.2. *Let V be a vector space with operations $\oplus$ and $\odot$ and let W be a nonempty subset of V. Then W is a subspace of V if and only if the following conditions hold:*
(a) *If α, β are any vectors in W, then $\alpha \oplus \beta$ is in W.*
(b) *If c is any real number and α is any vector in W, then $c \odot \alpha$ is in W.*

Proof: We first show that if W is a subspace of V, then (a) and (b) hold. This follows at once from the observation that if W is a subspace, then it is a vector space and so (a) and (b) of Definition 2.1 hold; these are precisely (a) and (b) of the theorem.

Conversely, suppose that (a) and (b) hold. We wish to show that W is a subspace of V. First, note that from (b) we have that $(-1) \odot \alpha$ is in W for any α in W. From (a) we have that $\alpha \oplus (-1) \odot \alpha$ is in W. But $\alpha \oplus (-1) \odot \alpha = \theta$, so θ is in W. Then $\alpha \oplus \theta = \alpha$ for any α in W. Finally, properties 1, 2, 5, 6, 7, and 8 hold in W because they hold in V. Hence W is a subspace of V.

EXAMPLE 12. Let W be the set of all vectors in R^3 of the form $\begin{bmatrix} a \\ b \\ a+b \end{bmatrix}$, where a and b are any real numbers. To verify Theorem 2.2(a) and (b), we let

$$\alpha = \begin{bmatrix} a_1 \\ b_1 \\ a_1 + b_1 \end{bmatrix} \quad \text{and} \quad \beta = \begin{bmatrix} a_2 \\ b_2 \\ a_2 + b_2 \end{bmatrix}$$

be two vectors in W. Then

$$\alpha \oplus \beta = \begin{bmatrix} a_1 + a_2 \\ b_1 + b_2 \\ (a_1 + b_1) + (a_2 + b_2) \end{bmatrix} = \begin{bmatrix} a_1 + a_2 \\ b_1 + b_2 \\ (a_1 + a_2) + (b_1 + b_2) \end{bmatrix}$$

is in W, for W consists of all those vectors whose third entry is the sum of the first two entries. Similarly,

$$c \odot \begin{bmatrix} a_1 \\ b_1 \\ a_1 + b_1 \end{bmatrix} = \begin{bmatrix} ca_1 \\ cb_1 \\ ca_1 + cb_1 \end{bmatrix}$$

is in W. Hence W is a subspace of R^3.

Henceforth we shall usually denote $\alpha \oplus \beta$ and $c \odot \alpha$ in a vector space V as $\alpha + \beta$ and $c\alpha$, respectively.

EXAMPLE 13. Let W be the set of all 2×3 matrices in $_2R_3$ of the form $\begin{bmatrix} a & b & 0 \\ 0 & c & d \end{bmatrix}$, where a, b, c, and d are real numbers. Then W is a subspace of $_2R_3$ (verify).

Another very important example of a subspace is provided by Example 14.

EXAMPLE 14. Consider the homogeneous system $AX = O$, where A is $m \times n$. A solution consists of a vector $X = \begin{bmatrix} x_1 \\ x_2 \\ \vdots \\ x_n \end{bmatrix}$, that is, a vector in R^n.

Thus the set of all solutions is a subset of R^n. We now show that it is a subspace of R^n (called the **solution space** of the homogeneous system) by verifying (a) and (b).

Let X_1 and X_2 be solutions. Then $X_1 + X_2$ is in W, that is, $X_1 + X_2$ is a solution, because $A(X_1 + X_2) = AX_1 + AX_2 = O + O = O$. Also, if X is in W, then cX is in W, that is, cX is a solution, because $A(cX) = c(AX) = cO = O$.

It should be noted that the set of all solutions to the linear system $AX = B$, $B \neq O$ is not a subspace of R^n (see Exercise 20).

Lines in R^3 (Optional)

In R^2 a line is determined by specifying its slope and one of its points. In R^3 a line is determined by specifying its direction and one of its points. Let

$$\delta = \begin{bmatrix} u \\ v \\ w \end{bmatrix}$$ be a nonzero vector in R^3. Then the line ℓ_0 through the origin and

parallel to δ consists of all the points $P(x, y, z)$ whose position vector

$$\alpha = \begin{bmatrix} x \\ y \\ z \end{bmatrix}$$ is of the form $\alpha = t\delta$, $-\infty < t < \infty$ (Figure 2.6a). It is easy to

verify that the line ℓ_0 is a subspace of R^3. Now let $P_0 = (x_0, y_0, z_0)$ be a point

in R^3, and let $\alpha_0 = \begin{bmatrix} x_0 \\ y_0 \\ z_0 \end{bmatrix}$ be the position vector of P_0. Then the line ℓ through

P_0 and parallel to δ consists of the points $P(x, y, z)$ whose position vector

$$\alpha = \begin{bmatrix} x \\ y \\ z \end{bmatrix}$$ satisfies (Figure 2.6b)

$$\alpha = \alpha_0 + t\delta \qquad -\infty < t < \infty. \tag{1}$$

Equation (1) is called the **parametric equation** of ℓ, since it contains the parameter t which can be assigned any real number. Equation (1) can also be written in terms of the components as

$$\begin{array}{ll} x = x_0 + tu \\ y = y_0 + tv & -\infty < t < \infty \\ z = z_0 + tw, \end{array} \tag{2}$$

which are called the **parametric equations** of ℓ.

EXAMPLE 15. The parametric equations of the line through the point

$P_0(-3, 2, 1)$, which is parallel to the vector $\delta = \begin{bmatrix} 2 \\ -3 \\ 4 \end{bmatrix}$, are

$$\begin{array}{ll} x = -3 + 2t \\ y = 2 - 3t & -\infty < t < \infty \\ z = 1 + 4t. \end{array}$$

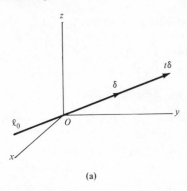

FIGURE 2.6

EXAMPLE 16. Find parametric equations of the line ℓ through the points $P_0(2, 3, -4)$ and $P_1(3, -2, 5)$.

Solution: The desired line is parallel to the vector $\delta = \overrightarrow{P_0P_1}$. Now

$$\delta = \begin{bmatrix} 3 - 2 \\ -2 - 3 \\ 5 - (-4) \end{bmatrix} = \begin{bmatrix} 1 \\ -5 \\ 9 \end{bmatrix}.$$

Since P_0 is on the line, we can write the parametric equations of ℓ as

$$\begin{aligned} x &= 2 + t \\ y &= 3 - 5t \qquad -\infty < t < \infty \\ z &= -4 + 9t. \end{aligned}$$

Of course, we could have used the point P_2 instead of P_1. In fact, we could use any point on the line in the parametric equation for ℓ. Thus a line can be represented in infinitely many ways in parametric form.

2.1. Exercises

1. Prove in detail that R^n is a vector space.
2. Show that P_2 is a vector space.

In Exercises 3 through 7 the given set together with the given operations is not a vector space. List the properties of Definition 2.1 that fail to hold.

3. The set of all positive real numbers with the operations of $\oplus$ as ordinary addition and $\odot$ as ordinary multiplication.

4. The set of all ordered pairs of real numbers with the operations $(x, y) \oplus (x', y')$ $= (x + x', y + y')$ and $r \odot (x, y) = (x, ry)$.

5. The set of all ordered triples of real numbers with the operations $(x, y, z) \oplus (x', y', z') = (x + x', y + y', z + z')$ and $r \odot (x, y, z) = (x, 1, z)$.

6. The set of all 2×1 matrices $\begin{bmatrix} x \\ y \end{bmatrix}$, where $x \le 0$, with the usual operations in R^2.

7. The set of all ordered pairs of real numbers with the operations $(x, y) \oplus (x', y') = (x + x', y + y')$ and $r \odot (x, y) = (0, 0)$.

8. Let V be the set of all positive real numbers; define $\oplus$ by $\alpha \oplus \beta = \alpha\beta$ ($\oplus$ is ordinary multiplication) and define $\odot$ by $c \odot \alpha = \alpha^c$. Prove that V is a vector space.

9. Let V be the set of all real-valued continuous functions. If f and g are in V, define $f \oplus g$ by $(f \oplus g)(t) = f(t) + g(t)$. If f is in V, define $c \odot f$ by $(c \odot f)(t) = cf(t)$. Prove that V is a vector space (this is the vector space defined in Example 6).

10. Let V be the set consisting of a single element θ. Let $\theta \oplus \theta = \theta$ and $c \odot \theta = \theta$. Prove that V is a vector space.

11. Consider the differential equation $y'' - y' + 2y = 0$. A solution is a real-valued function f satisfying the equation. Let V be the set of all solutions to the given differential equation; define $\oplus$ and $\odot$ as in Exercise 9. Prove that V is a vector space.

12. Prove Theorem 2.1(a)–(c).

13. Prove that P_2 is a subspace of P_3.

14. Which of the following subsets of the vector space, $_2R_3$, of all 2×3 matrices are subspaces? The set of all matrices of the form

(a) $\begin{bmatrix} a & b & c \\ d & 0 & 0 \end{bmatrix}$, where $b = a + c$.

(b) $\begin{bmatrix} a & b & c \\ d & 0 & 0 \end{bmatrix}$, where $c > 0$.

(c) $\begin{bmatrix} a & b & c \\ d & e & f \end{bmatrix}$, where $a = -2c$ and $f = 2e + d$.

15. Which of the following subsets of R^3 are subspaces? The set of all vectors of the form

(a) $\begin{bmatrix} a \\ b \\ 1 \end{bmatrix}$.   (b) $\begin{bmatrix} a \\ b \\ 0 \end{bmatrix}$.   (c) $\begin{bmatrix} a \\ b \\ a + 2b \end{bmatrix}$.

(d) $\begin{bmatrix} a \\ b \\ c \end{bmatrix}$, where $a > 0$.   (e) $\begin{bmatrix} a \\ 0 \\ 0 \end{bmatrix}$.   (f) $\begin{bmatrix} a \\ a \\ c \end{bmatrix}$.

16. Which of the following subsets of $_2R_2$ are subspaces? The set of all 2×2
    (a) Symmetric matrices.
    (b) Singular matrices.
    (c) Nonsingular matrices.

17. Show that P is a subspace of the vector space of all real-valued continuous functions introduced in Exercise 9.

18. Let V be the vector space of all real-valued continuous functions considered in Exercise 9. Which of the following subsets are subspaces of V? The set of all
    (a) Nonnegative functions.
    (b) Constant functions.
    (c) Functions f such that $f(0) = 0$.
    (d) Functions f such that $f(0) = 5$.
    (e) Differentiable functions.

19. Prove that the set V of all real-valued functions is a vector space under the operations defined as in Exercise 9. Show also that the space of Exercise 9 is a subspace of V.

20. Show that the set of all solutions to the linear system $AX = B$, $B \neq O$, is not a subspace of R^n.

21. Let V be a vector space and let α and β be vectors in V. Prove that the subset of V consisting of all elements of the form $a\alpha + b\beta$, where a and b are real numbers, is a subspace of V.

22. Prove that $-(-\alpha) = \alpha$.

23. Prove that if $\alpha \oplus \beta = \alpha \oplus \gamma$, then $\beta = \gamma$.

24. Prove that if $\alpha \neq \theta$ and $a\alpha = b\alpha$, then $a = b$.

25. (a) Show that a line ℓ_0 through the origin in R^3 is a subspace of R^3.
    (b) Show that a line ℓ in R^3 not passing through the origin is not a subspace of R^3.

26. State which of the following points are on the line

$$
\begin{aligned}
x &= 3 + 2t \\
y &= -2 + 3t \qquad -\infty < t < \infty \\
z &= 4 - 3t.
\end{aligned}
$$

(a) $(1, 1, 1)$.   (b) $(1, -1, 0)$.
(c) $(1, 0, -2)$.   (d) $(4, -\frac{1}{2}, \frac{5}{2})$.

27. State which of the following points are on the line

$$
\begin{aligned}
x &= 4 - 2t \\
y &= -3 + 2t \qquad -\infty < t < \infty \\
z &= 4 - 5t.
\end{aligned}
$$

(a) $(0, 1, -6)$.   (b) $(1, 2, 3)$.
(c) $(4, -3, 4)$.   (d) $(0, 1, -1)$.

28. Find the parametric equations of the line through $P_0 = (x_0, y_0, z_0)$ parallel to δ.

(a) $P_0 = (3, 4, -2)$, $\delta = \begin{bmatrix} 4 \\ -5 \\ 2 \end{bmatrix}$.   (b) $P_0 = (3, 2, 4)$, $\delta = \begin{bmatrix} -2 \\ 5 \\ 1 \end{bmatrix}$.

29. Find the parametric equations of the line through the given points.
    (a) $(2, -3, 1), (4, 2, 5)$.   (b) $(-3, -2, -2), (5, 5, 4)$.

2.2. Linear Independence and Basis

We have so far defined a mathematical system called a real vector space and noted some of its properties. We further observe that the only real vector space having a finite number of vectors in it is the vector space whose only vector is θ. For if $\alpha \neq \theta$ is in a vector space V, then $c \odot \alpha$ is in V, where c is any real number, and so V has infinitely many vectors in it. However, in this section we shall show that each vector space V studied here has a finite number of vectors that completely describe V. We now turn to a formulation of these ideas.

Definition 2.3. Let $S = \{\alpha_1, \alpha_2, \ldots, \alpha_k\}$ be a set of vectors in a vector space V. A vector α in V is called a **linear combination** of the vectors in S if

$$\alpha = a_1\alpha_1 + a_2\alpha_2 + \cdots + a_k\alpha_k$$

for some real numbers $a_1, a_2, \ldots, a_k$.

EXAMPLE 1. In R^3 let

$$\alpha_1 = \begin{bmatrix} 1 \\ 2 \\ 1 \end{bmatrix}, \qquad \alpha_2 = \begin{bmatrix} 1 \\ 0 \\ 2 \end{bmatrix}, \qquad \text{and} \qquad \alpha_3 = \begin{bmatrix} 1 \\ 1 \\ 0 \end{bmatrix}.$$

The vector

$$\alpha = \begin{bmatrix} 2 \\ 1 \\ 5 \end{bmatrix}$$

is a linear combination of α_1, α_2, and α_3 if we can find a_1, a_2, and a_3 so that $a_1\alpha_1 + a_2\alpha_2 + a_3\alpha_3 = \alpha$. Substituting for α, α_1, α_2, and α_3, we have

$$a_1 \begin{bmatrix} 1 \\ 2 \\ 1 \end{bmatrix} + a_2 \begin{bmatrix} 1 \\ 0 \\ 2 \end{bmatrix} + a_3 \begin{bmatrix} 1 \\ 1 \\ 0 \end{bmatrix} = \begin{bmatrix} 2 \\ 1 \\ 5 \end{bmatrix}.$$

Equating corresponding entries leads to the linear system

$$\begin{aligned} a_1 + a_2 + a_3 &= 2 \\ 2a_1 + a_3 &= 1 \\ a_1 + 2a_2 &= 5. \end{aligned}$$

Solving this linear system by the methods of Chapter 1 gives $a_1 = 1$, $a_2 = 2$, and $a_3 = -1$ (verify) which means that α is a linear combination of α_1, α_2, and α_3. Thus

$$\alpha = \alpha_1 + 2\alpha_2 - \alpha_3.$$

Definition 2.4. Let $S = \{\alpha_1, \alpha_2, \ldots, \alpha_k\}$ be a set of vectors in vector space V. The set S **spans** V, or V is **spanned** by S, if every vector in V is a linear combination of the vectors in S.

EXAMPLE 2. Let V be the vector space R^3. Let

$$\alpha_1 = \begin{bmatrix} 1 \\ 2 \\ 1 \end{bmatrix}, \qquad \alpha_2 = \begin{bmatrix} 1 \\ 0 \\ 2 \end{bmatrix}, \qquad \text{and} \qquad \alpha_3 = \begin{bmatrix} 1 \\ 1 \\ 0 \end{bmatrix}.$$

To find out whether $\{\alpha_1, \alpha_2, \alpha_3\}$ spans V, we pick any vector $\alpha = \begin{bmatrix} a \\ b \\ c \end{bmatrix}$ in V

(a, b, and c are arbitrary real numbers) and must find out whether there are constants a_1, a_2, and a_3 such that

$$a_1\alpha_1 + a_2\alpha_2 + a_3\alpha_3 = \alpha.$$

This leads to the linear system (verify)

$$\begin{aligned}
a_1 + a_2 + a_3 &= a \\
2a_1 + a_3 &= b \\
a_1 + 2a_2 &= c.
\end{aligned}$$

A solution is (verify)

$$a_1 = \frac{-2a + 2b + c}{3}, \qquad a_2 = \frac{a - b + c}{3}, \qquad a_3 = \frac{4a - b - 2c}{3}.$$

Thus $\{\alpha_1, \alpha_2, \alpha_3\}$ spans V.

EXAMPLE 3. Let V be P_2, the vector space consisting of all polynomials of degree ≤ 2 and the zero polynomial. Let $\alpha_1 = t^2 + 2t + 1$ and $\alpha_2 = t^2 + 2$. Does $\{\alpha_1, \alpha_2\}$ span V? Again let $\alpha = at^2 + bt + c$ be any vector in V, where

a, b, and c are any real numbers. We must find out whether there are constants a_1 and a_2 such that

$$\alpha = a_1\alpha_1 + a_2\alpha_2,$$

or

$$at^2 + bt + c = a_1(t^2 + 2t + 1) + a_2(t^2 + 2).$$

Thus

$$(a_1 + a_2)t^2 + (2a_1)t + (a_1 + 2a_2) = at^2 + bt + c.$$

Now two polynomials agree for all values of t only if the coefficients of respective powers of t agree. Thus we get the linear system

$$\begin{array}{rcr} a_1 + a_2 &=& a \\ 2a_1 &=& b \\ a_1 + 2a_2 &=& c. \end{array}$$

Transforming the augmented matrix of this linear system into reduced row echelon form, we obtain (verify)

$$\begin{bmatrix} 1 & 0 & \vdots & 2a - c \\ 0 & 1 & \vdots & c - a \\ 0 & 0 & \vdots & b - 4a + 2c \end{bmatrix}.$$

If $b - 4a + 2c \neq 0$, then there is no solution. Hence $\{\alpha_1, \alpha_2\}$ does not span V.

Definition 2.5. Let $S = \{\alpha_1, \alpha_2, \ldots, \alpha_k\}$ be a set of distinct vectors in a vector space V. Then S is said to be **linearly dependent** if there exist constants $a_1, a_2, \ldots, a_k$, not all zero, such that

$$a_1\alpha_1 + a_2\alpha_2 + \cdots + a_k\alpha_k = \theta. \tag{1}$$

Otherwise, S is called **linearly independent**. That is, S is linearly independent if (1) holds only when

$$a_1 = a_2 = \cdots = a_k = 0.$$

It should be emphasized that for any set of distinct vectors $S = \{\alpha_1, \alpha_2, \ldots, \alpha_k\}$, (1) always holds if we choose all the scalars $a_1, a_2, \ldots, a_k$ equal to zero. The important point in this definition is whether or not it is possible to satisfy (1) with at least one of the scalars different from zero.

EXAMPLE 4. Let V be R_3 and $\alpha_1 = [1 \ \ 0 \ \ 0]$, $\alpha_2 = [0 \ \ 1 \ \ 0]$, and $\alpha_3 = [0 \ \ 0 \ \ 1]$. To find out whether $S = \{\alpha_1, \alpha_2, \alpha_3\}$ is linearly dependent or linearly independent, we form (1), $a_1\alpha_1 + a_2\alpha_2 + a_3\alpha_3 = \theta$, and solve for $a_1, a_2,$ and a_3. Since $a_1, a_2,$ and a_3 turn out to be 0 (verify), we conclude that S is linearly independent.

EXAMPLE 5. Let V be P_2 and $\alpha_1 = t^2 + t + 2$, $\alpha_2 = 2t^2 + t$, $\alpha_3 = 3t^2 + 2t + 2$. To find out whether $S = \{\alpha_1, \alpha_2, \alpha_3\}$ is linearly dependent or linearly independent, we form (1), $a_1\alpha_1 + a_2\alpha_2 + a_3\alpha_3 = \theta$, and solve for $a_1, a_2,$ and a_3. The resulting homogeneous system is (verify)

$$
\begin{aligned}
a_1 + 2a_2 + 3a_3 &= 0 \\
a_1 + \ \ a_2 + 2a_3 &= 0 \\
2a_1 \ \ \ \ \ \ \ + 2a_3 &= 0,
\end{aligned}
$$

which has infinitely many solutions (verify). A particular solution is $a_1 = 1$, $a_2 = 1, a_3 = -1$, so

$$
\alpha_1 + \alpha_2 - \alpha_3 = \theta.
$$

Hence S is linearly dependent.

Theorem 2.3. *Let S_1 and S_2 be finite subsets of a vector space and let S_1 be a subset of S_2. Then:*
(a) *If S_1 is linearly dependent, so is S_2.*
(b) *If S_2 is linearly independent, so is S_1.*

Proof: Let $S_1 = \{\alpha_1, \alpha_2, \ldots, \alpha_k\}$ and $S_2 = \{\alpha_1, \alpha_2, \ldots, \alpha_m\}$. We first prove (a). Since S_1 is linearly dependent, there exist $a_1, a_2, \ldots, a_k$, not all zero, such that

$$
a_1\alpha_1 + a_2\alpha_2 + \cdots + a_k\alpha_k = \theta.
$$

Then

$$
a_1\alpha_1 + a_2\alpha_2 + \cdots + a_k\alpha_k + 0\alpha_{k+1} + \cdots + 0\alpha_m = \theta.
$$

Since not all the coefficients are zero, we conclude that S_2 is linearly dependent.
Next, we prove (b). Let S_2 be linearly independent. If S_1 is assumed as

linearly dependent, then S_2 is linearly dependent, by (a), a contradiction. Hence S_1 must be linearly independent.

We now note that the set $S = \{\theta\}$ consisting only of θ is linearly dependent, for $5\theta = \theta$ and $5 \neq 0$. From this it follows that if S is any set of vectors that contains θ, then S must be linearly dependent. Also, a set consisting of a single nonzero vector is linearly independent (verify).

We now consider the meaning of linear dependence in R^2 and R^3. Suppose that $\{\alpha_1, \alpha_2\}$ is linearly dependent in R^2. Then there exist scalars a_1 and a_2 not both zero such that $a_1\alpha_1 + a_2\alpha_2 = \theta$. If $a_1 \neq 0$, then $\alpha_1 = -(a_2/a_1)\alpha_2$. If $a_2 \neq 0$, then $\alpha_2 = -(a_1/a_2)\alpha_1$. Thus one of the vectors is a multiple of the other. Conversely, suppose that $\alpha_1 = a\alpha_2$. Then $\alpha_1 - a\alpha_2 = \theta$, and since the coefficients of α_1 and α_2 are not both zero, it follows that $\{\alpha_1, \alpha_2\}$ is linearly dependent. Thus $\{\alpha_1, \alpha_2\}$ is linearly dependent in R^2 if and only if one of the vectors is a multiple of the other (Figure 2.7a). Hence two vectors in R^2 are linearly dependent if and only if they both lie on the same line passing through the origin (Figure 2.7a and b).

Suppose now that $\{\alpha_1, \alpha_2, \alpha_3\}$ is a linearly dependent set of vectors in R^3. Then we can write

$$a_1\alpha_1 + a_2\alpha_2 + a_3\alpha_3 = \theta,$$

where a_1, a_2, and a_3 are not all zero. If $a_2 \neq 0$, then

$$\alpha_2 = -\frac{a_1}{a_2}\alpha_1 - \frac{a_3}{a_2}\alpha_3,$$

which means that α_2 is in the subspace of R^3 spanned by the vectors α_1 and α_3, that is, the plane through the origin determined by α_1 and α_3. Conversely, if α_2 is in the plane through the origin spanned by α_1 and α_3, then α_2 is a linear combination of α_1 and α_3:

$$\alpha_2 = a_1\alpha_1 + a_3\alpha_3.$$

Then

$$a_1\alpha_1 - 1\alpha_2 + a_3\alpha_3 = \theta,$$

FIGURE 2.7

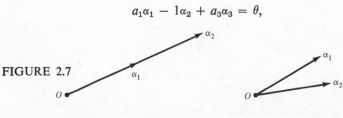

(a) Linearly dependent.   (b) Linearly independent.

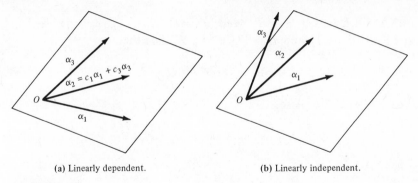

(a) Linearly dependent.   (b) Linearly independent.

FIGURE 2.8

which means that $\{\alpha_1, \alpha_2, \alpha_3\}$ is linearly dependent. Thus three vectors in R^3 are linearly dependent if and only if they all lie in the same plane passing through the origin (Figure 2.8a and b).

Theorem 2.4. Let $S = \{\alpha_1, \alpha_2, \ldots, \alpha_n\}$ be a set of nonzero vectors in a vector space V. Then S is linearly dependent if and only if one of the vectors α_j is a linear combination of the preceding vectors in S.

Proof: If α_j is a linear combination of the preceding vectors,

$$\alpha_j = a_1\alpha_1 + a_2\alpha_2 + \cdots + a_{j-1}\alpha_{j-1},$$

then

$$a_1\alpha_1 + a_2\alpha_2 + \cdots + a_{j-1}\alpha_{j-1} + (-1)\alpha_j + 0\alpha_{j+1} + \cdots + 0\alpha_n = \theta.$$

Since at least one coefficient, -1, is nonzero, we conclude that S is linearly dependent.

Conversely, let S be linearly dependent. Then there exist scalars $a_1, a_2, \ldots, a_n$, not all zero, such that $a_1\alpha_1 + a_2\alpha_2 + \cdots + a_n\alpha_n = \theta$. Now let j be the largest subscript for which $a_j \neq 0$. If $j > 1$, then

$$\alpha_j = -\left(\frac{a_1}{a_j}\right)\alpha_1 - \left(\frac{a_2}{a_j}\right)\alpha_2 - \cdots - \left(\frac{a_{j-1}}{a_j}\right)\alpha_{j-1}.$$

If $j = 1$, then $a_j\alpha_j = \theta$, which implies that $\alpha_j = \theta$, a contradiction to the hypothesis that none of the vectors in S is the zero vector. Thus one of the vectors in S is a linear combination of the preceding vectors in S.

EXAMPLE 6. Let $V = R_3$ and $\alpha_1 = [1 \quad 2 \quad -1]$, $\alpha_2 = [1 \quad -2 \quad 1]$, $\alpha_3 = [-3 \quad 2 \quad -1]$, and $\alpha_4 = [2 \quad 0 \quad 0]$. We find that

$$\alpha_1 + 2\alpha_2 + \alpha_3 + 0\alpha_4 = \theta,$$

so $\alpha_3 = -\alpha_1 - 2\alpha_2$. Also,

$$\alpha_1 + \alpha_2 + 0\alpha_3 - \alpha_4 = \theta,$$

so $\alpha_4 = \alpha_1 + \alpha_2 + 0\alpha_3$.

We can also prove that if $S = \{\alpha_1, \alpha_2, \ldots, \alpha_n\}$ is a set of vectors in a vector space V, then S is linearly dependent if and only if one of the vectors in S is a linear combination of all the other vectors in S (see Exercise 4). Thus, in Example 6, $\alpha_1 = -\alpha_3 - 2\alpha_2$, $\alpha_2 = -\frac{1}{2}\alpha_1 - \frac{1}{2}\alpha_3$, $\alpha_1 = \alpha_4 - \alpha_2$, and $\alpha_2 = \alpha_4 - a_1$.

Basis

Definition 2.6. A set of vectors $S = \{\alpha_1, \alpha_2, \ldots, \alpha_k\}$ in a vector space V is called a **basis** for V if S spans V and S is linearly independent.

EXAMPLE 7. Let $V = R^3$ and $S = \left\{ \begin{bmatrix} 1 \\ 0 \\ 0 \end{bmatrix}, \begin{bmatrix} 0 \\ 1 \\ 0 \end{bmatrix}, \begin{bmatrix} 0 \\ 0 \\ 1 \end{bmatrix} \right\}$. Then S is a basis for R^3, called the **natural basis** for R^3. One can easily see how to generalize this to obtain the natural basis for R^n. Similarly,

$$S = \{[1 \quad 0 \quad 0], [0 \quad 1 \quad 0], [0 \quad 0 \quad 1]\}$$

is the natural basis for R_3.

The natural basis for R^n is denoted by $\varepsilon_1, \varepsilon_2, \ldots, \varepsilon_n$, where

$$\varepsilon_i = \begin{bmatrix} 0 \\ 0 \\ \vdots \\ 1 \\ 0 \\ \vdots \\ 0 \end{bmatrix} \leftarrow ith \text{ row};$$

that is, ε_i is an $n \times 1$ matrix with a 1 in the ith row and zeros elsewhere.
The natural basis for R^3 is also often denoted by

$$i = \begin{bmatrix} 1 \\ 0 \\ 0 \end{bmatrix}, \quad j = \begin{bmatrix} 0 \\ 1 \\ 0 \end{bmatrix}, \quad \text{and} \quad k = \begin{bmatrix} 0 \\ 0 \\ 1 \end{bmatrix}.$$

These vectors are shown in Figure 2.9. Thus any vector $\alpha = \begin{bmatrix} a_1 \\ a_2 \\ a_3 \end{bmatrix}$ in R^3 can

be written as

$$a_1 i + a_2 j + a_3 k.$$

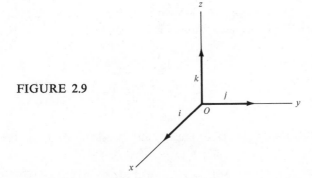

FIGURE 2.9

EXAMPLE 8. The set $S = \{t^2 + 1, t - 1, 2t + 2\}$ is a basis for the vector
space P_2. To show this, we must show that S spans V and is linearly independent. To show that it spans V, we take any vector in V, that is, a polynomial
$at^2 + bt + c$, and must find constants a_1, a_2, and a_3 such that

$$at^2 + bt + c = a_1(t^2 + 1) + a_2(t - 1) + a_3(2t + 2)$$
$$= a_1 t^2 + (a_2 + 2a_3)t + (a_1 - a_2 + 2a_3).$$

Since two polynomials agree for all values of t only if the coefficients of
respective powers of t agree, we get the linear system

$$\begin{aligned} a_1 \qquad\qquad\quad &= a \\ a_2 + 2a_3 &= b \\ a_1 - a_2 + 2a_3 &= c. \end{aligned}$$

Solving, we have

$$a_1 = a, \qquad a_2 = \frac{a + b - c}{2}, \qquad a_3 = \frac{c + b - a}{4}.$$

Hence S spans V. To show that S is linearly independent, we form

$$a_1(t^2 + 1) + a_2(t - 1) + a_3(2t + 2) = 0.$$

Then

$$a_1 t^2 + (a_2 + 2a_3)t + (a_1 - a_2 + 2a_3) = 0.$$

Again, this can hold for all values of t only if

$$
\begin{aligned}
a_1 \phantom{{}+ a_2 + 2a_3} &= 0 \\
a_2 + 2a_3 &= 0 \\
a_1 - a_2 + 2a_3 &= 0.
\end{aligned}
$$

The only solution to this homogeneous system is $a_1 = a_2 = a_3 = 0$, which implies that S is linearly independent. Thus S is a basis for P_2.

EXAMPLE 9. The set

$$S = \left\{ \begin{bmatrix} 1 & 0 \\ 0 & 0 \end{bmatrix}, \begin{bmatrix} 0 & 1 \\ 0 & 0 \end{bmatrix}, \begin{bmatrix} 0 & 0 \\ 1 & 0 \end{bmatrix}, \begin{bmatrix} 0 & 0 \\ 0 & 1 \end{bmatrix} \right\}$$

is a basis for $_2R_2$. To verify that S is linearly independent, we form the following equation:

$$a_1 \begin{bmatrix} 1 & 0 \\ 0 & 0 \end{bmatrix} + a_2 \begin{bmatrix} 0 & 1 \\ 0 & 0 \end{bmatrix} + a_3 \begin{bmatrix} 0 & 0 \\ 1 & 0 \end{bmatrix} + a_4 \begin{bmatrix} 0 & 0 \\ 0 & 1 \end{bmatrix} = \begin{bmatrix} 0 & 0 \\ 0 & 0 \end{bmatrix}.$$

This gives

$$\begin{bmatrix} a_1 & a_2 \\ a_3 & a_4 \end{bmatrix} = \begin{bmatrix} 0 & 0 \\ 0 & 0 \end{bmatrix},$$

which implies that $a_1 = a_2 = a_3 = a_4 = 0$. Hence S is linearly independent. To verify that S spans $_2R_2$, we take any vector $\begin{bmatrix} a & b \\ c & d \end{bmatrix}$ in $_2R_2$ and we must find scalars $a_1, a_2, a_3,$ and a_4 such that

$$\begin{bmatrix} a & b \\ c & d \end{bmatrix} = a_1 \begin{bmatrix} 1 & 0 \\ 0 & 0 \end{bmatrix} + a_2 \begin{bmatrix} 0 & 1 \\ 0 & 0 \end{bmatrix} + a_3 \begin{bmatrix} 0 & 0 \\ 1 & 0 \end{bmatrix} + a_4 \begin{bmatrix} 0 & 0 \\ 0 & 1 \end{bmatrix}.$$

We find that $a_1 = a, a_2 = b, a_3 = c,$ and $a_4 = d,$ so that S spans $_2R_2$.

We now establish some results that will tell about the number of vectors in a basis, compare two different bases, and give properties of bases. First, we observe that if $\{\alpha_1, \alpha_2, \ldots, \alpha_k\}$ is a basis for a vector space V, then $\{c\alpha_1, \alpha_2, \ldots, \alpha_k\}$ is also a basis when $c \neq 0$. Thus a real vector space always has infinitely many bases.

Theorem 2.5. *If $S = \{\alpha_1, \alpha_2, \ldots, \alpha_n\}$ is a basis for a vector space V, then every vector in V can be written in one and only one way as a linear combination of the vectors in S.*

Proof: First, every vector α in V can be written as a linear combination of the vectors in S, because S spans V. Now let

$$\alpha = a_1\alpha_1 + a_2\alpha_2 + \cdots + a_n\alpha_n$$

and

$$\alpha = b_1\alpha_1 + b_2\alpha_2 + \cdots + b_n\alpha_n.$$

We must show that $a_i = b_i$ for $i = 1, 2, \ldots, n$. We have

$$\theta = \alpha - \alpha = (a_1 - b_1)\alpha_1 + (a_2 - b_2)\alpha_2 + \cdots + (a_n - b_n)\alpha_n.$$

Since S is linearly independent, we conclude that

$$a_i - b_i = 0 \quad \text{for } i = 1, 2, \ldots, n.$$

Theorem 2.6. *If $S = \{\alpha_1, \alpha_2, \ldots, \alpha_n\}$ is a set of nonzero vectors spanning a vector space V, then S contains a basis T for V.*

Proof: If S itself is linearly independent, then S is a basis for V. If S is linearly dependent, then some α_j is a linear combination of the preceding vectors in S (Theorem 2.4). We now delete α_j from S, getting a subset S_1 of S. Then $S_1 = \{\alpha_1, \alpha_2, \ldots, \alpha_{j-1}, \alpha_{j+1}, \ldots, \alpha_n\}$ also spans V, for if α is any vector in V, then, since S spans V, we can find scalars $a_1, a_2, \ldots, a_n$ such that

$$\alpha = a_1\alpha_1 + a_2\alpha_2 + \cdots + a_{j-1}\alpha_{j-1} + a_j\alpha_j + a_{j+1}\alpha_{j+1} + \cdots + a_n\alpha_n.$$

Now if

$$\alpha_j = b_1\alpha_1 + b_2\alpha_2 + \cdots + b_{j-1}\alpha_{j-1},$$

then

$$\alpha = a_1\alpha_1 + a_2\alpha_2 + \cdots + a_{j-1}\alpha_{j-1} + a_j(b_1\alpha_1 + b_2\alpha_2 + \cdots + b_{j-1}\alpha_{j-1})$$
$$+ a_{j+1}\alpha_{j+1} + \cdots + a_n\alpha_n = c_1\alpha_1 + c_2\alpha_2$$
$$+ \cdots + c_{j-1}\alpha_{j-1} + c_{j+1}\alpha_{j+1} + \cdots + c_n\alpha_n,$$

which means that S_1 spans V. If S_1 is linearly independent, then S_1 is a basis. If S_1 is linearly dependent, delete a vector that is a linear combination of the preceding vectors of S_1 and get a new set S_2 which spans V. Continuing, since S is a finite set, we find a subset T of S that is linearly independent and spans V. The set T is a basis for V.

EXAMPLE 10. Let $V = R_3$ and $S = \{\alpha_1, \alpha_2, \alpha_3, \alpha_4, \alpha_5\}$, where $\alpha_1 = [1 \quad 0 \quad 1]$, $\alpha_2 = [0 \quad 1 \quad 1]$, $\alpha_3 = [1 \quad 1 \quad 2]$, $\alpha_4 = [1 \quad 2 \quad 1]$, and $\alpha_5 = [-1 \quad 1 \quad -2]$. We find that S spans R_3 and now wish to find a subset of R_3 that is a basis for R_3. First, S is linearly dependent and

$$2\alpha_1 + \alpha_2 - \alpha_4 + \alpha_5 = \theta$$

(verify). Hence $\alpha_5 = -2\alpha_1 - \alpha_2 + \alpha_4$. If $\alpha = a_1\alpha_1 + a_2\alpha_2 + \cdots + a_5\alpha_5$ is any vector in R_3, then

$$\alpha = a_1\alpha_1 + \cdots + a_4\alpha_4 + a_5(-2\alpha_1 - \alpha_2 + \alpha_4).$$

Thus α is a linear combination of α_1, α_2, α_3, and α_4, which means that $S_1 = \{\alpha_1, \alpha_2, \alpha_3, \alpha_4\}$ also spans R_3. We now find that S_1 is linearly dependent (verify) and that $\alpha_1 + \alpha_2 - \alpha_3 = \theta$, so $\alpha_3 = \alpha_1 + \alpha_2$. If we delete α_3 from S_1, we get $S_2 = \{\alpha_1, \alpha_2, \alpha_4\}$, which also spans R_3. Since S_2 is linearly independent (verify), it is a basis for R_3.

We are now about to establish a major result (Corollary 2.1) of this section, which will tell us about the number of vectors in two different bases.

Theorem 2.7. *If* $S = \{\alpha_1, \alpha_2, \ldots, \alpha_n\}$ *is a basis for a vector space V and T =* $\{\beta_1, \beta_2, \ldots, \beta_r\}$ *is a linearly independent set of vectors in V, then* $n \geq r$.

Proof: Suppose that $r > n$. Since S spans V, β_1 is a linear combination of the vectors in S:

$$\beta_1 = a_1\alpha_1 + a_2\alpha_2 + \cdots + a_n\alpha_n. \tag{2}$$

Observe that none of the vectors in T can be zero, since T is linearly independent. Thus $\beta_1 \neq \theta$, so in Equation (2) some $a_i \neq 0$. Suppose that

$a_n \neq 0$ (if this is not so, we can renumber the elements of T to make it so). Then

$$\alpha_n = \frac{1}{a_n} \beta_1 - \frac{1}{a_n} (a_1\alpha_1 + a_2\alpha_2 + \cdots + a_{n-1}\alpha_{n-1}).$$

This means that $S_1 = \{\alpha_1, \alpha_2, \ldots, \alpha_{n-1}, \beta_1\}$ spans V. We can then write

$$\beta_2 = b_1\alpha_1 + b_2\alpha_2 + \cdots + b_{n-1}\alpha_{n-1} + c\beta_1. \tag{3}$$

Since $\beta_2 \neq \theta$, not all the coefficients in (3) can be zero. If $b_1 = b_2 = \cdots = b_{n-1} = 0$, then $\beta_2 = c\beta_1$, which would imply by Theorem 2.4 that T is linearly dependent, a contradiction. Thus some $b_i \neq 0$. Suppose that $b_{n-1} \neq 0$ (if this is not so, we can again renumber $\beta_1, \beta_2, \ldots, \beta_{n-1}$). Then

$$\alpha_{n-1} = \frac{1}{b_{n-1}} \beta_2 - \frac{1}{b_{n-1}} (b_1\alpha_1 + b_2\alpha_2 + \cdots + b_{n-2}\alpha_{n-2} + c\beta_1),$$

which implies that $S_2 = \{\alpha_1, \alpha_2, \ldots, \alpha_{n-2}, \beta_1, \beta_2\}$ spans V. Repeating this process n times, we find that the set $S_n = \{\beta_1, \beta_2, \ldots, \beta_n\}$ spans V. Since $r > n$, we can also write

$$\beta_{n+1} = d_1\beta_1 + d_2\beta_2 + \cdots + d_n\beta_n,$$

which implies that T is linearly dependent, a contradiction. Hence $n \geq r$.

Corollary 2.1. *If $S = \{\alpha_1, \alpha_2, \ldots, \alpha_n\}$ and $T = \{\beta_1, \beta_2, \ldots, \beta_m\}$ are bases for a vector space V, then $n = m$.*

Proof: Since S is a basis and T is linearly independent, Theorem 2.7 implies that $n \geq m$. Similarly, we obtain $m \geq n$ because T is a basis and S is linearly independent. Hence $n = m$.

Thus, although a vector space has many bases, all bases have the same number of vectors. We can then make the following definition.

Definition 2.7. The **dimension** of a nonzero vector space V is the number of vectors in a basis for V. We often write **dim** V for the dimension of V. We also define the dimension of $\{\theta\}$ to be zero.

EXAMPLE 11. The set $S = \{t^2, t, 1\}$ is a basis for P_2, so dim $P_2 = 3$.

EXAMPLE 12. Let V be the subspace of R_3 spanned by $S = \{\alpha_1, \alpha_2, \alpha_3\}$, where $\alpha_1 = [0 \quad 1 \quad 1]$, $\alpha_2 = [1 \quad 0 \quad 1]$, and $\alpha_3 = [1 \quad 1 \quad 2]$. Thus every vector in V is of the form $a_1\alpha_1 + a_2\alpha_2 + a_3\alpha_3$, where a_1, a_2, and a_3 are arbitrary real numbers. We find that S is linearly dependent, and $\alpha_3 = \alpha_1 + \alpha_2$ (verify). Thus $S_1 = \{\alpha_1, \alpha_2\}$ also spans V. Since S_1 is linearly independent (verify), we conclude that it is a basis for V. Hence dim $V = 2$.

If V has dimension n, then any set of $n + 1$ vectors in V is necessarily linearly dependent, and also a set of $n - 1$ vectors in V cannot span V. For if $S = \{\alpha_1, \alpha_2, \ldots, \alpha_{n-1}\}$ spans V, then S contains a subset T which is a basis for V, implying that the dimension of V is $< n$, a contradiction. It is also easy to see that since the set $\{\theta\}$ is linearly dependent, then it is natural to say that the vector space $\{\theta\}$ has dimension zero.

Thus R^3 has dimension 3, R_2 has dimension 2, and R^n and R_n both have dimension n. Similarly, P_3 has dimension 4 because $\{t^3, t^2, t, 1\}$ is a basis for P_3. In general, P_n has dimension $n + 1$. All vector spaces considered henceforth in this book are finite-dimensional; that is, their dimension is a finite number. However, we point out that there are vector spaces of infinite dimension which are extremely important in mathematics and physics; their study lies beyond the scope of this book. The vector space P of all polynomials is an infinite-dimensional vector space (Exercise 18).

It is easy to show (Exercise 28) that if W is a nonzero subspace of a finite-dimensional vector space V, then dim $W \leq$ dim V.

We might also consider all the subspaces of R^2, the (x, y)-plane. First, we have $\{\theta\}$ and R^2, the trivial subspaces of dimension 0 and 2, respectively. The subspace V of R^2 spanned by a vector $\alpha \neq \theta$ is a one-dimensional subspace of R^2; V is a line through the origin. Thus the subspaces of R^2 are $\{\theta\}$, R^2, and all the lines through the origin.

We now prove a theorem that we shall have occasion to use several times in constructing a basis containing a given set of linearly independent vectors.

Theorem 2.8. *If S is a linearly independent set of vectors in a finite-dimensional vector space V, then there is a basis T for V which contains S.*

Proof: Let $S = \{\alpha_1, \alpha_2, \ldots, \alpha_m\}$ be a linearly independent set of vectors in the n-dimensional vector space V, where $m < n$. Now let $\{\gamma_1, \gamma_2, \ldots, \gamma_n\}$ be a basis for V and let $S_1 = \{\alpha_1, \alpha_2, \ldots, \alpha_m, \gamma_1, \gamma_2, \ldots, \gamma_n\}$. Since S_1 spans V, it contains, by Theorem 2.6, a basis T for V. Recall that T is obtained by deleting from S_1 every vector that is a linear combination of the preceding vectors. Since S is linearly independent, none of the α_i can be linear combinations of other α_j and thus are not deleted. Hence T will contain S.

EXAMPLE 13. To find a basis for R_3 that contains the vector $\alpha = [1 \quad 0 \quad 1]$, we use Theorem 2.8 as follows. First, let $\{\varepsilon_1', \varepsilon_2', \varepsilon_3'\}$ be the natural basis for R_3, where $\varepsilon_1' = [1 \quad 0 \quad 0]$, $\varepsilon_2' = [0 \quad 1 \quad 0]$, and $\varepsilon_3' = [0 \quad 0 \quad 1]$. Let $S_1 = \{[1 \quad 0 \quad 1], [1 \quad 0 \quad 0], [0 \quad 1 \quad 0], [0 \quad 0 \quad 1]\}$. Since ε_1' is not a linear combination of α, we retain ε_1'. Now we check whether $\varepsilon_2' = [0 \quad 1 \quad 0]$ is a linear combination of α and ε_1'. Since the answer is no, we retain ε_2'. Since ε_3' is a linear combination of α, ε_1', and ε_2', we delete it. Thus $\{\alpha, \varepsilon_1', \varepsilon_2'\}$ is a basis for R_3.

As defined earlier, a given set S of vectors in a vector space V is a basis for V if it spans V and is linearly independent. However, if we are given the *additional* information that the dimension of V is n, we need only verify one of the two conditions. This is the content of the following theorem.

Theorem 2.9. *Let V be an n-dimensional vector space.*
   (a) *If $S = \{\alpha_1, \alpha_2, \ldots, \alpha_n\}$ is a linearly independent set of vectors in V, then S is a basis for V.*
   (b) *If $S = \{\alpha_1, \alpha_2, \ldots, \alpha_n\}$ spans V, then S is a basis for V.*

Proof: Exercise.

We introduce another useful property of bases.

Definition 2.8. Let S be a set of vectors in a vector space V. A subset T of S is called a **maximal independent subset** of S if T is a linearly independent set of vectors of S, and if there is no linearly independent subset of S having more vectors than T does.

EXAMPLE 14. Let V be R^3 and consider the set $S = \{\alpha_1, \alpha_2, \alpha_3, \alpha_4\}$, where

$$\alpha_1 = \begin{bmatrix} 1 \\ 0 \\ 0 \end{bmatrix}, \quad \alpha_2 = \begin{bmatrix} 0 \\ 1 \\ 0 \end{bmatrix}, \quad \alpha_3 = \begin{bmatrix} 0 \\ 0 \\ 1 \end{bmatrix}, \quad \text{and} \quad \alpha_4 = \begin{bmatrix} 1 \\ 1 \\ 1 \end{bmatrix}.$$

Maximal independent subsets of S are

$$\{\alpha_1, \alpha_2, \alpha_3\}, \quad \{\alpha_1, \alpha_2, \alpha_4\}, \quad \{\alpha_1, \alpha_3, \alpha_4\}, \quad \text{and} \quad \{\alpha_2, \alpha_3, \alpha_4\}.$$

Theorem 2.10. *Let S be a finite subset of the vector space V that spans V. A maximal independent subset T of S is a basis for V.*

Proof: Exercise.

We conclude this section with one more important example of the properties of the solution space of the homogeneous system $AX = O$.

EXAMPLE 15. Suppose that we wish to find a basis for the solution space V of the homogeneous system

$$\begin{bmatrix} 1 & 2 & 0 & 3 & 1 \\ 2 & 3 & 0 & 3 & 1 \\ 1 & 1 & 2 & 2 & 1 \\ 3 & 5 & 0 & 6 & 2 \\ 2 & 3 & 2 & 5 & 2 \end{bmatrix} \begin{bmatrix} x_1 \\ x_2 \\ x_3 \\ x_4 \\ x_5 \end{bmatrix} = \begin{bmatrix} 0 \\ 0 \\ 0 \\ 0 \\ 0 \end{bmatrix}.$$

Gauss–Jordan reduction leads to the equivalent system

$$\begin{bmatrix} 1 & 0 & 0 & -3 & -1 \\ 0 & 1 & 0 & 3 & 1 \\ 0 & 0 & 1 & 1 & \frac{1}{2} \\ 0 & 0 & 0 & 0 & 0 \\ 0 & 0 & 0 & 0 & 0 \end{bmatrix} \begin{bmatrix} x_1 \\ x_2 \\ x_3 \\ x_4 \\ x_5 \end{bmatrix} = \begin{bmatrix} 0 \\ 0 \\ 0 \\ 0 \\ 0 \end{bmatrix},$$

whose general solution is (verify)

$$X = \begin{bmatrix} 3s + t \\ -3s - t \\ -s - \frac{1}{2}t \\ s \\ t \end{bmatrix},$$

where s and t are any real numbers. We can write

$$X = s \begin{bmatrix} 3 \\ -3 \\ -1 \\ 1 \\ 0 \end{bmatrix} + t \begin{bmatrix} 1 \\ -1 \\ -\frac{1}{2} \\ 0 \\ 1 \end{bmatrix}. \tag{4}$$

Since s and t can take on any values, letting them first be 1 and 0, and then 0 and 1, we get as solutions

$$
X_1 = \begin{bmatrix} 3 \\ -3 \\ -1 \\ 1 \\ 0 \end{bmatrix} \quad \text{and} \quad X_2 = \begin{bmatrix} 1 \\ -1 \\ -\frac{1}{2} \\ 0 \\ 1 \end{bmatrix}.
$$

From (4) it is clear that $\{X_1, X_2\}$ spans V, and since $\{X_1, X_2\}$ is linearly independent (verify), it is a basis for V.

2.2. Exercises

1. Does the set $\left\{ \begin{bmatrix} 1 \\ 2 \\ 3 \end{bmatrix}, \begin{bmatrix} 0 \\ 0 \\ 1 \end{bmatrix} \right\}$ span R^3?

2. Does the set $\{t^2 + 1, t + 2\}$ span P_2?

3. Prove that the set $\{[1 \quad 0 \quad 0], [0 \quad 1 \quad 0], [0 \quad 0 \quad 1], [1 \quad 1 \quad 1]\}$ spans R_3.

4. Let $S = \{\alpha_1, \alpha_2, \ldots, \alpha_k\}$ be a set of vectors in a vector space V. Prove that S is linearly dependent if and only if one of the vectors is a linear combination of the others.

5. Which of the following sets of vectors in R_3 are linearly dependent? For those which are, express one vector as a linear combination of the rest:
   (a) $\{[1 \quad 1 \quad 0], [0 \quad 2 \quad 3], [1 \quad 2 \quad 3], [3 \quad 6 \quad 6]\}$.
   (b) $\{[1 \quad 1 \quad 0], [3 \quad 4 \quad 2]\}$.
   (c) $\{[1 \quad 1 \quad 0], [0 \quad 2 \quad 3], [1 \quad 2 \quad 3], [0 \quad 0 \quad 0]\}$.

6. Consider the vector space $_2R_2$. Follow the directions of Exercise 5.
   (a) $\left\{ \begin{bmatrix} 1 & 1 \\ 2 & 1 \end{bmatrix}, \begin{bmatrix} 1 & 0 \\ 0 & 2 \end{bmatrix}, \begin{bmatrix} 0 & 3 \\ 2 & 1 \end{bmatrix}, \begin{bmatrix} 4 & 6 \\ 8 & 6 \end{bmatrix} \right\}$.
   (b) $\left\{ \begin{bmatrix} 1 & 1 \\ 1 & 1 \end{bmatrix}, \begin{bmatrix} 1 & 0 \\ 0 & 2 \end{bmatrix}, \begin{bmatrix} 0 & 1 \\ 0 & 2 \end{bmatrix} \right\}$,
   (c) $\left\{ \begin{bmatrix} 1 & 1 \\ 1 & 1 \end{bmatrix}, \begin{bmatrix} 2 & 3 \\ 1 & 2 \end{bmatrix}, \begin{bmatrix} 3 & 1 \\ 2 & 1 \end{bmatrix}, \begin{bmatrix} 2 & 2 \\ 1 & 1 \end{bmatrix} \right\}$.

7. Consider the vector space P_2. Follow the directions of Exercise 5.
   (a) $\{t^2 + 1, t - 2, t + 3\}$.
   (b) $\{2t^2 + t, t^2 + 3, t\}$.
   (c) $\{2t^2 + t + 1, 3t^2 + t - 5, t + 13\}$.

8. Let V be the vector space of all real-valued continuous functions. Follow the directions of Exercise 5.
   (a) $\{\cos t, \sin t, e^t\}$.
   (b) $\{t, e^t, \sin t\}$.
   (c) $\{t^2, t, e^t\}$.   (d) $\{\cos^2 t, \sin^2 t, \cos 2t\}$.

9. Let A be an $m \times n$ matrix in reduced row echelon form. Prove that the non-zero rows of A, viewed as vectors in R_n, form a linearly independent set of vectors.

10. Which of the following subsets form a basis for R^3? Express the vector

$$\begin{bmatrix} 2 \\ 1 \\ 3 \end{bmatrix}$$ as a linear combination of the vectors in each subset that is a basis.

(a) $\left\{ \begin{bmatrix} 1 \\ 1 \\ 1 \end{bmatrix}, \begin{bmatrix} 1 \\ 2 \\ 3 \end{bmatrix}, \begin{bmatrix} 0 \\ 1 \\ 0 \end{bmatrix} \right\}$. (b) $\left\{ \begin{bmatrix} 1 \\ 2 \\ 2 \end{bmatrix}, \begin{bmatrix} 2 \\ 1 \\ 3 \end{bmatrix}, \begin{bmatrix} 0 \\ 0 \\ 0 \end{bmatrix} \right\}$.

(c) $\left\{ \begin{bmatrix} 2 \\ 1 \\ 3 \end{bmatrix}, \begin{bmatrix} 1 \\ 2 \\ 1 \end{bmatrix}, \begin{bmatrix} 1 \\ 1 \\ 4 \end{bmatrix}, \begin{bmatrix} 1 \\ 5 \\ 1 \end{bmatrix} \right\}$.

11. Find a basis for R^3 that includes

(a) The vector $\begin{bmatrix} 1 \\ 0 \\ 2 \end{bmatrix}$. (b) The vectors $\begin{bmatrix} 1 \\ 0 \\ 2 \end{bmatrix}$ and $\begin{bmatrix} 0 \\ 1 \\ 3 \end{bmatrix}$.

12. Find a basis for the solution space V of the homogeneous system

$$\begin{bmatrix} 1 & 2 & 1 & 2 & 1 \\ 1 & 2 & 2 & 1 & 2 \\ 2 & 4 & 3 & 3 & 3 \\ 0 & 0 & 1 & -1 & -1 \end{bmatrix} \begin{bmatrix} x_1 \\ x_2 \\ x_3 \\ x_4 \\ x_5 \end{bmatrix} = \begin{bmatrix} 0 \\ 0 \\ 0 \\ 0 \end{bmatrix}.$$

What is the dimension of V?

13. Find a basis for P_3 that includes the vectors $t^3 + t$ and $t^2 - t$.

14. Find a basis for $_2R_3$. What is the dimension of $_2R_3$? Generalize to $_mR_n$.

15. Find a basis for the subspace of R^3 spanned by

$$\left\{ \begin{bmatrix} 1 \\ 2 \\ 2 \end{bmatrix}, \begin{bmatrix} 3 \\ 2 \\ 1 \end{bmatrix}, \begin{bmatrix} 11 \\ 10 \\ 7 \end{bmatrix}, \begin{bmatrix} 7 \\ 6 \\ 4 \end{bmatrix} \right\}.$$

What is the dimension of W?

16. Prove Theorem 2.9.

17. Prove Theorem 2.10.

18. Prove that the vector space P of all polynomials is not finite-dimensional. (*Hint:* Suppose that $\{p_1(t), p_2(t), \ldots, p_k(t)\}$ is a finite basis for P. Let $d_j = $ degree $p_j(t)$. Establish a contradiction.)

19. Find the dimension of the subspace W of P_3 spanned by $\{t^3 + t^2 + 1, t^2 + 1, t^3 - 1\}$.

20. Let W be the subspace of the space of all continuous real-valued functions spanned by $\{\cos^2 t, \sin^2 t, \cos 2t\}$. Find a basis for W. What is the dimension of W?

21. Classify all subspaces of R^3.

22. Does the set $S = \left\{ \begin{bmatrix} 1 & 1 \\ 0 & 0 \end{bmatrix}, \begin{bmatrix} 0 & 0 \\ 1 & 1 \end{bmatrix}, \begin{bmatrix} 1 & 0 \\ 0 & 1 \end{bmatrix}, \begin{bmatrix} 0 & 1 \\ 1 & 1 \end{bmatrix} \right\}$ span $_2R_2$?

23. Which of the following subsets form a basis for P_2? Express $5t^2 - 3t + 8$ as a linear combination of the vectors in each subset that is a basis.
    (a) $\{t^2 + t, t - 1, t + 1\}$.   (b) $\{t^2 + 1, t - 1\}$.
    (c) $\{t^2 + t, t^2, t^2 + 1\}$.   (d) $\{t^2 + 1, t^2 - t + 1\}$.

24. Which of the subsets of Exercise 6 form a basis for $_2R_2$? Express $\begin{bmatrix} 3 & -2 \\ 4 & -3 \end{bmatrix}$
    as a linear combination of the vectors in each subset that is a basis.

25. Let V be an n-dimensional vector space. Show that any $n + 1$ vectors in V form a linearly dependent set.

26. Find the dimension of the solution space of the homogeneous system

$$\begin{bmatrix} 1 & 0 & 2 \\ 2 & 1 & 3 \\ 3 & 1 & 2 \end{bmatrix} \begin{bmatrix} x_1 \\ x_2 \\ x_3 \end{bmatrix} = \begin{bmatrix} 0 \\ 0 \\ 0 \end{bmatrix}.$$

27. Find a basis for the solution space V of the homogeneous system

$$\begin{bmatrix} 1 & 2 & 2 & -1 & 1 \\ 0 & 2 & 2 & -2 & -1 \\ 2 & 6 & 2 & -4 & 1 \\ 1 & 4 & 0 & -3 & 0 \end{bmatrix} \begin{bmatrix} x_1 \\ x_2 \\ x_3 \\ x_4 \\ x_5 \end{bmatrix} = \begin{bmatrix} 0 \\ 0 \\ 0 \\ 0 \end{bmatrix}.$$

   What is the dimension of V?

28. Show that if W is a nonzero subspace of a finite-dimensional vector space V, then dim $W \leq$ dim V.

29. Show that if W is a subspace of a finite-dimensional vector space V and dim $W =$ dim V, then $W = V$.

2.3. Coordinates and Isomorphisms

Coordinates

If V is an n-dimensional vector space, we know that V has a basis S with n vectors in it; so far we have not paid any attention to the order of the vectors in S. However, in the discussion in this section we shall speak of an **ordered** basis $S = \{\alpha_1, \alpha_2, \ldots, \alpha_n\}$ for V; thus $S_1 = \{\alpha_2, \alpha_1, \ldots, \alpha_n\}$ is a different ordered basis for V.

If $S = \{\alpha_1, \alpha_2, \ldots, \alpha_n\}$ is an ordered basis for the n-dimensional vector space V, then every vector α in V can be uniquely expressed in the form

$$\alpha = a_1\alpha_1 + a_2\alpha_2 + \cdots + a_n\alpha_n,$$

where $a_1, a_2, \ldots, a_n$ are real numbers. We shall refer to

$$[\alpha]_S = \begin{bmatrix} a_1 \\ a_2 \\ \vdots \\ a_n \end{bmatrix}$$

as the **coordinate vector of** α **with respect to the ordered basis** S. The entries of $[\alpha]_S$ are called the **coordinates of** α **with respect to** S. Theorem 2.5 implies that $[\alpha]_S$ is unique.

EXAMPLE 1. Consider the vector space P_1 of all polynomials of degree ≤ 1 and the zero polynomial and let $S = \{\alpha_1, \alpha_2\}$ be an ordered basis for P_1, where $\alpha_1 = t$ and $\alpha_2 = 1$. If $\alpha = p(t) = 5t - 2$, then $[\alpha]_S = \begin{bmatrix} 5 \\ -2 \end{bmatrix}$ is the coordinate vector of α with respect to the ordered basis S. On the other hand, if $T = \{t + 1, t - 1\}$ is the ordered basis, we have $5t - 2 = \frac{3}{2}(t + 1) + \frac{7}{2}(t - 1)$, which implies that

$$[\alpha]_T = \begin{bmatrix} \frac{3}{2} \\ \frac{7}{2} \end{bmatrix}.$$

The choice of an ordered basis and the consequent assignment of a co-ordinate vector for every α in V enables us to "picture" the vector space. We illustrate this notion by using Example 1. Choose a fixed point O in the plane R^2 and draw any two arrows β_1 and β_2 from O which depict the basis vectors t and 1 in the ordered basis $S = \{t, 1\}$ for P_2 (see Figure 2.10). The directions of β_1 and β_2 determine two lines, which we call the x_1- **and** x_2-**axes**, respectively. The positive direction on the x_1-axis is in the direction of β_1; the negative direction on the x_1-axis is along $-\beta_1$. Similarly, the positive direction on the x_2-axis is in the direction of β_2; the negative direction on the x_2-axis is along $-\beta_2$. The lengths of β_1 and β_2 determine the scales on the x_1- and x_2-axes, respectively. If α is a vector in P_1, we can write α, uniquely, as $\alpha = a_1\alpha_1 + $

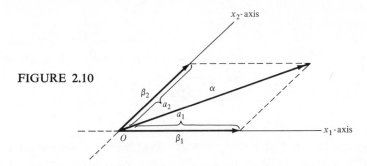

FIGURE 2.10

$a_2\alpha_2$. We now mark off a segment of length $|a_1|$ on the x_1-axis (in the positive direction if a_1 is positive and in the negative direction if a_1 is negative) and draw a line through the end point of this segment parallel to β_2. Similarly, mark off a segment of length $|a_2|$ on the x_2-axis (in the positive direction if a_2 is positive and in the negative direction if a_2 is negative) and draw a line through the end point of this segment parallel to β_1. We draw a directed line segment from O to the point of intersection of these two lines. This directed line segment represents α.

If α and β are vectors in an n-dimensional vector space V with an ordered basis $S = \{\alpha_1, \alpha_2, \ldots, \alpha_n\}$, then we can write α and β uniquely, as

$$\alpha = a_1\alpha_1 + a_2\alpha_2 + \cdots + a_n\alpha_n, \qquad \beta = b_1\alpha_1 + b_2\alpha_2 + \cdots + b_n\alpha_n.$$

Then $\alpha + \beta = (a_1 + b_1)\alpha_1 + (a_2 + b_2)\alpha_2 + \cdots + (a_n + b_n)\alpha_n$, which means that the coordinate vector $[\alpha + \beta]_S$ of $\alpha + \beta$ with respect to the basis S is

$$[\alpha + \beta]_S = \begin{bmatrix} (a_1 + b_1) \\ (a_2 + b_2) \\ \vdots \\ (a_n + b_n) \end{bmatrix} = [\alpha]_S + [\beta]_S.$$

Also, if c is a real number, then $c\alpha = (ca_1)\alpha_1 + (ca_2)\alpha_2 + \cdots + (ca_n)\alpha_n$, which implies that

$$[c\alpha]_S = \begin{bmatrix} ca_1 \\ ca_2 \\ \vdots \\ ca_n \end{bmatrix} = c[\alpha]_S.$$

This discussion suggests that, from an algebraic point of view, V and R^n behave "rather similarly." We now clarify this notion.

Isomorphism

Definition 2.9. Let V be a real vector space with operations $\oplus$ and $\odot$, and let W be a real vector space with operations $\boxplus$ and $\boxdot$. A one-one function L mapping V onto W is called an **isomorphism** of V onto W if:
(a) $L(\alpha \oplus \beta) = L(\alpha) \boxplus L(\beta)$ for α, β in V.
(b) $L(c \odot \alpha) = c \boxdot L(\alpha)$ for α in V, c a real number.

As a result of Theorem 2.12, we can replace the expressions "V is isomorphic to W" and "W is isomorphic to V" by "V and W are isomorphic."

Recall that L is **one-one** if $L(\alpha_1) = L(\alpha_2)$, for α_1, α_2 in V, implies that $\alpha_1 = \alpha_2$. Also, L is **onto** if for each β in W there is at least one α in V for which $L(\alpha) = \beta$. Thus the mapping $L: R^3 \to R^2$ defined by

$$L\left(\begin{bmatrix} a_1 \\ a_2 \\ a_3 \end{bmatrix}\right) = \begin{bmatrix} a_1 + a_2 \\ a_1 \end{bmatrix}$$

is onto. To see this, suppose that $\beta = \begin{bmatrix} b_1 \\ b_2 \end{bmatrix}$; we seek an $\alpha = \begin{bmatrix} a_1 \\ a_2 \\ a_3 \end{bmatrix}$ such that

$$L(\alpha) = \begin{bmatrix} a_1 + a_2 \\ a_1 \end{bmatrix} = \beta = \begin{bmatrix} b_1 \\ b_2 \end{bmatrix}.$$

Thus we obtain the solution: $a_1 = b_2$, $a_2 = b_1 - b_2$, and a_3 is arbitrary. However, L is not one-one, for if $\alpha_1 = \begin{bmatrix} 1 \\ 2 \\ 3 \end{bmatrix}$ and $\alpha_2 = \begin{bmatrix} 1 \\ 2 \\ 4 \end{bmatrix}$, then

$$L(\alpha_1) = L(\alpha_2) = \begin{bmatrix} 3 \\ 1 \end{bmatrix}.$$

Isomorphic vector spaces differ only in the nature of their elements; their algebraic properties are identical. That is, if the vector spaces V and W are isomorphic, under the isomorphism L, then for each α in V there is a unique β in W so that $L(\alpha) = \beta$ and, conversely, for each β in W there is a unique α in V so that $L(\alpha) = \beta$. If we now replace each element of V by its image under L and replace the operations $\oplus$ and $\odot$ of V by $\boxplus$ and $\boxdot$, respectively, we get precisely W. The most important example of isomorphic vector spaces is given in the following theorem.

Theorem 2.11. *If V is an n-dimensional real vector space, then V is isomorphic to R^n.*

Proof: Let $S = \{\alpha_1, \alpha_2, \ldots, \alpha_n\}$ be an ordered basis for V, and let $L: V \to R^n$ be defined by

$$L(\alpha) = [\alpha]_S = \begin{bmatrix} a_1 \\ a_2 \\ \vdots \\ a_n \end{bmatrix},$$

where $\alpha = a_1\alpha_1 + a_2\alpha_2 + \cdots + a_n\alpha_n$. We show that L is an isomorphism.

First, L is one-one. For let

$$[\alpha]_S = \begin{bmatrix} a_1 \\ a_2 \\ \vdots \\ a_n \end{bmatrix} \quad \text{and} \quad [\beta]_S = \begin{bmatrix} b_1 \\ b_2 \\ \vdots \\ b_n \end{bmatrix}$$

and suppose that $L(\alpha) = L(\beta)$.

Then $\begin{bmatrix} a_1 \\ a_2 \\ \vdots \\ a_n \end{bmatrix} = \begin{bmatrix} b_1 \\ b_2 \\ \vdots \\ b_n \end{bmatrix}$, which implies that $a_i = b_i$ for $i = 1, 2, \ldots, n$. Hence

$\alpha = \beta$, by Theorem 2.5.

Next, L is onto, for if $\beta = \begin{bmatrix} b_1 \\ b_2 \\ \vdots \\ b_n \end{bmatrix}$ is a given vector in R^n and $\alpha = b_1\alpha_1 +$

$b_2\alpha_2 + \cdots + b_n\alpha_n$, then $L(\alpha) = \beta$.

Finally, L satisfies Definition 2.9(a) and (b). Let α and β be vectors in V

such that $[\alpha]_S = \begin{bmatrix} a_1 \\ a_2 \\ \vdots \\ a_n \end{bmatrix}$ and $[\beta]_S = \begin{bmatrix} b_1 \\ b_2 \\ \vdots \\ b_n \end{bmatrix}$. Then $L(\alpha + \beta) = [\alpha + \beta]_S = [\alpha]_S$

$+ [\beta]_S = L(\alpha) + L(\beta)$ and $L(c\alpha) = [c\alpha]_S = c[\alpha]_S = cL(\alpha)$, as we saw before. Hence V and R^n are isomorphic.

Another example of isomorphism is given by the vector spaces discussed in the review section at the beginning of this chapter: R^2, the vector space of directed line segments emanating from a point in the plane and the vector space of all ordered pairs of real numbers.

Some important properties of isomorphisms are given in Theorem 2.12.

Theorem 2.12

(a) *Every vector space V is isomorphic to itself.*

(b) *If V is isomorphic to W, then W is isomorphic to V.*

(c) *If U is isomorphic to V and V is isomorphic to W, then U is isomorphic to W.*

Proof: Exercise. [Parts (a) and (b) are not difficult to show; (c) is slightly harder and will essentially be proved in Theorem 4.6.]

Corollary 2.2. *Any two n-dimensional vector spaces are isomorphic.*

Proof: Use Theorems 2.11 and 2.12.

This corollary means that all vector spaces of the same dimension are, algebraically speaking, alike.

Theorem 2.13. *Two finite-dimensional vector spaces are isomorphic if and only if their dimensions are equal.*

Proof: Let V and W be n-dimensional vector spaces. Then V and R^n are isomorphic and W and R^n are isomorphic. From Theorem 2.12 it follows that V and W are isomorphic.

Conversely, let V and W be isomorphic finite-dimensional vector spaces; let $L: V \to W$ be an isomorphism. Assume that dim $V = n$, and let $S = \{\alpha_1, \alpha_2, \ldots, \alpha_n\}$ be a basis for V.

We now prove that the set $T = \{L(\alpha_1), L(\alpha_2), \ldots, L(\alpha_n)\}$ is a basis for W. First, T spans W. For, if β is any vector in W, then $\beta = L(\alpha)$ for some α in V. Now $\alpha = a_1\alpha_1 + a_2\alpha_2 + \cdots + a_n\alpha_n$, where the a_i are uniquely determined real numbers, so

$$L(\alpha) = L(a_1\alpha_1 + a_2\alpha_2 + \cdots + a_n\alpha_n) = L(a_1\alpha_1) + L(a_2\alpha_2)$$
$$+ \cdots + L(a_n\alpha_n) = a_1L(\alpha_1) + a_2L(\alpha_2) + \cdots + a_nL(\alpha_n).$$

Thus T spans W.

Now suppose that

$$a_1L(\alpha_1) + a_2L(\alpha_2) + \cdots + a_nL(\alpha_n) = \theta_W.$$

Then $L(a_1\alpha_1 + a_2\alpha_2 + \cdots + a_n\alpha_n) = \theta_W$. Now, from Exercise 2, $L(\theta_V) = \theta_W$. Since L is one-one, we get $a_1\alpha_1 + a_2\alpha_2 + \cdots + a_n\alpha_n = \theta_V$. Since S is linearly independent, we conclude that $a_1 = a_2 = \cdots = a_n = 0$, which means that T is linearly independent. Hence T is a basis for W and dim $W = n$.

It is easy to show, as a consequence of Theorem 2.13, that the spaces R^n and R^m are isomorphic if and only if $n = m$ (see Exercise 3).

We can now establish the converse of Theorem 2.11, as follows.

Corollary 2.3. *If V is a finite-dimensional vector space that is isomorphic to R^n, then dim $V = n$.*

Proof: This result follows from Theorem 2.13.

If $L: V \to W$ is an isomorphism, then since L is a one-one onto mapping, it has an inverse L^{-1}. (This will be shown in Theorem 4.6.) It is not difficult to show that $L^{-1}: W \to V$ is also an isomorphism. (This will also be essentially shown in Theorem 4.6.) Moreover, if $S = \{\alpha_1, \alpha_2, \ldots, \alpha_n\}$ is a basis for V, then one can prove that $T = L(S) = \{L(\alpha_1), L(\alpha_2), \ldots, L(\alpha_n)\}$ is a basis for W, as we have seen in the proof of Theorem 2.13.

We have shown in this section that the idea of a finite-dimensional vector space, which at first seemed fairly abstract, is not so mysterious. In fact, such a vector space does not differ much from R^n.

We now look at the relationship between two coordinate vectors for the same vector α with respect to different bases. Thus let $S = \{\alpha_1, \alpha_2, \ldots, \alpha_n\}$ and $T = \{\beta_1, \beta_2, \ldots, \beta_n\}$ be two ordered bases for the n-dimensional vector space V. If α is any vector in V, then $\alpha = c_1\alpha_1 + c_2\alpha_2 + \ldots + c_n\alpha_n$ and $\alpha = d_1\beta_1 + d_2\beta_2 + \cdots + d_n\beta_n$. Now let

$$\begin{aligned}
\alpha_1 &= a_{11}\beta_1 + a_{21}\beta_2 + \cdots + a_{n1}\beta_n \\
\alpha_2 &= a_{12}\beta_1 + a_{22}\beta_2 + \cdots + a_{n2}\beta_n \\
&\vdots \\
\alpha_n &= a_{1n}\beta_1 + a_{2n}\beta_2 + \cdots + a_{nn}\beta_n.
\end{aligned}$$

Then the coordinate vector of α_j with respect to T is

$$[\alpha_j]_T = \begin{bmatrix} a_{1j} \\ a_{2j} \\ \vdots \\ a_{nj} \end{bmatrix}.$$

Let

$$P = \begin{bmatrix} a_{11} & a_{12} & \cdots & a_{1n} \\ a_{21} & a_{22} & \cdots & a_{2n} \\ \vdots & & & \vdots \\ a_{n1} & a_{n2} & \cdots & a_{nn} \end{bmatrix};$$

P is called the **transition matrix from the S-basis to the T-basis**. Now

$$\begin{aligned}
\alpha &= c_1\alpha_1 + c_2\alpha_2 + \cdots + c_n\alpha_n \\
&= c_1(a_{11}\beta_1 + a_{21}\beta_2 + \cdots + a_{n1}\beta_n) \\
&\quad + c_2(a_{12}\beta_1 + a_{22}\beta_2 + \cdots + a_{n2}\beta_n) + \cdots \\
&\quad + c_n(a_{1n}\beta_1 + a_{2n}\beta_2 + \cdots + a_{nn}\beta_n) \\
&= \beta_1(a_{11}c_1 + a_{12}c_2 + \cdots + a_{1n}c_n) \\
&\quad + \beta_2(a_{21}c_1 + a_{22}c_2 + \cdots + a_{2n}c_n) + \cdots \\
&\quad + \beta_n(a_{n1}c_1 + a_{n2}c_2 + \cdots + a_{nn}c_n) \\
&= d_1\beta_1 + d_2\beta_2 + \cdots + d_n\beta_n.
\end{aligned}$$

We can write this in matrix form as

$$[\alpha]_T = P[\alpha]_S \quad \text{or} \quad \begin{bmatrix} d_1 \\ d_2 \\ \vdots \\ d_n \end{bmatrix} = P \begin{bmatrix} c_1 \\ c_2 \\ \vdots \\ c_n \end{bmatrix}.$$

Now if $P[\alpha]_S = \theta_{R^n} = [\alpha]_T$, then $\alpha = \theta_V$, which means that $[\alpha]_S = \theta_{R^n}$. Thus the homogeneous system $PX = 0$ has only the trivial solution; it then follows from Theorem 1.26 that P is nonsingular. Of course, we then also have $[\alpha]_S = P^{-1}[\alpha]_T$. That is, P^{-1} is then the **transition matrix from the T-basis to the S-basis**; the jth column of P^{-1} is $[\beta_j]_S$.

EXAMPLE 2. Let V be R^3 and let

$$S = \left\{ \begin{bmatrix} 6 \\ 3 \\ 3 \end{bmatrix}, \begin{bmatrix} 4 \\ -1 \\ 3 \end{bmatrix}, \begin{bmatrix} 5 \\ 5 \\ 2 \end{bmatrix} \right\} \quad \text{and} \quad T = \left\{ \begin{bmatrix} 2 \\ 0 \\ 1 \end{bmatrix}, \begin{bmatrix} 1 \\ 2 \\ 0 \end{bmatrix}, \begin{bmatrix} 1 \\ 1 \\ 1 \end{bmatrix} \right\}$$

be ordered bases for R^3. To obtain the transition matrix P from the S-basis to the T-basis, we note that

$$\alpha_1 = \begin{bmatrix} 6 \\ 3 \\ 3 \end{bmatrix} = 2\begin{bmatrix} 2 \\ 0 \\ 1 \end{bmatrix} + 1\begin{bmatrix} 1 \\ 2 \\ 0 \end{bmatrix} + 1\begin{bmatrix} 1 \\ 1 \\ 1 \end{bmatrix}, \text{ so } [\alpha_1]_T = \begin{bmatrix} 2 \\ 1 \\ 1 \end{bmatrix}$$

$$\alpha_2 = \begin{bmatrix} 4 \\ -1 \\ 3 \end{bmatrix} = 2\begin{bmatrix} 2 \\ 0 \\ 1 \end{bmatrix} - 1\begin{bmatrix} 1 \\ 2 \\ 0 \end{bmatrix} + 1\begin{bmatrix} 1 \\ 1 \\ 1 \end{bmatrix}, \text{ so } [\alpha_2]_T = \begin{bmatrix} 2 \\ -1 \\ 1 \end{bmatrix}$$

and

$$\alpha_3 = \begin{bmatrix} 5 \\ 5 \\ 2 \end{bmatrix} = 1\begin{bmatrix} 2 \\ 0 \\ 1 \end{bmatrix} + 2\begin{bmatrix} 1 \\ 2 \\ 0 \end{bmatrix} + 1\begin{bmatrix} 1 \\ 1 \\ 1 \end{bmatrix}, \text{ so } [\alpha_3]_T = \begin{bmatrix} 1 \\ 2 \\ 1 \end{bmatrix}.$$

Then

$$P = \begin{bmatrix} 2 & 2 & 1 \\ 1 & -1 & 2 \\ 1 & 1 & 1 \end{bmatrix}.$$

If $\alpha = \begin{bmatrix} 4 \\ -9 \\ 5 \end{bmatrix}$, then to find $[\alpha]_T$ we can first express α in terms of S as

$$\alpha = \begin{bmatrix} 4 \\ -9 \\ 5 \end{bmatrix} = 1\begin{bmatrix} 6 \\ 3 \\ 3 \end{bmatrix} + 2\begin{bmatrix} 4 \\ -1 \\ 3 \end{bmatrix} - 2\begin{bmatrix} 5 \\ 5 \\ 2 \end{bmatrix}.$$

Thus $[\alpha]_S = \begin{bmatrix} 1 \\ 2 \\ -2 \end{bmatrix}$. Then

$$[\alpha]_T = P[\alpha]_S = \begin{bmatrix} 2 & 2 & 1 \\ 1 & -1 & 2 \\ 1 & 1 & 1 \end{bmatrix}\begin{bmatrix} 1 \\ 2 \\ -2 \end{bmatrix} = \begin{bmatrix} 4 \\ -5 \\ 1 \end{bmatrix}.$$

If we compute $[\alpha]_T$ directly, we find that

$$\alpha = \begin{bmatrix} 4 \\ -9 \\ 5 \end{bmatrix} = 4\begin{bmatrix} 2 \\ 0 \\ 1 \end{bmatrix} - 5\begin{bmatrix} 1 \\ 2 \\ 0 \end{bmatrix} + 1\begin{bmatrix} 1 \\ 1 \\ 1 \end{bmatrix} \quad \text{so } [\alpha]_T = \begin{bmatrix} 4 \\ -5 \\ 1 \end{bmatrix}.$$

The transition matrix from the T-basis to the S-basis is

$$P^{-1} = \begin{bmatrix} \frac{3}{2} & \frac{1}{2} & -\frac{5}{2} \\ -\frac{1}{2} & -\frac{1}{2} & \frac{3}{2} \\ -1 & 0 & 2 \end{bmatrix}.$$

To check that this is so, we compute the transition matrix Q from the T-basis to the S-basis, directly. We have

$$\begin{bmatrix} 2 \\ 0 \\ 1 \end{bmatrix} = \tfrac{3}{2}\begin{bmatrix} 6 \\ 3 \\ 3 \end{bmatrix} - \tfrac{1}{2}\begin{bmatrix} 4 \\ -1 \\ 3 \end{bmatrix} - 1\begin{bmatrix} 5 \\ 5 \\ 2 \end{bmatrix}.$$

Similarly,

$$\begin{bmatrix} 1 \\ 2 \\ 0 \end{bmatrix} = \tfrac{1}{2}\begin{bmatrix} 6 \\ 3 \\ 3 \end{bmatrix} - \tfrac{1}{2}\begin{bmatrix} 4 \\ -1 \\ 3 \end{bmatrix} + 0\begin{bmatrix} 5 \\ 5 \\ 2 \end{bmatrix}.$$

and

$$\begin{bmatrix} 1 \\ 1 \\ 1 \end{bmatrix} = -\tfrac{5}{2}\begin{bmatrix} 6 \\ 3 \\ 3 \end{bmatrix} + \tfrac{3}{2}\begin{bmatrix} 4 \\ -1 \\ 3 \end{bmatrix} + 2\begin{bmatrix} 5 \\ 5 \\ 2 \end{bmatrix}.$$

Hence $Q = P^{-1}$.

2.3. Exercises

1. Prove Theorem 2.12:
   (b) Every vector space is isomorphic to itself.
   (b) If V and W are vector spaces and V is isomorphic to W, then W is isomorphic to V. (This is essentially proved in Theorem 4.6.)

    (c) If U, V, and W are vector spaces such that U is isomorphic to V, and V is isomorphic to W, then U is isomorphic to W.

2. Let $L: V \rightarrow W$ be an isomorphism of vector space V onto vector space W.
    (a) Prove that $L(\theta_V) = \theta_W$.
    (b) Show that $L(\alpha - \beta) = L(\alpha) - L(\beta)$.

3. Prove that R^n and R^m are isomorphic if and only if $n = m$.

4. Show that R_n and R^n are isomorphic.

5. Show that P_2 and R^3 are isomorphic.

6. (a) Show that $_2R_2$ is isomorphic to R^4.
    (b) What is dim $_2R_2$?

7. Let $S = \left\{ \begin{bmatrix} 1 \\ 1 \end{bmatrix}, \begin{bmatrix} 2 \\ 3 \end{bmatrix} \right\}$ and $T = \left\{ \begin{bmatrix} 1 \\ 2 \end{bmatrix}, \begin{bmatrix} 0 \\ 1 \end{bmatrix} \right\}$ be ordered bases for R^2. Let $\alpha = \begin{bmatrix} 1 \\ 5 \end{bmatrix}$ and $\beta = \begin{bmatrix} 5 \\ 4 \end{bmatrix}$.
    (a) Find the coordinate vectors of α and β with respect to the basis S.
    (b) What is the transition matrix P from the S to the T-basis?
    (c) Find the coordinate vectors of α and β with respect to T using P.
    (d) Find the coordinate vectors of α and β with respect to T directly.
    (e) Find the transition matrix Q from the T to the S-basis.
    (f) Find the coordinate vectors of α and β with respect to S using Q. Compare the answers with those of (a).

8. Let $S = \{2t^2 + t, t^2 + 3, t\}$ and $T = \{t^2 + 1, t - 2, t + 3\}$ be ordered bases for P_2. Let $\alpha = 8t^2 - 4t + 6$ and $\beta = 7t^2 - t + 9$. Follow the directions of Exercise 7.

9. Let $S = \{[0 \quad 1 \quad 1], [1 \quad 0 \quad 0], [1 \quad 0 \quad 1]\}$ and $T = \{[1 \quad 1 \quad 1], [1 \quad 2 \quad 3], [1 \quad 0 \quad 1]\}$ be ordered bases for R_3. Let $\alpha = [-1 \quad 4 \quad 5]$ and $\beta = [2 \quad 0 \quad -6]$. Follow the directions of Exercise 7.

10. Let

$$S = \left\{ \begin{bmatrix} 1 & 1 \\ 0 & 0 \end{bmatrix}, \begin{bmatrix} 0 & 0 \\ 1 & 0 \end{bmatrix}, \begin{bmatrix} 0 & 0 \\ 0 & 1 \end{bmatrix}, \begin{bmatrix} 1 & 0 \\ 0 & 0 \end{bmatrix} \right\}$$

and

$$T = \left\{ \begin{bmatrix} 1 & 0 \\ 0 & 0 \end{bmatrix}, \begin{bmatrix} 0 & 1 \\ 1 & 0 \end{bmatrix}, \begin{bmatrix} 0 & 2 \\ 0 & 1 \end{bmatrix}, \begin{bmatrix} 0 & 0 \\ 1 & 1 \end{bmatrix} \right\}$$

be ordered bases for $_2R_2$. Let $\alpha = \begin{bmatrix} 1 & 1 \\ 1 & 1 \end{bmatrix}$ and $\beta = \begin{bmatrix} 1 & 2 \\ -2 & 1 \end{bmatrix}$. Follow the directions of Exercise 7.

11. Let V be the subspace of the vector space of all real-valued continuous functions that has basis $S = \{e^t, e^{-t}\}$. Show that V and R^2 are isomorphic.

12. Let V be the subspace of the vector space of all real-valued functions that is *spanned* by the set $S = \{\cos^2 t, \sin^2 t, \cos 2t\}$. Show that V and R_2 are isomorphic.

13. The set $S = \left\{ \begin{bmatrix} 1 \\ -1 \\ 1 \end{bmatrix}, \begin{bmatrix} 0 \\ 0 \\ 1 \end{bmatrix}, \begin{bmatrix} 1 \\ 1 \\ 0 \end{bmatrix} \right\}$ is an ordered basis for R^3. Find α if $[\alpha]_S$ is

(a) $\begin{bmatrix} 2 \\ -1 \\ 3 \end{bmatrix}$. (b) $\begin{bmatrix} 0 \\ 0 \\ 1 \end{bmatrix}$. (c) $\begin{bmatrix} 0 \\ 2 \\ 1 \end{bmatrix}$.

14. The set $S = \{t^2 + 1, t + 1, t^2 + t\}$ is an ordered basis for P_2. Find α if $[\alpha]_S$ is

(a) $\begin{bmatrix} 3 \\ -1 \\ -2 \end{bmatrix}$. (b) $\begin{bmatrix} 1 \\ 0 \\ 0 \end{bmatrix}$. (c) $\begin{bmatrix} 2 \\ 0 \\ -1 \end{bmatrix}$.

15. If the vector α in P_2 has the coordinate vector $\begin{bmatrix} 1 \\ 2 \\ 3 \end{bmatrix}$ with respect to the ordered basis $S = \{t^2, t - 1, 1\}$, what is $[\alpha]_T$ if $T = \{t^2 + t + 1, t + 1, 1\}$?

16. Let V and W be isomorphic vector spaces. Prove that if V_1 is a subspace of V, then V_1 is isomorphic to a subspace W_1 of W.

2.4. The Rank of a Matrix

In this section we obtain a good method for finding a basis for a vector space V spanned by a given set $S = \{\alpha_1, \alpha_2, \ldots, \alpha_k\}$ of vectors. Of course, we could do this by finding a maximal independent subset T of S, but this trial-and-error method could be very lengthy. For example, if S has 20 vectors and T has 5 vectors, there are 15,504 different ways of choosing 5 vectors out of 20 vectors, and we do not know at which point in this search we will find a suitable T. We shall also attach a unique number to a matrix that we later show gives us information about the dimension of the solution space of a homogeneous system.

Definition 2.10. Let

$$A = \begin{bmatrix} a_{11} & a_{12} & \cdots & a_{1n} \\ a_{21} & a_{22} & \cdots & a_{2n} \\ & & \vdots & \\ a_{m1} & a_{m2} & \cdots & a_{mn} \end{bmatrix}$$

be an $m \times n$ matrix. The rows of A, considered as vectors in R_n, span a subspace of R_n, called the **row space** of A. Similarly, the columns of A, considered as vectors in R^m, span a subspace of R^m called the **column space** of A.

Theorem 2.14. *If A and B are two m × n row (column) equivalent matrices, then the row (column) spaces of A and B are identical.*

Proof: If *A* and *B* are row equivalent, then the rows of *B* are obtained from the rows of *A* by a finite number of the three elementary row operations. Thus each row of *B* is a linear combination of the rows of *A*. Hence the row space of *B* is contained in the row space of *A*. If we apply the inverse elementary row operations to *B*, we get *A*, so the row space of *A* is contained in the row space of *B*. Hence the row spaces of *A* and *B* are identical. The proof for the column spaces is similar.

We can use this theorem to find a basis for a subspace spanned by a given set of vectors. We illustrate this method with the following example.

EXAMPLE 1. Find a basis for the subspace V of R_4 spanned by $S = \{\alpha_1, \alpha_2, \alpha_3, \alpha_4, \alpha_5\}$, where $\alpha_1 = [1 \quad 2 \quad 1 \quad 2]$, $\alpha_2 = [2 \quad 1 \quad 2 \quad 1]$, $\alpha_3 = [3 \quad 2 \quad 3 \quad 2]$, $\alpha_4 = [3 \quad 3 \quad 3 \quad 3,]$ and $\alpha_5 = [5 \quad 3 \quad 5 \quad 3]$.

Solution: Note that V is the row space of the matrix

$$A = \begin{bmatrix} 1 & 2 & 1 & 2 \\ 2 & 1 & 2 & 1 \\ 3 & 2 & 3 & 2 \\ 3 & 3 & 3 & 3 \\ 5 & 3 & 5 & 3 \end{bmatrix}.$$

Using elementary row operations, we find that A is row equivalent to the matrix (verify)

$$B = \begin{bmatrix} 1 & 0 & 1 & 0 \\ 0 & 1 & 0 & 1 \\ 0 & 0 & 0 & 0 \\ 0 & 0 & 0 & 0 \\ 0 & 0 & 0 & 0 \end{bmatrix},$$

which is in reduced row echelon form. The row spaces of A and B are identical and a basis for the row space of B consists of $\beta_1 = [1 \quad 0 \quad 1 \quad 0]$ and $\beta_2 = [0 \quad 1 \quad 0 \quad 1]$. (See Exercise 9 in Section 2.2.) Hence $\{[1 \quad 0 \quad 1 \quad 0], [0 \quad 1 \quad 0 \quad 1]\}$ is a basis for V.

It is not necessary to find a matrix B in reduced row echelon form which is row equivalent to A. All that is required is that we have a matrix B that is row equivalent to A and such that we can easily obtain a basis for the row space of B. Often one does not have to reduce A all the way to reduced row echelon form to get such a matrix B. For it is easy to show that if A is row equivalent to a matrix B that is in row echelon form, then the nonzero rows of B form a basis for the row space of A (Exercise 6).

EXAMPLE 2. Let V be the subspace of P_4 spanned by $S = \{\alpha_1, \alpha_2, \alpha_3, \alpha_4\}$, where $\alpha_1 = t^4 + t^2 + 2t + 1$, $\alpha_2 = t^4 + t^2 + 2t + 2$, $\alpha_3 = 2t^4 + t^3 + t + 2$, and $\alpha_4 = t^4 + t^3 - t^2 - t$. To find a basis for V, we proceed as follows.

Since P_4 is isomorphic to R_5 under the isomorphism L defined by $L(at^4 + bt^3 + ct^2 + dt + e) = [a \quad b \quad c \quad d \quad e]$, then $L(V)$ is isomorphic to a subspace W of R_5 (see Exercise 16 of Section 2.3). W is spanned by $\{L(\alpha_1), L(\alpha_2), L(\alpha_3), L(\alpha_4)\}$, as we have seen in the proof of Theorem 2.13. We now find a basis for W by proceeding as in Example 1. Thus W is the row space of the matrix

$$A = \begin{bmatrix} 1 & 0 & 1 & 2 & 1 \\ 1 & 0 & 1 & 2 & 2 \\ 2 & 1 & 0 & 1 & 2 \\ 1 & 1 & -1 & -1 & 0 \end{bmatrix},$$

and A is row equivalent to (verify)

$$B = \begin{bmatrix} 1 & 0 & 1 & 2 & 0 \\ 0 & 1 & -2 & -3 & 0 \\ 0 & 0 & 0 & 0 & 1 \\ 0 & 0 & 0 & 0 & 0 \end{bmatrix}.$$

A basis for W is therefore $T = \{\beta_1, \beta_2, \beta_3\}$, where $\beta_1 = [1 \quad 0 \quad 1 \quad 2 \quad 0]$, $\beta_2 = [0 \quad 1 \quad -2 \quad -3 \quad 0]$, and $\beta_3 = [0 \quad 0 \quad 0 \quad 0 \quad 1]$. A basis for V is then

$$\{L^{-1}(\beta_1), L^{-1}(\beta_2), L^{-1}(\beta_3)\} = \{t^4 + t^2 + 2t, t^3 - 2t^2 - 3t, 1\}.$$

Definition 2.11. The dimension of the row (column) space of A is called the **row (column) rank** of A.

We note that if A and B are row equivalent, then row rank A = row rank B and if A and B are column equivalent, then column rank A = column rank B.

Therefore, if we start out with an $m \times n$ matrix A and find a matrix B in reduced row echelon form that is row equivalent to A, then A and B have equal row ranks. But the row rank of B is clearly the number of nonzero rows. Thus we have a good method for finding the row rank of a given matrix A.

EXAMPLE 3. To determine the row rank of the matrix

$$A = \begin{bmatrix} 1 & 2 & 3 & 1 & 2 \\ 2 & 1 & 2 & 3 & 1 \\ 3 & 3 & 5 & 4 & 3 \\ 1 & -1 & -1 & 2 & -1 \end{bmatrix},$$

we find that A is row equivalent to (verify)

$$B = \begin{bmatrix} 1 & 0 & \frac{1}{3} & \frac{5}{3} & 0 \\ 0 & 1 & \frac{4}{3} & -\frac{1}{3} & 1 \\ 0 & 0 & 0 & 0 & 0 \\ 0 & 0 & 0 & 0 & 0 \end{bmatrix},$$

which means that row rank A = row rank B = 2.

We may also conclude that if A is an $m \times n$ matrix and P is a nonsingular $m \times m$ matrix, then row rank(PA) = row rank A, for A and PA are row equivalent (Exercise 17 in Section 1.6). Similarly, if Q is a nonsingular $n \times n$ matrix, then column rank(AQ) = column rank A. Moreover, since dimension $R_n = n$, we see that row rank $A \leq n$. Also, since the row space of A is spanned by m vectors, row rank $A \leq m$. Thus row rank $A \leq$ minimum$\{m, n\}$.

We now prove the following important theorem.

Theorem 2.15. *The row rank and column rank of the $m \times n$ matrix $A = [a_{ij}]$ are equal.*

Proof: Let $\alpha_1, \alpha_2, \ldots, \alpha_m$ be the row vectors of A, where

$$\alpha_i = [a_{i1} \quad a_{i2} \quad \cdots \quad a_{in}], \qquad i = 1, 2, \ldots, m.$$

Let row rank $A = k$ and let the set of vectors $\{\beta_1, \beta_2, \ldots, \beta_k\}$ form a basis for the row space of A, where $\beta_i = [b_{i1} \quad b_{i2} \quad \cdots \quad b_{in}]$ for $i = 1, 2, \ldots, k$.

Now each of the row vectors is a linear combination of $\beta_1, \beta_2, \ldots, \beta_k$:

$$
\begin{aligned}
\alpha_1 &= r_{11}\beta_1 + r_{12}\beta_2 + \cdots + r_{1k}\beta_k \\
\alpha_2 &= r_{21}\beta_1 + r_{22}\beta_2 + \cdots + r_{2k}\beta_k \\
&\vdots \\
\alpha_m &= r_{m1}\beta_1 + r_{m2}\beta_2 + \cdots + r_{mk}\beta_k,
\end{aligned}
$$

where the r_{ij} are uniquely determined real numbers. Recalling that two matrices are equal if and only if the corresponding entries are equal, we equate the entries of these vector equations to get

$$
\begin{aligned}
a_{1j} &= r_{11}b_{1j} + r_{12}b_{2j} + \cdots + r_{1k}b_{kj} \\
a_{2j} &= r_{21}b_{1j} + r_{22}b_{2j} + \cdots + r_{2k}b_{kj} \\
&\vdots \\
a_{mj} &= r_{m1}b_{1j} + r_{m2}b_{2j} + \cdots + r_{mk}b_{kj}
\end{aligned}
$$

or

$$
\begin{bmatrix} a_{1j} \\ a_{2j} \\ \vdots \\ a_{mj} \end{bmatrix} = b_{1j} \begin{bmatrix} r_{11} \\ r_{21} \\ \vdots \\ r_{m1} \end{bmatrix} + b_{2j} \begin{bmatrix} r_{12} \\ r_{22} \\ \vdots \\ r_{m2} \end{bmatrix} + \cdots + b_{kj} \begin{bmatrix} r_{1k} \\ r_{2k} \\ \vdots \\ r_{mk} \end{bmatrix}
$$

for $j = 1, 2, \ldots, n$.

Since every column of A is a linear combination of k vectors, the dimension of the column space of A is at most k, or column rank $A \le k = $ row rank A. Similarly, we get row rank $A \le$ column rank A. Hence the row and column ranks of A are equal.

EXAMPLE 4. Let A be the matrix of Example 3. The column space of A is the subspace V of R^4 spanned by the columns of A:

$$
\alpha_1 = \begin{bmatrix} 1 \\ 2 \\ 3 \\ 1 \end{bmatrix}, \quad \alpha_2 = \begin{bmatrix} 2 \\ 1 \\ 3 \\ -1 \end{bmatrix}, \quad \alpha_3 = \begin{bmatrix} 3 \\ 2 \\ 5 \\ -1 \end{bmatrix},
$$

$$
\alpha_4 = \begin{bmatrix} 1 \\ 3 \\ 4 \\ 2 \end{bmatrix}, \quad \text{and} \quad \alpha_5 = \begin{bmatrix} 2 \\ 1 \\ 3 \\ -1 \end{bmatrix}.
$$

Now column rank $A = \dim V$. Using the method of Example 10 of Section 2.2, we find (verify) that $\{\alpha_1, \alpha_2\}$ is a basis for V. Hence column rank $A = 2$.

Since the row and column ranks of a matrix are equal, we shall now merely refer to the **rank** of a matrix. Note that rank $I_n = n$. Theorem 1.30 states that A is equivalent to B if and only if there exist nonsingular matrices P and Q such that $B = PAQ$. If A is equivalent to B, then rank $A = $ rank B, for rank $B = $ rank $(PAQ) = $ rank $(PA) = $ rank A.

We also recall from Section 1.7 that if A is an $m \times n$ matrix, then A is equivalent to a matrix $C = \begin{bmatrix} I_k & O \\ O & O \end{bmatrix}$. Now rank $A = $ rank $C = k$. We use these facts to establish the result that if A and B are $m \times n$ matrices of equal rank, then A and B are equivalent. Thus let rank $A = k = $ rank B. Then there exist nonsingular matrices P_1, Q_1, P_2, and Q_2 such that

$$P_1 A Q_1 = \begin{bmatrix} I_k & O \\ O & O \end{bmatrix} = P_2 B Q_2.$$

Then $P_2^{-1} P_1 A Q_1 Q_2^{-1} = B$. Letting $P = P_2^{-1} P_1$ and $Q = Q_1 Q_2^{-1}$, we find that P and Q are nonsingular and $B = PAQ$. Hence A and B are equivalent.

At this point we can show that an $n \times n$ matrix is nonsingular if and only if its rank is n. We first show Theorem 2.16.

Theorem 2.16. *If A is an $n \times n$ matrix, then rank $A = n$ if and only if A is row equivalent to I_n.*

Proof: If rank $A = n$, then A is row equivalent to a matrix B in reduced row echelon form, and rank $B = n$. Since rank $B = n$, we conclude that B has no zero rows, and this implies (by Exercise 3 of Section 1.5) that $B = I_n$. Hence A is row equivalent to I_n.

Conversely, if A is row equivalent to I_n, then rank $A = $ rank $I_n = n$.

Corollary 2.4. *A is nonsingular if and only if rank $A = n$.*

Proof: This follows from Theorem 2.16 and Corollary 1.3.

From a practical point of view, this result is not too useful, since most of the time we want to know not only whether A is nonsingular but also its inverse. The method developed in Chapter 1 enables us to find A^{-1}, if it exists, and tells us if it does not exist. Thus we do not have to learn first if A^{-1} exists and then go through another procedure to obtain it.

Corollary 2.5. *The homogeneous system $AX = O$, where A is $n \times n$, has a nontrivial solution if and only if rank $A < n$.*

Proof: This follows from Corollary 2.4 and from the fact that $AX = O$ has a nontrivial solution if and only if A is singular (Theorem 1.26). $\blacksquare$

EXAMPLE 5. Let $A = \begin{bmatrix} 1 & 2 & 0 \\ 0 & 1 & 3 \\ 2 & 1 & 3 \end{bmatrix}$. Then if we transform A to reduced row echelon form B, we find that $B = I_3$. Thus rank $A = 3$ and A is nonsingular. Moreover, the homogeneous system $AX = O$ has only the trivial solution.

EXAMPLE 6. Let $A = \begin{bmatrix} 1 & 2 & 0 \\ 1 & 1 & -3 \\ 1 & 3 & 3 \end{bmatrix}$. Then A is row equivalent to $\begin{bmatrix} 1 & 0 & -6 \\ 0 & 1 & 3 \\ 0 & 0 & 0 \end{bmatrix}$, a matrix in reduced row echelon form. Hence rank $A < 3$, and A is singular. Moreover, the homogeneous system $AX = O$ has a nontrivial solution.

Our final application of rank is to linear systems. If we consider the linear system $AX = B$ of m equations in n unknowns, where $A = [a_{ij}]$, then we observe that the system can also be written as the equation (verify)

$$x_1 \begin{bmatrix} a_{11} \\ a_{21} \\ \vdots \\ a_{m1} \end{bmatrix} + x_2 \begin{bmatrix} a_{12} \\ a_{22} \\ \vdots \\ a_{m2} \end{bmatrix} + \cdots + x_n \begin{bmatrix} a_{1n} \\ a_{2n} \\ \vdots \\ a_{mn} \end{bmatrix} = \begin{bmatrix} b_1 \\ b_2 \\ \vdots \\ b_m \end{bmatrix}.$$

Thus $AX = B$ is consistent (has a solution) if and only if B is a linear combination of the columns of A; that is, if and only if B belongs to the column space of A. This means that if $AX = B$ has a solution, then rank $A = $ rank $[A \vdots B]$, where $[A \vdots B]$ is the augmented matrix of the linear system. Conversely, if rank $[A \vdots B] = $ rank A, then B is in the column space of A, which means that the linear system has a solution. Thus $AX = B$ is consistent if and only if rank $A = $ rank $[A \vdots B]$. This result, although of interest, is not of great computational value, since we usually are interested in finding a solution rather than in knowing whether or not a solution exists.

EXAMPLE 7. Consider the linear system

$$\begin{bmatrix} 2 & 1 & 3 \\ 1 & -2 & 2 \\ 0 & 1 & 3 \end{bmatrix} \begin{bmatrix} x_1 \\ x_2 \\ x_3 \end{bmatrix} = \begin{bmatrix} 1 \\ 2 \\ 3 \end{bmatrix}.$$

Since rank A = rank $[A \mathbin{\vdots} B] = 3$, the linear system has a solution.

EXAMPLE 8. The linear system

$$\begin{bmatrix} 1 & 2 & 3 \\ 1 & -3 & 4 \\ 2 & -1 & 7 \end{bmatrix} \begin{bmatrix} x_1 \\ x_2 \\ x_3 \end{bmatrix} = \begin{bmatrix} 4 \\ 5 \\ 6 \end{bmatrix}.$$

has no solution, because rank $A = 2$ and rank $[A \mathbin{\vdots} B] = 3$.

2.4. Exercises

1. Find a basis for the subspace V of R^3 spanned by

$$S = \left\{ \begin{bmatrix} 1 \\ 2 \\ 3 \end{bmatrix}, \begin{bmatrix} 2 \\ 1 \\ 4 \end{bmatrix}, \begin{bmatrix} -1 \\ -1 \\ 2 \end{bmatrix}, \begin{bmatrix} 0 \\ 1 \\ 2 \end{bmatrix}, \begin{bmatrix} 1 \\ 1 \\ 1 \end{bmatrix} \right\}.$$

2. Find a basis for the subspace of P_3 spanned by

$$S = \{t^3 + t^2 + 2t + 1, t^3 - 3t + 1, t^2 + t + 2, t + 1, t^3 + 1\}.$$

3. Find a basis for the subspace of $_2R_2$ spanned by

$$S = \left\{ \begin{bmatrix} 1 & 2 \\ 1 & 1 \end{bmatrix}, \begin{bmatrix} 2 & 1 \\ 3 & 1 \end{bmatrix}, \begin{bmatrix} 0 & 2 \\ 1 & 2 \end{bmatrix}, \begin{bmatrix} 3 & 2 \\ 1 & 4 \end{bmatrix}, \begin{bmatrix} 5 & 0 \\ 0 & -1 \end{bmatrix} \right\}.$$

4. Find a basis for the subspace of R_2 spanned by

$$S = \{[1 \quad 2], [2 \quad 3], [3 \quad 1], [-4 \quad 3]\}.$$

5. Find the row and column ranks of the following matrices.

(a) $\begin{bmatrix} 1 & 2 & 3 & 2 & 1 \\ 3 & 1 & -5 & -2 & 1 \\ 7 & 8 & -1 & 2 & 5 \end{bmatrix}$.

(b) $\begin{bmatrix} 1 & 3 & 2 & 0 & 0 & 1 \\ 2 & 1 & -5 & 1 & 2 & 0 \\ 3 & 2 & 5 & 1 & -2 & 1 \\ 5 & 8 & 9 & 1 & -2 & 2 \\ 9 & 9 & 4 & 2 & 0 & 2 \end{bmatrix}$.

(c) $\begin{bmatrix} 1 & 2 & 3 & 2 & 1 \\ 0 & 5 & 4 & 0 & -1 \\ 2 & -1 & 2 & 4 & 3 \end{bmatrix}$.

√ **6.** Let A be an $m \times n$ matrix in row echelon form. Prove that rank A = the number of nonzero rows of A.

7. For each of the following matrices, verify Theorem 2.15 by computing the row and column ranks.

(a) $\begin{bmatrix} 1 & 2 & 3 \\ -1 & 2 & 1 \\ 3 & 1 & 2 \end{bmatrix}$.

(b) $\begin{bmatrix} 1 & -2 & -1 \\ 2 & -1 & 3 \\ 7 & -8 & 3 \end{bmatrix}$.

(c) $\begin{bmatrix} 1 & -2 & -1 \\ 2 & -1 & 3 \\ 7 & -8 & 3 \\ 5 & -7 & 0 \end{bmatrix}$.

8. Find the rank of each matrix.

(a) $\begin{bmatrix} 1 & -1 & 2 & 3 \\ 2 & 6 & -8 & 1 \\ 5 & 3 & -2 & 10 \end{bmatrix}$.

(b) $\begin{bmatrix} 1 & 2 & 0 & 3 \\ 3 & 2 & -1 & 0 \\ 2 & -1 & 0 & 1 \end{bmatrix}$.

(c) $\begin{bmatrix} 1 & 3 & -2 & 4 \\ -1 & 4 & -5 & 10 \\ 3 & 2 & 1 & -2 \\ 3 & -5 & 8 & -16 \end{bmatrix}$.

9. Which of the following matrices are equivalent?

$A = \begin{bmatrix} 1 & 2 & 1 & 3 \\ 2 & 1 & -4 & -5 \\ 7 & 8 & -5 & -1 \\ 10 & 14 & -2 & 8 \end{bmatrix}$,
$B = \begin{bmatrix} 1 & 2 & 1 & 3 \\ 2 & 1 & -4 & -5 \\ 1 & 1 & 0 & 0 \\ 0 & 0 & 1 & 1 \end{bmatrix}$,

$C = \begin{bmatrix} 1 & 5 & 1 & 3 \\ 2 & 1 & 2 & 1 \\ -3 & 0 & 1 & 0 \\ 4 & 7 & -4 & 3 \end{bmatrix}$,
$D = \begin{bmatrix} 1 & 2 & -4 & 3 \\ 4 & 7 & -4 & 1 \\ 7 & 12 & -4 & -1 \\ 2 & 3 & 4 & -5 \end{bmatrix}$,

$E = \begin{bmatrix} 4 & 3 & -1 & -5 \\ -2 & -6 & -7 & 10 \\ -2 & -3 & -2 & 5 \\ 0 & -6 & -10 & 10 \end{bmatrix}$.

10. By comparing the ranks of the coefficient and augmented matrices, determine which of the following linear systems are consistent.

(a) $\begin{bmatrix} 1 & 2 & 5 & -2 \\ 2 & 3 & -2 & 4 \\ 5 & 1 & 0 & 2 \end{bmatrix} \begin{bmatrix} x_1 \\ x_2 \\ x_3 \\ x_4 \end{bmatrix} = \begin{bmatrix} 0 \\ 0 \\ 0 \end{bmatrix}$.

(b) $\begin{bmatrix} 1 & 2 & 5 & -2 \\ 2 & 3 & -2 & 4 \\ 5 & 1 & 0 & 2 \end{bmatrix} \begin{bmatrix} x_1 \\ x_2 \\ x_3 \\ x_4 \end{bmatrix} = \begin{bmatrix} -1 \\ -13 \\ 3 \end{bmatrix}$.

(c) $\begin{bmatrix} 1 & -2 & -3 & 4 \\ 4 & -1 & -5 & 6 \\ 2 & 3 & 1 & -2 \end{bmatrix} \begin{bmatrix} x_1 \\ x_2 \\ x_3 \\ x_4 \end{bmatrix} = \begin{bmatrix} 1 \\ 2 \\ 2 \end{bmatrix}$.

(d) $\begin{bmatrix} 1 & 1 & 1 \\ 1 & -1 & 1 \\ 5 & 1 & 5 \end{bmatrix} \begin{bmatrix} x_1 \\ x_2 \\ x_3 \end{bmatrix} = \begin{bmatrix} 6 \\ 2 \\ 5 \end{bmatrix}.$

11. Use Corollary 2.4 to find which of the following matrices are nonsingular.

(a) $\begin{bmatrix} 1 & 2 & -3 \\ -1 & 2 & 3 \\ 0 & 8 & 0 \end{bmatrix}.$ (b) $\begin{bmatrix} 1 & 2 & -3 \\ -1 & 2 & 3 \\ 0 & 1 & 1 \end{bmatrix}.$

(c) $\begin{bmatrix} 1 & 1 & 4 & -1 \\ 1 & 2 & 3 & 2 \\ -1 & 3 & 2 & 1 \\ -2 & 6 & 12 & -4 \end{bmatrix}.$

12. Use Corollary 2.5 to find which of the following homogeneous systems have a nontrivial solution.

(a) $\begin{bmatrix} 1 & 1 & 2 & -1 \\ 1 & 3 & -1 & 2 \\ 1 & 1 & 1 & 3 \\ 1 & 2 & 1 & 1 \end{bmatrix} \begin{bmatrix} x_1 \\ x_2 \\ x_3 \\ x_4 \end{bmatrix} = \begin{bmatrix} 0 \\ 0 \\ 0 \\ 0 \end{bmatrix}.$

(b) $\begin{bmatrix} 1 & 2 & 3 \\ 0 & 1 & 0 \\ 1 & 0 & 3 \end{bmatrix} \begin{bmatrix} x_1 \\ x_2 \\ x_3 \end{bmatrix} = \begin{bmatrix} 0 \\ 0 \\ 0 \end{bmatrix}.$ (c) $\begin{bmatrix} 1 & 2 & -1 \\ 2 & -1 & 3 \\ 5 & -4 & 3 \end{bmatrix} \begin{bmatrix} x_1 \\ x_2 \\ x_3 \end{bmatrix} = \begin{bmatrix} 0 \\ 0 \\ 0 \end{bmatrix}.$

13. For each of the following matrices A, find rank A by obtaining a matrix of the form $\begin{bmatrix} I_k & O \\ O & O \end{bmatrix}$ which is equivalent to A.

(a) $\begin{bmatrix} 1 & 1 & -2 \\ 1 & 2 & 3 \\ 0 & 1 & 3 \end{bmatrix}.$ (b) $\begin{bmatrix} 1 & 1 & -2 & 0 & 0 \\ 1 & 2 & 3 & 6 & 7 \\ 2 & 1 & 3 & 6 & 5 \end{bmatrix}.$

(c) $\begin{bmatrix} 1 & 1 & -2 \\ 1 & 2 & 3 \\ 3 & 4 & -1 \end{bmatrix}.$

CHAPTER 3

Inner Product Spaces[*]

As we noted in Chapter 2, when physicists talk about vectors in R^3, they usually refer to objects that have magnitude and direction. However, thus far in our study of vector spaces we have refrained from discussing these notions. In this chapter we deal with magnitude and direction in a vector space.

3.1. The Standard Inner Product on R^3

In this section we review the usual notions of magnitude and direction in R^3 and, in the next section, generalize these to R^n. We consider R^3 with the usual Cartesian coordinate system. Let $\alpha = \begin{bmatrix} a_1 \\ a_2 \\ a_3 \end{bmatrix}$ be a vector in R^3. Using the Pythagorean theorem twice (see Figure 3.1), we can obtain the **length** of α, $|\alpha|$, as

$$|\alpha| = \sqrt{a^2 + a_3^2} = \sqrt{(\sqrt{a_1^2 + a_2^2})^2 + a_3^2} = \sqrt{a_1^2 + a_2^2 + a_3^2}.$$

[*] This chapter may also be covered after Section 6.1 and before Section 6.2, which is where it is used.

116

If $\alpha = \begin{bmatrix} a_1 \\ a_2 \\ a_3 \end{bmatrix}$ and $\beta = \begin{bmatrix} b_1 \\ b_2 \\ b_3 \end{bmatrix}$, then $\alpha - \beta = \begin{bmatrix} a_1 - b_1 \\ a_2 - b_2 \\ a_3 - b_3 \end{bmatrix}$. Thus

$$|\alpha - \beta| = \sqrt{(a_1 - b_1)^2 + (a_2 - b_2)^2 + (a_3 - b_3)^2}. \tag{1}$$

This is the **distance** between α and β.

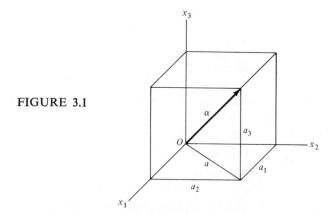

FIGURE 3.1

Of course, (1) also gives the distance between two points P and Q in R^3 with respective coordinates (a_1, a_2, a_3) and (b_1, b_2, b_3)

EXAMPLE 1. The length of $\alpha = \begin{bmatrix} 1 \\ 2 \\ 3 \end{bmatrix}$ is $|\alpha| = \sqrt{1^2 + 2^2 + 3^2} = \sqrt{14}$.

EXAMPLE 2. Let $\alpha = \begin{bmatrix} 1 \\ 2 \\ 3 \end{bmatrix}$ and $\beta = \begin{bmatrix} -4 \\ 3 \\ 5 \end{bmatrix}$. Then

$$|\alpha - \beta| = \sqrt{(1 + 4)^2 + (2 - 3)^2 + (3 - 5)^2} = \sqrt{30}.$$

The direction of a vector α in R^3 is specified by giving the cosines of the angles that the vector α makes with the positive x_1-, x_2-, and x_3-axes (see Figure 3.2); these are called **direction cosines**. Instead of dealing with the special problem of finding the cosines of these angles, we look at the problem

of determining the **angle** ϕ between two nonzero vectors $\alpha = \begin{bmatrix} a_1 \\ a_2 \\ a_3 \end{bmatrix}$ and

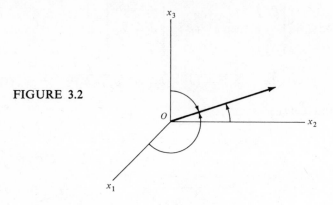

FIGURE 3.2

$$\beta = \begin{bmatrix} b_1 \\ b_2 \\ b_3 \end{bmatrix}$$ (see Figure 3.3), which satisfies $0 \le \phi \le \pi$. Using the law of cosines, we have

$$|PQ|^2 = |\alpha|^2 + |\beta|^2 - 2|\alpha|\,|\beta|\cos\phi,$$

where $|PQ|$ is the length of the line segment PQ. It is easy to see from geometrical considerations that $|PQ| = |\beta - \alpha|$. Hence

$$\cos\phi = \frac{|\alpha|^2 + |\beta|^2 - |\beta - \alpha|^2}{2|\alpha|\,|\beta|} = \frac{(a_1^2 + a_2^2 + a_3^2) + (b_1^2 + b_2^2 + b_3^2)}{2|\alpha|\,|\beta|}$$

$$- \frac{(a_1 - b_1)^2 + (a_2 - b_2)^2 + (a_3 - b_3)^2}{2|\alpha|\,|\beta|}$$

$$= \frac{a_1 b_1 + a_2 b_2 + a_3 b_3}{|\alpha|\,|\beta|}.$$

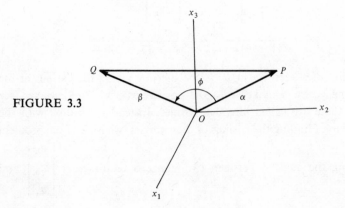

FIGURE 3.3

EXAMPLE 3. Let $\alpha = \begin{bmatrix} 1 \\ 1 \\ 0 \end{bmatrix}$ and $\beta = \begin{bmatrix} 0 \\ 1 \\ 1 \end{bmatrix}$. The angle ϕ between α and

β is determined by

$$\cos \phi = \frac{(1)(0) + (1)(1) + (0)(1)}{\sqrt{1^2 + 1^2 + 0^2} \sqrt{0^2 + 1^2 + 1^2}} = \frac{1}{2}.$$

Thus $\phi = 60°$.

We can express $\cos \phi$ in terms of a very useful function, which we now define.

Definition 3.1. Let $\alpha = \begin{bmatrix} a_1 \\ a_2 \\ a_3 \end{bmatrix}$ and $\beta = \begin{bmatrix} b_1 \\ b_2 \\ b_3 \end{bmatrix}$ be vectors in R^3. The **standard inner product on R^3** is defined as the real number

$$a_1 b_1 + a_2 b_2 + a_3 b_3$$

and is denoted by (α, β).

We now see that if $\alpha = \begin{bmatrix} a_1 \\ a_2 \\ a_3 \end{bmatrix}$ is a vector in R^3, then $|\alpha| = \sqrt{(\alpha, \alpha)}$ and if

$\beta = \begin{bmatrix} b_1 \\ b_2 \\ b_3 \end{bmatrix}$, then the cosine of the angle ϕ between nonzero vectors α and β is

$$\cos \phi = \frac{(\alpha, \beta)}{|\alpha| \, |\beta|}, \qquad 0 \le \phi \le \pi.$$

We also observe that two vectors α and β in R^3 are **orthogonal**, or **perpendicular**, if and only if $(\alpha, \beta) = 0$.

We note the following properties of the standard inner product on R^3 that will motivate our next section.

Theorem 3.1. *The standard inner product on R^3 has the following properties:*
(a) $(\alpha, \alpha) > 0$ *if* $\alpha \ne \theta$.
(b) $(\alpha, \beta) = (\beta, \alpha)$.
(c) $(\alpha + \beta, \gamma) = (\alpha, \gamma) + (\beta, \gamma)$.
(d) $(c\alpha, \beta) = c(\alpha, \beta)$.

Proof: Exercise.

The standard inner product (α, β) on R^3 is usually called the **dot product** of α and β, and it is denoted by $\alpha \cdot \beta$. Thus if $\alpha = \begin{bmatrix} a_1 \\ a_2 \\ a_3 \end{bmatrix}$ and $\beta = \begin{bmatrix} b_1 \\ b_2 \\ b_3 \end{bmatrix}$, then $\alpha \cdot \beta = a_1 b_1 + a_2 b_2 + a_3 b_3$.

Cross Product (Optional)

Suppose that $\alpha = a_1\mathbf{i} + a_2\mathbf{j} + a_3\mathbf{k}$ and $\beta = b_1\mathbf{i} + b_2\mathbf{j} + b_3\mathbf{k}$ and that we want to find a vector $\gamma = \begin{bmatrix} u \\ v \\ w \end{bmatrix}$ orthogonal (perpendicular) to both α and β.

Thus we want $\alpha \cdot \gamma = 0$ and $\beta \cdot \gamma = 0$, which leads to the linear system

$$\begin{aligned} a_1 u + a_2 v + a_3 w &= 0 \\ b_1 u + b_2 v + b_3 w &= 0. \end{aligned} \tag{2}$$

It can be shown that

$$\gamma = \begin{bmatrix} a_2 b_3 - a_3 b_2 \\ a_3 b_1 - a_1 b_3 \\ a_1 b_2 - a_2 b_1 \end{bmatrix}$$

is a solution to (2) (verify). Of course, we can also write γ as

$$\gamma = (a_2 b_3 - a_3 b_2)\mathbf{i} + (a_3 b_1 - a_1 b_3)\mathbf{j} + (a_1 b_2 - a_2 b_1)\mathbf{k}. \tag{3}$$

This vector is called the **cross product** of α and β and is denoted by $\alpha \times \beta$. Note that the cross product, $\alpha \times \beta$, is a vector, while the dot product, $\alpha \cdot \beta$, is a number. Although the cross product is not defined on R^n if $n \neq 3$, it has many applications; we shall use it when we study planes in R^3.

EXAMPLE 4. Let $\alpha = 2\mathbf{i} + \mathbf{j} + 2\mathbf{k}$ and $\beta = 3\mathbf{i} - \mathbf{j} - 3\mathbf{k}$. From (3),

$$\alpha \times \beta = -\mathbf{i} + 12\mathbf{j} - 5\mathbf{k}.$$

Let α, β, and γ be vectors in R^3 and c a scalar. The cross-product operation satisfies the following properties, whose verification we leave to the reader:

(a) $\alpha \times \beta = -(\beta \times \alpha)$.

(b) $\alpha \times (\beta + \gamma) = \alpha \times \beta + \alpha \times \gamma$.

(c) $(\alpha + \beta) \times \gamma = \alpha \times \gamma + \beta \times \gamma$.

(d) $c(\alpha \times \beta) = (c\alpha) \times \beta = \alpha \times (c\beta)$.

(e) $\alpha \times \alpha = \theta$.

(f) $\theta \times \alpha = \alpha \times \theta = \theta$.

(g) $\alpha \times (\beta \times \gamma) = (\alpha \cdot \gamma)\beta - (\alpha \cdot \beta)\gamma$.

(h) $(\alpha \times \beta) \times \gamma = (\gamma \cdot \alpha)\beta - (\gamma \cdot \beta)\alpha$.

EXAMPLE 5. It follows from (3) that

$$\mathbf{i} \times \mathbf{i} = \mathbf{j} \times \mathbf{j} = \mathbf{k} \times \mathbf{k} = \theta,$$
$$\mathbf{i} \times \mathbf{j} = \mathbf{k}, \quad \mathbf{j} \times \mathbf{k} = \mathbf{i}, \quad \mathbf{k} \times \mathbf{i} = \mathbf{j}.$$

Also,

$$\mathbf{j} \times \mathbf{i} = -\mathbf{k}, \quad \mathbf{k} \times \mathbf{j} = -\mathbf{i}, \quad \mathbf{i} \times \mathbf{k} = -\mathbf{j}.$$

These can be remembered by the method indicated in Figure 3.4. Moving around the circle in a clockwise direction, we see that the cross product of two vectors taken in the indicated order is the third vector; moving in a counterclockwise direction, we see that the cross product taken in the indicated order is the negative of the third vector. The cross product of a vector with itself is the zero vector.

FIGURE 3.4

Although many of the familiar properties of the real numbers hold for the cross product, it should be noted that two important properties do not hold. The commutative law does not hold, since $\alpha \times \beta = -(\beta \times \alpha)$. Also, the associative law does not hold, since $\mathbf{i} \times (\mathbf{i} \times \mathbf{j}) = \mathbf{i} \times \mathbf{k} = -\mathbf{j}$ while $(\mathbf{i} \times \mathbf{i}) \times \mathbf{j} = \theta \times \mathbf{j} = \theta$.

Thus far we know that the vector $\alpha \times \beta$ is orthogonal to both α and β. The direction of $\alpha \times \beta$ can be determined as follows. Since $\mathbf{i} \times \mathbf{j} = \mathbf{k}$, the direction of $\alpha \times \beta$ is such that α, β and $\alpha \times \beta$ form a right-handed coordinate system. That is, the direction of $\alpha \times \beta$ is that in which a right-hand screw

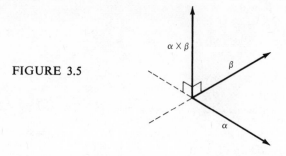

FIGURE 3.5

perpendicular to the plane of α and β will move if we rotate it through an acute angle from α to β (Figure 3.5).

Planes (Optional)

A plane in R^3 can be determined by specifying a point in it and a **normal** to it, that is, a vector perpendicular to it.

To obtain an equation of the plane passing through the point $P_0(x_0, y_0, z_0)$ and having the nonzero vector $\gamma = a\mathbf{i} + b\mathbf{j} + c\mathbf{k}$ as a normal, we proceed as follows. A point $P(x, y, z)$ lies in the plane if and only if the vector $\overrightarrow{P_0P}$ is perpendicular to γ (Figure 3.6). Thus $P(x, y, z)$ lies in the plane if and only if

$$\gamma \cdot \overrightarrow{P_0P} = 0. \tag{4}$$

Since

$$\overrightarrow{P_0P} = (x - x_0)\mathbf{i} + (y - y_0)\mathbf{j} + (z - z_0)\mathbf{k},$$

we can write (4) as

$$a(x - x_0) + b(y - y_0) + c(z - z_0) = 0. \tag{5}$$

EXAMPLE 6. Find an equation of the plane passing through the point $(3, 4, -3)$ and perpendicular to the vector $\gamma = 5\mathbf{i} - 2\mathbf{j} + 4\mathbf{k}$.

Solution: Substituting in (5), we obtain the equation of the plane as

$$5(x - 3) - 2(y - 4) + 4(z + 3) = 0.$$

A plane is also determined by three noncollinear points in it, as we show in the following example.

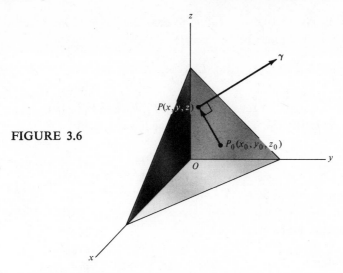

FIGURE 3.6

EXAMPLE 7. Find an equation of the plane passing through the points $P_1(2, -2, 1)$, $P_2(-1, 0, 3)$, and $P_3(5, -3, 4)$.

Solution: The nonparallel vectors $\overrightarrow{P_1P_2} = -3\mathbf{i} + 2\mathbf{j} + 2\mathbf{k}$ and $\overrightarrow{P_1P_3} = 3\mathbf{i} - \mathbf{j} + 3\mathbf{k}$ lie in the plane, since the points P_1, P_2, and P_3 lie in the plane. The vector

$$\delta = \overrightarrow{P_1P_2} \times \overrightarrow{P_1P_3} = 8\mathbf{i} + 15\mathbf{j} - 3\mathbf{k}$$

is then perpendicular to both $\overrightarrow{P_1P_2}$ and $\overrightarrow{P_1P_3}$ and is thus a normal to the plane. Using the vector δ and the point $P_1(2, -2, 1)$ in (5), we obtain

$$8(x - 2) + 15(y + 2) - 3(z - 1) = 0$$

as an equation of the plane.

If we multiply out and simplify, (5) can be rewritten as

$$ax + by + cz + d = 0. \tag{6}$$

EXAMPLE 8. Equation (6) for the plane in Example 7 can be rewritten as

$$8x + 15y - 3z + 17 = 0. \tag{7}$$

It is not difficult to show (Exercise 32) that the graph of an equation of the form given in (6), where a, b, c, and d are constants, is a plane with normal $\gamma = a\mathbf{i} + b\mathbf{j} + c\mathbf{k}$; moreover, if $d = 0$, it is a two-dimensional subspace of R^3.

EXAMPLE 9. An alternative solution to Example 7 is as follows. Let the equation of the desired plane be

$$ax + by + cz + d = 0, \tag{8}$$

where a, b, c, and d are to be determined. Since P_1, P_2, and P_3 lie in the plane, their coordinates satisfy (8). Thus we obtain the linear system

$$
\begin{aligned}
2a - 2b + c + d &= 0 \\
-a + 3c + d &= 0 \\
5a - 3b + 4c + d &= 0.
\end{aligned}
$$

Solving this system, we have

$$a = \tfrac{8}{17}r, \qquad b = \tfrac{15}{17}r, \qquad c = -\tfrac{3}{17}r, \qquad \text{and} \qquad d = r,$$

where r is any real number. Letting $r = 17$, we obtain

$$a = 8, \qquad b = 15, \qquad c = -3, \qquad \text{and} \qquad d = 17,$$

which yields (7) as in the first solution.

EXAMPLE 10. Find the parametric equations of the line of intersection of the planes

$$\pi_1: \quad 2x + 3y - 2z + 4 = 0 \qquad \text{and} \qquad \pi_2: \quad x - y + 2z + 3 = 0.$$

Solution: Solving the linear system consisting of the equations for π_1 and π_2, we obtain (verify)

$$
\begin{aligned}
x &= -\tfrac{13}{5} - \tfrac{4}{5}t \\
y &= \tfrac{2}{5} + \tfrac{6}{5}t \qquad -\infty < t < \infty \\
z &= 0 + t
\end{aligned}
$$

as the parametric equations for the line l of intersection of the planes (see Figure 3.7).

As we have indicated, the cross product cannot be generalized to R^n. However, we can generalize the notions of length, direction, and standard inner product to R^n in the natural manner, but there are some things to be checked. For example, if we define the cosine of the angle ϕ between two

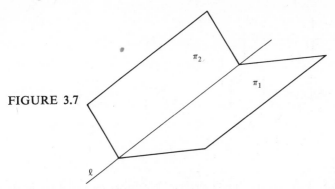

FIGURE 3.7

nonzero vectors α and β in R^n as $\cos \phi = (\alpha, \beta)/(|\alpha|\, |\beta|)$, we must check that $-1 \le \cos \phi \le 1$; otherwise, it would be misleading to call this fraction $\cos \phi$. Rather than verify this property for R^n at this point, we shall obtain this result in our next section, where we formulate the notion of inner product in any real vector space.

3.1. Exercises

1. Find the length of each of the following vectors.

   (a) $\begin{bmatrix} 1 \\ 0 \\ 0 \end{bmatrix}$.
   (b) $\begin{bmatrix} 0 \\ 0 \\ 0 \end{bmatrix}$.
   (c) $\begin{bmatrix} 1 \\ 2 \\ 3 \end{bmatrix}$.
   (d) $\begin{bmatrix} -1 \\ -3 \\ -4 \end{bmatrix}$.
   (e) $\begin{bmatrix} 1 \\ -2 \\ 4 \end{bmatrix}$.

2. Find $|\alpha - \beta|$.

   (a) $\alpha = \begin{bmatrix} 1 \\ 0 \\ 0 \end{bmatrix}$, $\beta = \begin{bmatrix} 0 \\ 0 \\ 0 \end{bmatrix}$.
   (b) $\alpha = \begin{bmatrix} 0 \\ 0 \\ 0 \end{bmatrix}$, $\beta = \begin{bmatrix} 1 \\ 0 \\ 0 \end{bmatrix}$.

   (c) $\alpha = \begin{bmatrix} 1 \\ 2 \\ 3 \end{bmatrix}$, $\beta = \begin{bmatrix} 4 \\ 5 \\ 6 \end{bmatrix}$.
   (d) $\alpha = \begin{bmatrix} -1 \\ -2 \\ -3 \end{bmatrix}$, $\beta = \begin{bmatrix} -4 \\ -5 \\ -6 \end{bmatrix}$.

3. Find the distance between α and β.

   (a) $\alpha = \begin{bmatrix} 1 \\ 2 \\ 3 \end{bmatrix}$, $\beta = \begin{bmatrix} -4 \\ -5 \\ -6 \end{bmatrix}$.
   (b) $\alpha = \begin{bmatrix} 1 \\ 2 \\ 3 \end{bmatrix}$, $\beta = \begin{bmatrix} 4 \\ -5 \\ 6 \end{bmatrix}$.

   (c) $\alpha = \begin{bmatrix} -1 \\ -2 \\ -3 \end{bmatrix}$, $\beta = \begin{bmatrix} 4 \\ 5 \\ 6 \end{bmatrix}$.

4. Find (α, β).

   (a) $\alpha = \begin{bmatrix} 0 \\ 0 \\ 0 \end{bmatrix}$, $\beta = \begin{bmatrix} 0 \\ 0 \\ 0 \end{bmatrix}$.
   (b) $\alpha = \begin{bmatrix} 0 \\ 0 \\ 0 \end{bmatrix}$, $\beta = \begin{bmatrix} 1 \\ 2 \\ 3 \end{bmatrix}$.

(c) $\alpha = \begin{bmatrix} 2 \\ -3 \\ 4 \end{bmatrix}$, $\beta = \begin{bmatrix} 0 \\ 0 \\ 0 \end{bmatrix}$.

(d) $\alpha = \begin{bmatrix} 1 \\ 2 \\ -3 \end{bmatrix}$, $\beta = \begin{bmatrix} -4 \\ 5 \\ 6 \end{bmatrix}$.

(e) $\alpha = \begin{bmatrix} 1 \\ 3 \\ 2 \end{bmatrix}$, $\beta = \begin{bmatrix} 2 \\ 3 \\ 4 \end{bmatrix}$.

5. For each pair of vectors α and β in Exercise 3, find the cosine of the angle ϕ between α and β.

6. For each of the following vectors α, find the direction cosines (the cosines of the angle between α and the positive x_1-, x_2-, and x_3-axes).

(a) $\alpha = \begin{bmatrix} 1 \\ 0 \\ 0 \end{bmatrix}$.   (b) $\alpha = \begin{bmatrix} 1 \\ 3 \\ 2 \end{bmatrix}$.   (c) $\alpha = \begin{bmatrix} -1 \\ -2 \\ -3 \end{bmatrix}$.

(d) $\alpha = \begin{bmatrix} 4 \\ -3 \\ 2 \end{bmatrix}$.   (e) $\alpha = \begin{bmatrix} 1 \\ 1 \\ 0 \end{bmatrix}$.   (f) $\alpha = \begin{bmatrix} 1 \\ 1 \\ 1 \end{bmatrix}$.

7. Prove Theorem 3.1.

8. Verify Theorem 3.1 for

$$\alpha = \begin{bmatrix} 1 \\ 2 \\ 3 \end{bmatrix}, \quad \beta = \begin{bmatrix} -2 \\ 4 \\ 3 \end{bmatrix}, \quad \gamma = \begin{bmatrix} 0 \\ 3 \\ -2 \end{bmatrix}, \quad \text{and} \quad c = -3.$$

9. Let P and Q be the points in R^3 with respective coordinates $(3, -1, 2)$ and $(4, 2, -3)$. Find the length of the segment PQ.

10. For each pair of vectors α and β in Exercise 2, find (α, β).

11. (a) Let α, β, and γ be vectors in R^3 such that α and β are orthogonal and α and γ are orthogonal. Prove that α and any linear combination of β and γ, $r\beta + s\gamma$, are orthogonal.

   (b) Let α be a vector in R^3. Prove that the set W of all vectors β in R^3 such that α and β are orthogonal is a subspace of R^3.

12. Prove that if $(\alpha, \beta) = 0$ for all β in R^3, then $\alpha = 0$.

13. Let points A, B, C, and D in R^3 have respective coordinates $(1, 2, 3)$, $(-2, 3, 5)$, $(0, 3, 6)$, and $(3, 2, 4)$. Prove that $ABCD$ is a parallelogram.

14. Let $S = \{\alpha_1, \alpha_2, \alpha_3\}$ be a set of nonzero vectors in R^3 such that any two vectors in S are orthogonal. Prove that S is linearly independent.

15. Prove that for any vectors α, β, and γ in R^3 we have $(\alpha, \beta + \gamma) = (\alpha, \beta) + (\alpha, \gamma)$.

16. Prove that for any vectors α, β, and γ in R^3 and any scalar c we have
   (a) $(\alpha + c\beta, \gamma) = (\alpha, \gamma) + c(\beta, \gamma)$.

(b) $(\alpha, c\beta) = c(\alpha, \beta)$.

(c) $(\alpha + \beta, c\gamma) = c(\alpha, \gamma) + c(\beta, \gamma)$.

17. Which of the following pairs of lines are perpendicular?

$$
\begin{array}{llll}
x = & 2 + 2t & x = 2 + t & x = 3 - t & x = & 2t \\
\text{(a) } y = & -3 - 3t & \text{and} \quad y = 4 - t & \text{(b) } y = 4 + t & \text{and} \quad y = 3 - 2t \\
z = & 4 + 4t & z = 5 - t. & z = 2 + 2t & z = 4 + 2t.
\end{array}
$$

18. Find parametric equations for the line passing through $(3, -1, -3)$ and perpendicular to the line passing through $(3, -2, 4)$ and $(0, 3, 5)$.

19. Compute $\alpha \times \beta$.

(a) $\alpha = 2\mathbf{i} + 3\mathbf{j} + 4\mathbf{k}, \beta = -\mathbf{i} + 3\mathbf{j} - \mathbf{k}$.

(b) $\alpha = \mathbf{i} + \mathbf{k}, \beta = 2\mathbf{i} + 3\mathbf{j} - \mathbf{k}$.

(c) $\alpha = \mathbf{i} - \mathbf{j} + 2\mathbf{k}, \beta = 3\mathbf{i} - 4\mathbf{j} + \mathbf{k}$.

(d) $\alpha = (2\mathbf{i} - \mathbf{j} + \mathbf{k}), \beta = -2\alpha$.

20. Compute $\alpha \times \beta$.

(a) $\alpha = \mathbf{i} - \mathbf{j} + 2\mathbf{k}, \beta = 3\mathbf{i} + \mathbf{j} + 2\mathbf{k}$.

(b) $\alpha = 2\mathbf{i} + \mathbf{j} - 2\mathbf{k}, \beta = \mathbf{i} + 3\mathbf{k}$.

(c) $\alpha = 2\mathbf{j} + \mathbf{k}, \beta = 3\alpha$.

(d) $\alpha = 4\mathbf{i} - 2\mathbf{k}, \beta = 2\mathbf{j} - \mathbf{k}$.

21. Let $\alpha = \mathbf{i} + 2\mathbf{j} - 3\mathbf{k}$, $\beta = 2\mathbf{i} + 3\mathbf{j} + \mathbf{k}$, $\gamma = 2\mathbf{i} - \mathbf{j} + 2\mathbf{k}$, and $c = -3$. Verify properties (a) through (d) for the cross product operation.

22. Verify that each of the cross products $\alpha \times \beta$ in Exercise 1 is orthogonal to both α and β.

23. (a) Show that $(\alpha \times \beta) \cdot \gamma = \alpha \cdot (\beta \times \gamma)$.

(b) Show that $(\alpha \times \beta) \times \gamma = (\gamma \cdot \alpha)\beta - (\gamma \cdot \beta)\alpha$.

24. Prove the *Jacobi identity*

$$(\alpha \times \beta) \times \gamma + (\beta \times \gamma) \times \alpha + (\gamma \times \alpha) \times \beta = \theta.$$

25. State which of the following points are on the plane

$$3(x - 2) + 2(y + 3) - 4(z - 4) = 0.$$

(a) $(0, -2, 3)$.   (b) $(1, -2, 3)$.

26. Find an equation of the plane passing through the given point and perpendicular to the given vector.

(a) $(0, 2, -3), 3\mathbf{i} - 2\mathbf{j} + 4\mathbf{k}$.   (b) $(-1, 3, 2), \mathbf{j} - 3\mathbf{k}$.

27. Find an equation of the plane passing through the given points.

(a) $(0, 1, 2), (3, -2, 5), (2, 3, 4)$.   (b) $(2, 3, 4), (-1, -2, 3), (-5, -4, 2)$.

28. Find parametric equations of the line of intersection of the given planes.

(a) $2x + 3y - 4z + 5 = 0$ and $-3x + 2y + 5z + 6 = 0$.

(b) $3x - 2y - 5z + 4 = 0$ and $2x + 3y + 4z + 8 = 0$.

29. Find an equation of the plane through $(-2, 3, 4)$ and perpendicular to the line through $(4, -2, 5)$ and $(0, 2, 4)$.

30. Find the point of intersection of the line

$$
\begin{aligned}
x &= 2 - 3t \\
y &= 4 + 2t \\
z &= 3 - 5t
\end{aligned}
$$

and the plane $2x + 3y + 4z + 8 = 0$.

31. Find a line passing through $(-2, 5, -3)$ and perpendicular to the plane $2x - 3y + 4z + 7 = 0$.
32. (a) Show that the graph of an equation of the form given in (6) is a plane with normal $\gamma = ai + bj + ck$.
    (b) Show that the set of all points on the plane $ax + by + cz = 0$ is a subspace of R^3.
    (c) Find a basis for the plane $2x - 3y + 4z = 0$.

3.2. Inner Product Spaces

In this section we shall use the properties of the standard inner product on R^3 listed in Theorem 3.1 as our foundation for generalizing the notion of the inner product to any real vector space. Here V is an arbitrary real vector space, not necessarily finite dimensional.

Definition 3.2. Let V be any real vector space. An **inner product** on V is a function that assigns to each ordered pair of vectors α, β of V a real number (α, β) satisfying:
  (a) $(\alpha, \alpha) > 0$ for $\alpha \neq \theta_V$.
  (b) $(\beta, \alpha) = (\alpha, \beta)$ for any α, β in V.
  (c) $(\alpha + \beta, \gamma) = (\alpha, \gamma) + (\beta, \gamma)$ for any α, β, γ in V.
  (d) $(c\alpha, \beta) = c(\alpha, \beta)$ for α, β in V and c a scalar.

From these properties it follows that $(\alpha, c\beta) = c(\alpha, \beta)$ for $(\alpha, c\beta) = (c\beta, \alpha)$ $= c(\beta, \alpha) = c(\alpha, \beta)$.

EXAMPLE 1. In Section 3.1 we have defined the standard inner product on R^3.

EXAMPLE 2. We can define the **standard inner product** on R^n by defining

$$(\alpha, \beta) \text{ for } \alpha = \begin{bmatrix} a_1 \\ a_2 \\ \vdots \\ a_n \end{bmatrix} \text{ and } \beta = \begin{bmatrix} b_1 \\ b_2 \\ \vdots \\ b_n \end{bmatrix} \text{ in } R^n, \text{ as}$$

$$(\alpha, \beta) = a_1 b_1 + a_2 b_2 + \cdots + a_n b_n.$$

Of course, one has to verify that this function satisfies the properties of Definition 3.2.

Later we shall need the following interesting result dealing with the standard inner product on R^n. If $A = [a_{ij}]$ is an $n \times n$ matrix and γ and δ are vectors in R^n, then

$$(A\gamma, \delta) = (\gamma, A'\delta). \tag{1}$$

To see this, let

$$\gamma = \begin{bmatrix} c_1 \\ c_2 \\ \vdots \\ c_n \end{bmatrix} \quad \text{and} \quad \delta = \begin{bmatrix} d_1 \\ d_2 \\ \vdots \\ d_n \end{bmatrix}.$$

Then

$$A\gamma = \begin{bmatrix} \sum_{j=1}^{n} a_{1j}c_j \\ \sum_{j=1}^{n} a_{2j}c_j \\ \vdots \\ \sum_{j=1}^{n} a_{nj}c_j \end{bmatrix}.$$

Thus

$$(A\gamma, \delta) = d_1 \left(\sum_{j=1}^{n} a_{1j}c_j \right) + d_2 \left(\sum_{j=1}^{n} a_{2j}c_j \right) + \cdots + d_n \left(\sum_{j=1}^{n} a_{nj}c_j \right),$$

which we can rewrite as

$$c_1 \left(\sum_{i=1}^{n} a_{i1}d_i \right) + c_2 \left(\sum_{i=1}^{n} a_{i2}d_i \right) + \cdots + c_n \left(\sum_{i=1}^{n} a_{in}d_i \right).$$

However, this last expression is $(\gamma, A'\delta)$.

EXAMPLE 3. Let $\alpha = \begin{bmatrix} a_1 \\ a_2 \end{bmatrix}$ and $\beta = \begin{bmatrix} b_1 \\ b_2 \end{bmatrix}$ be vectors in R^2. We define $(\alpha, \beta) = a_1b_1 - a_2b_1 - a_1b_2 + 3a_2b_2$. This is an inner product on R^2. We have $(\alpha, \alpha) = a_1^2 - 2a_1a_2 + a_2^2 + 2a_2^2 = (a_1 - a_2)^2 + 2a_2^2 > 0$ if $\alpha \neq \theta$. We can also verify (see Exercise 2) the remaining three properties of Definition 3.2.

Example 3 shows that on one vector space we may have many different inner products, since we also have the standard inner product on R^2.

EXAMPLE 4. Let V be the vector space of all continuous real-valued functions on the unit interval $[0, 1]$. For f and g in V, we let $(f, g) = \int_0^1 f(t)g(t)\, dt$. We now verify that this is an inner product on V, that is, that the properties of Definition 3.2 are satisfied.

Using results from the calculus, we have

$$(f, f) = \int_0^1 (f(t))^2\, dt > 0 \qquad \text{if } f \neq 0,$$

the zero function, so (a) holds. Also,

$$(f, g) = \int_0^1 f(t)g(t)\, dt = \int_0^1 g(t)f(t)\, dt = (g, f).$$

Next,

$$(f + g, h) = \int_0^1 (f(t) + g(t))h(t)\, dt = \int_0^1 f(t)h(t)\, dt + \int_0^1 g(t)h(t)\, dt$$

$$= (f, h) + (g, h).$$

Finally,

$$(cf, g) = \int_0^1 (cf(t))g(t)\, dt = c \int_0^1 f(t)g(t)\, dt = c(f, g).$$

Thus, if f and g are the functions defined by $f(t) = t + 1$, $g(t) = 2t + 3$, then

$$(f, g) = \int_0^1 (t + 1)(2t + 3)\, dt = \int_0^1 (2t^2 + 5t + 3)\, dt = \tfrac{37}{6}.$$

EXAMPLE 5. Let $V = R_2$; if $\alpha = [a_1 \quad a_2]$ and $\beta = [b_1 \quad b_2]$ are vectors in V, we define $(\alpha, \beta) = a_1 b_1 - a_2 b_1 - a_1 b_2 + 5a_2 b_2$. The verification that this function is an inner product is entirely analogous to the verification required in Example 3.

EXAMPLE 6. Let $V = P$; if $p(t)$ and $q(t)$ are polynomials in P, we define $(p(t), q(t)) = \int_0^1 p(t)q(t)\, dt$. The verification that this function is an inner product is identical to the verification given for Example 4.

EXAMPLE 7. Let V be any finite-dimensional vector space and let $S = \{\alpha_1, \alpha_2, \ldots, \alpha_n\}$ be an ordered basis for V. We define an inner product on V as follows: If $\alpha = a_1\alpha_1 + a_2\alpha_2 + \cdots + a_n\alpha_n$ and $\beta = b_1\alpha_1 + b_2\alpha_2 + \cdots + b_n\alpha_n$, then

$$(\alpha, \beta) = a_1b_1 + a_2b_2 + \cdots + a_nb_n.$$

It is not difficult to verify that this defines an inner product on V.

Example 7 shows that we can define an inner product on any finite-dimensional vector space. Of course, if we change the basis in Example 7, we obtain a different inner product.

We now show that every inner product on a finite-dimensional vector space is completely determined, in terms of a given basis, by a certain matrix. Let V be a finite-dimensional vector space and $S = \{\alpha_1, \alpha_2, \ldots, \alpha_n\}$ be an ordered basis for V. Assume that we are given an inner product on V. Now let $c_{ij} = (\alpha_i, \alpha_j)$, and let $C = [c_{ij}]$; C is a symmetric matrix that determines (α, β) for any vectors α and β in V. For if α and β are in V, then $\alpha = a_1\alpha_1 + a_2\alpha_2 + \cdots + a_n\alpha_n$ and $\beta = b_1\alpha_1 + b_2\alpha_2 + \cdots + b_n\alpha_n$, where $a_1, a_2, \ldots, a_n, b_1, b_2, \ldots, b_n$ are uniquely determined real numbers. Then

$$(\alpha, \beta) = (a_1\alpha_1 + a_2\alpha_2 + \cdots + a_n\alpha_n, \beta) = \sum_{i=1}^{n} a_i(\alpha_i, \beta)$$

$$= \sum_{i=1}^{n} a_i(\alpha_i, b_1\alpha_1 + b_2\alpha_2 + \cdots + b_n\alpha_n)$$

$$= \sum_{i=1}^{n} a_i \sum_{j=1}^{n} b_j(\alpha_i, \alpha_j) = \sum_{j=1}^{n} \sum_{i=1}^{n} b_j c_{ij} a_i$$

$$= [\alpha]_S' C [\beta]_S = X'CY,$$

where $X = [\alpha]_S$ and $Y = [\beta]_S$. Thus C determines (α, β) for every α and β in V. The matrix C is called the **matrix of the inner product with respect to the ordered basis** S. Since $(\alpha, \alpha) > 0$ for $\alpha \neq \theta$, we must have $X'CX > 0$ for any nonzero vector X in R^n. An $n \times n$ matrix C with the property that $X'CX > 0$ for any nonzero vector X in R^n is called **positive definite**. Such a matrix is nonsingular, for if C is singular, then the homogeneous system $CX = O$ has a nontrivial solution X_0. Then $X_0'CX_0 = 0$, contradicting the requirement that $X'CX > 0$ for nonzero X.

Conversely, if $C = [c_{ij}]$ is any $n \times n$ symmetric matrix that is positive definite (that is, $X'CX > 0$ if X is a nonzero vector in R^n), then we define (α, β), for $\alpha = a_1\alpha_1 + a_2\alpha_2 + \cdots + a_n\alpha_n$ and $\beta = b_1\alpha_1 + b_2\alpha_2 + \cdots + b_n\alpha_n$ in V by $(\alpha, \beta) = \sum_{j=1}^{n} \sum_{i=1}^{n} b_j c_{ij} a_i$. It is not difficult to show that this defines an inner product on V. The only gap in the above discussion is that we still do not know when a real symmetric matrix is positive definite. In Section 6.3 (see Theorem 6.12) we provide a characterization of a positive definite matrix.

EXAMPLE 8. Let $C = \begin{bmatrix} 2 & 1 \\ 1 & 2 \end{bmatrix}$. In this case we may verify that C is positive definite as follows.

$$
\begin{aligned}
X'CX &= [x_1 \quad x_2] \begin{bmatrix} 2 & 1 \\ 1 & 2 \end{bmatrix} \begin{bmatrix} x_1 \\ x_2 \end{bmatrix} \\
&= 2x_1^2 + 2x_1x_2 + 2x_2^2 \\
&= x_1^2 + x_2^2 + (x_1 + x_2)^2 > 0 \qquad \text{if } X \neq O.
\end{aligned}
$$

We now define an inner product on P_1 whose matrix with respect to the ordered basis $S = \{t, 1\}$ is C. Thus let $p(t) = a_1 t + a_2$ and $q(t) = b_1 t + b_2$ be any two vectors in P_1. Let $(p(t), q(t)) = 2a_1 b_1 + a_2 b_1 + a_1 b_2 + 2a_2 b_2$. We must verify that if $p(t)$ is not the zero polynomial, then $(p(t), p(t)) > 0$; that is, if not both a_1 and a_2 are zero, then $2a_1^2 + 2a_1 a_2 + 2a_2^2 > 0$. We now have

$$
2a_1^2 + 2a_1 a_2 + 2a_2^2 = a_1^2 + a_2^2 + (a_1 + a_2)^2 > 0.
$$

The remaining properties are easy to verify.

Definition 3.3. A real vector space that has an inner product defined on it is called an **inner product space**. If the space is finite-dimensional, it is called a **Euclidean space**.

If V is an inner product space, then by **dimension** of V we mean the dimension of V as a real vector space, and a set S is a **basis** for V if S is a basis for the real vector space V. Examples 4 and 6 are inner product spaces. In an inner product space we define the **length** of a vector α by

$$
|\alpha| = \sqrt{(\alpha, \alpha)}.
$$

This definition of length seems reasonable because at least we have $|\alpha| > 0$ if $\alpha \neq \theta$. We can show (see Exercise 7) that $|\theta| = 0$.

We shall now prove a result that will enable us to give a worthwhile definition for the cosine of an angle between two nonzero vectors α and β in an inner product space V. This result, called the **Cauchy–Schwarz inequality**, has many important applications in mathematics. The proof, although not difficult, is one that is not too natural and does call for a clever start.

Theorem 3.2 (Cauchy–Schwarz Inequality). *If α and β are any two vectors in an inner product space V, then*

$$(\alpha, \beta)^2 \leq |\alpha|^2 \, |\beta|^2.$$

Proof: If $\alpha = \theta$, then $|\alpha| = 0$ and $(\alpha, \beta) = 0$, so the inequality holds. Now suppose that α is nonzero. Let r be a scalar and consider the vector $r\alpha + \beta$. Since the inner product of a vector with itself is always nonnegative, we have

$$0 \leq (r\alpha + \beta, r\alpha + \beta) = (\alpha, \alpha)r^2 + 2r(\alpha, \beta) + (\beta, \beta) = ar^2 + 2br + c,$$

where $a = (\alpha, \alpha)$, $b = (\alpha, \beta)$, and $c = (\beta, \beta)$. If we fix α and β, then $ar^2 + 2br + c = p(r)$ is a quadratic polynomial in r that is nonnegative for all values of r. This means that $p(r)$ has at most one real root, for if it had two distinct roots, r_1 and r_2, it would be negative between r_1 and r_2 (Figure 3.8).

From the quadratic formula, the roots of $p(r)$ are given by

$$\frac{-b + \sqrt{b^2 - ac}}{a} \quad \text{and} \quad \frac{-b - \sqrt{b^2 - ac}}{a}$$

($a \neq 0$, since $\alpha \neq \theta$). Thus we must have $b^2 - ac \leq 0$, which means that $b^2 \leq ac$, the desired inequality.

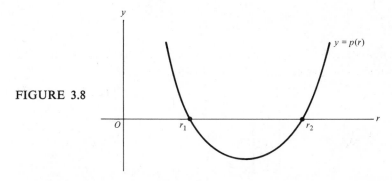

FIGURE 3.8

EXAMPLE 9. Let $\alpha = \begin{bmatrix} 1 \\ 2 \\ -3 \end{bmatrix}$ and $\beta = \begin{bmatrix} -3 \\ 2 \\ 2 \end{bmatrix}$ be in the Euclidean

space R^3 with the standard inner product. Then $(\alpha, \beta) = -5$, $|\alpha| = \sqrt{14}$, and $|\beta| = \sqrt{17}$. Therefore, $(\alpha, \beta)^2 \leq |\alpha|^2 |\beta|^2$.

If α and β are any two nonzero vectors in an inner product space V, the Cauchy–Schwarz inequality can be written as

$$\frac{(\alpha, \beta)^2}{|\alpha|^2 |\beta|^2} \leq 1 \qquad \text{or} \qquad -1 \leq \frac{(\alpha, \beta)}{|\alpha| |\beta|} \leq 1.$$

It then follows that there is one and only one angle ϕ such that

$$\cos \phi = \frac{(\alpha, \beta)}{|\alpha| |\beta|}, \qquad 0 \leq \phi \leq \pi.$$

We define this angle to be the **angle between** α and β.

The triangle inequality is an easy consequence of the Cauchy–Schwarz inequality.

Corollary 3.1 (Triangle Inequality). *If α and β are any vectors in an inner product space V, then $|\alpha + \beta| \leq |\alpha| + |\beta|$.*

Proof: We have

$$|\alpha + \beta|^2 = (\alpha + \beta, \alpha + \beta) = (\alpha, \alpha) + 2(\alpha, \beta) + (\beta, \beta) = |\alpha|^2 + 2(\alpha, \beta) + |\beta|^2.$$

The Cauchy–Schwarz inequality states that $(\alpha, \beta) \leq |(\alpha, \beta)| \leq |\alpha| |\beta|$, so we get

$$|\alpha + \beta|^2 \leq |\alpha|^2 + 2|\alpha| |\beta| + |\beta|^2 = (|\alpha| + |\beta|)^2.$$

Taking square roots, we obtain

$$|\alpha + \beta| \leq |\alpha| + |\beta|.$$

We now state the Cauchy–Schwarz inequality for the inner product spaces introduced in several of our examples. In Example 2, if

$$\alpha = \begin{bmatrix} a_1 \\ a_2 \\ \vdots \\ a_n \end{bmatrix} \qquad \text{and} \qquad \beta = \begin{bmatrix} b_1 \\ b_2 \\ \vdots \\ b_n \end{bmatrix},$$

then

$$(\alpha, \beta)^2 = \left(\sum_{i=1}^{n} a_i b_i\right)^2 \leq \left(\sum_{i=1}^{n} a_i^2\right)\left(\sum_{i=1}^{n} b_i^2\right) = |\alpha|^2 \, |\beta|^2.$$

In Example 4, if f and g are continuous functions on $[0, 1]$, then

$$(f, g)^2 = \left(\int_0^1 f(t)g(t) \, dt\right)^2 \leq \left(\int_0^1 f^2(t) \, dt\right)\left(\int_0^1 g^2(t) \, dt\right).$$

EXAMPLE 10. Let V be the Euclidean space P_2 with inner product defined as in Example 6. If $p(t) = t + 2$, then the length of $p(t)$ is

$$|p(t)| = \sqrt{(p(t), p(t))} = \sqrt{\int_0^1 (t + 2)^2 \, dt} = \sqrt{\frac{19}{3}}.$$

If $q(t) = 2t - 3$, then to find the cosine of the angle between $p(t)$ and $q(t)$, we proceed as follows. First,

$$|q(t)| = \sqrt{\int_0^1 (2t - 3)^2 \, dt} = \sqrt{\frac{13}{3}}.$$

Next,

$$(p(t), q(t)) = \int_0^1 (t + 2)(2t - 3) \, dt = \int_0^1 (2t^2 + t - 6) \, dt = -\frac{29}{6}.$$

Then

$$\cos \phi = \frac{(p(t), q(t))}{|p(t)| \, |q(t)|} = \frac{-29/6}{\sqrt{19/3}\sqrt{13/3}} = \frac{-29}{2\sqrt{(19)(13)}}.$$

Definition 3.4. If V is an inner product space, we define the **distance** between two vectors α and β in V as $d(\alpha, \beta) = |\alpha - \beta|$.

Definition 3.5. Let V be an inner product space. Vectors α and β in V are **orthogonal** if $(\alpha, \beta) = 0$.

EXAMPLE 11. Let V be the Euclidean space P_2 considered in Example 10. The vectors t and $t - \frac{2}{3}$ are orthogonal, for

$$\left(t, t - \frac{2}{3}\right) = \int_0^1 t\left(t - \frac{2}{3}\right) dt = \int_0^1 \left(t^2 - \frac{2t}{3}\right) dt = 0.$$

Of course, the vector θ_V in an inner product space V is orthogonal to every vector in V, and two nonzero vectors in V are orthogonal if the angle ϕ between them is $\pi/2$. Also, the subset of vectors in V orthogonal to a fixed vector in V is a subspace of V (see Exercise 17).

We know from calculus that we can work with any set of coordinate axes for R^3, but that the work becomes less burdensome when we deal with Cartesian coordinates. The comparable notion in an inner product space is that of a basis whose vectors are mutually orthogonal. We now proceed to formulate this idea.

Definition 3.6. Let V be an inner product space. A set S of vectors in V is called **orthogonal** if any two distinct vectors in S are orthogonal. If, in addition, each vector in S is of unit length, then S is called **orthonormal**.

We note here that if α is a nonzero vector in an inner product space, then we can always find a vector of unit length (called a **unit vector**) in the same direction as α; we let $\beta = \alpha/|\alpha|$. Then

$$|\beta| = \sqrt{(\beta, \beta)} = \sqrt{\left(\frac{\alpha}{|\alpha|}, \frac{\alpha}{|\alpha|}\right)} = \sqrt{\frac{(\alpha, \alpha)}{|\alpha|\,|\alpha|}} = \sqrt{\frac{|\alpha|^2}{|\alpha|\,|\alpha|}} = 1,$$

and the cosine of the angle between α and β is 1, so α and β have the same direction.

EXAMPLE 12. The natural bases for R^n and R_n are orthonormal sets with respect to the standard inner products on these vector spaces.

An important result about orthogonal sets of vectors in an inner product space is the following.

Theorem 3.3. Let $S = \{\alpha_1, \alpha_2, \ldots, \alpha_n\}$ be a finite orthogonal set of nonzero vectors in an inner product space V. Then S is linearly independent.

Proof: Suppose that $a_1\alpha_1 + a_2\alpha_2 + \cdots + a_n\alpha_n = \theta$. Then taking the inner product of both sides with α_i, we have

$$(a_1\alpha_1 + a_2\alpha_2 + \cdots + a_i\alpha_i + \cdots + a_n\alpha_n, \alpha_i) = (\theta, \alpha_i) = 0.$$

The left-hand side is

$$a_1(\alpha_1, \alpha_i) + a_2(\alpha_2, \alpha_i) + \cdots + a_i(\alpha_i, \alpha_i) + \cdots + a_n(\alpha_n, \alpha_i),$$

and since S is orthogonal, this is $a_i(\alpha_i, \alpha_i)$. Thus $a_i(\alpha_i, \alpha_i) = 0$. Since $\alpha_i \neq \theta$, $(\alpha_i, \alpha_i) \neq 0$, so $a_i = 0$. Repeating this for $i = 1, 2, \ldots, n$, we find that $a_1 = a_2 = \cdots = a_n = 0$, so S is linearly independent.

3.2. Exercises

1. Verify that the standard inner product on R^n satisfies the properties of Definition 3.2.

2. Verify that the function in Example 3 satisfies the remaining three properties of Definition 3.2.

3. Let V be the real vector space of all $n \times n$ matrices. If A and B are in V, we define $(A, B) = \text{Tr}(B'A)$, where Tr is the trace function defined in Exercise 12 of Section 1.2. Prove that this function is an inner product on V.

4. Verify that the function defined on R_2 in Example 5 is an inner product.

5. Verify that the function defined on P_n in Example 6 is an inner product.

6. Verify that the function defined on V in Example 7 is an inner product.

7. Let V be an inner product space. Prove the following:
   (a) $(\theta, \theta) = 0$. (*Hint:* Use $\theta + \theta = \theta$.) Thus $(\alpha, \alpha) = 0$ if and only if $\alpha = \theta$.
   (b) $|\theta| = 0$.
   (c) $(\alpha, \theta) = (\theta, \alpha) = 0$ for any α in V.
   (d) If $(\alpha, \beta) = 0$ for all β in V, then $\alpha = \theta$.
   (e) If $(\alpha, \gamma) = (\beta, \gamma)$ for all γ in V, then $\alpha = \beta$.
   (f) If $(\gamma, \alpha) = (\gamma, \beta)$ for all γ in V, then $\alpha = \beta$.

8. Let V be Euclidean space R_4 with the standard inner product. Find (α, β) for the following pairs of vectors in R_4.
   (a) $\alpha = [1 \quad 3 \quad -1 \quad 2], \beta = [-1 \quad 2 \quad 0 \quad 1]$.
   (b) $\alpha = [0 \quad 0 \quad 1 \quad 1], \beta = [1 \quad 1 \quad 0 \quad 0]$.
   (c) $\alpha = [-2 \quad 1 \quad 3 \quad 4], \beta = [3 \quad 2 \quad 1 \quad -2]$.
   (d) $\alpha = [1 \quad 2 \quad 3 \quad 4], \beta = [-1 \quad 0 \quad -1 \quad -1]$.
   (e) $\alpha = [1 \quad 2 \quad 1 \quad 2], \beta = [2 \quad -1 \quad -1 \quad 2]$.

9. For the inner product space of continuous functions on $[0, 1]$ defined in Example 4, find (f, g) for the following:
   (a) $f(t) = 1 + t, g(t) = 2 - t$.   (b) $f(t) = 1, g(t) = 3$.
   (c) $f(t) = 1, g(t) = 3 + 2t$.   (d) $f(t) = 3t, g(t) = 2t^2$.
   (e) $f(t) = t, g(t) = e^t$.

10. Let V be the Euclidean space of Example 3. Find the length of the following vectors.
   (a) $\begin{bmatrix} 1 \\ 3 \end{bmatrix}$.   (b) $\begin{bmatrix} 3 \\ -1 \end{bmatrix}$.   (c) $\begin{bmatrix} 1 \\ 0 \end{bmatrix}$.   (d) $\begin{bmatrix} 0 \\ -2 \end{bmatrix}$.   (e) $\begin{bmatrix} 2 \\ -4 \end{bmatrix}$.

11. Let V be the inner product space of Example 6. Find the cosine of the angle between each of the following pairs of vectors in V.
   (a) $p(t) = t, q(t) = t - 1$.   (b) $p(t) = t, q(t) = t$.

(c) $p(t) = 1, q(t) = 2t + 3.$ (d) $p(t) = 1, q(t) = 1.$

(e) $p(t) = t^2, q(t) = 2t^3 - \frac{4}{3}t.$

12. Prove the **parallelogram law** for any two vectors in an inner product space:

$$|\alpha + \beta|^2 + |\alpha - \beta|^2 = 2|\alpha|^2 + 2|\beta|^2.$$

13. Let V be an inner product space. Show that $|a\alpha| = |a| \, |\alpha|$ for any vector α and any scalar a (here $|a|$ is the absolute value of a and $|\alpha|$ is the length of α).

14. State the Cauchy–Schwarz inequality for the inner product spaces defined in Example 3, Example 5, and Exercise 3.

15. Let V be an inner product space. Prove that if α and β are any vectors in V, then $|\alpha + \beta|^2 = |\alpha|^2 + |\beta|^2$ if and only if $(\alpha, \beta) = 0$, that is, if and only if α and β are orthogonal. This result is known as the **Pythagorean theorem**.

16. Let V be the Euclidean space R_4 considered in Exercise 8. Find which of the pairs of vectors listed there are orthogonal.

17. Let V be an inner product space and α a fixed vector in V. Prove that the set of all vectors in V orthogonal to α is a subspace of V.

18. For each of the inner products defined in Examples 1, 3, and 5, choose an ordered basis S for the vector space and find the matrix of the inner product with respect to S.

19. Let $C = \begin{bmatrix} 3 & -2 \\ -2 & 3 \end{bmatrix}$. Define an inner product on R_2 whose matrix with respect to the natural ordered basis is $C = \begin{bmatrix} 3 & -2 \\ -2 & 3 \end{bmatrix}$.

20. If V is an inner product space, prove that the distance function of Definition 3.4 satisfies the following properties for all vectors α, β, and γ in V.

(a) $d(\alpha, \beta) \geq 0.$ (b) $d(\alpha, \beta) = 0$ if and only if $\alpha = \beta.$

(c) $d(\alpha, \beta) = d(\beta, \alpha).$ (d) $d(\alpha, \beta) \leq d(\alpha, \gamma) + d(\gamma, \beta).$

21. If V is the inner product space of Example 4, find the distance between each of the following pairs of vectors α and β in V. (a) $\sin t, \cos t.$ (b) $t, t^2.$ (c) $2t + 3,$ $3t^2 - t.$ (d) $3t + 1, 1.$

22. Which of the following sets of vectors in R^3, with the standard inner product, are orthogonal? Orthonormal?

(a) $\left\{ \begin{bmatrix} 1/\sqrt{2} \\ 0 \\ 1/\sqrt{2} \end{bmatrix}, \begin{bmatrix} -1/\sqrt{2} \\ 0 \\ 1/\sqrt{2} \end{bmatrix}, \begin{bmatrix} 0 \\ 1 \\ 0 \end{bmatrix} \right\}.$ (b) $\left\{ \begin{bmatrix} 1 \\ 1 \\ 0 \end{bmatrix}, \begin{bmatrix} 0 \\ 0 \\ 1 \end{bmatrix}, \begin{bmatrix} 0 \\ 1 \\ 0 \end{bmatrix} \right\}.$

(c) $\left\{ \begin{bmatrix} 0 \\ 1 \\ -1 \end{bmatrix}, \begin{bmatrix} 0 \\ 1 \\ 1 \end{bmatrix}, \begin{bmatrix} 2 \\ 0 \\ 0 \end{bmatrix} \right\}.$

23. Let V be the inner product space of Example 6.

(a) Let $p(t) = 3t + 1$ and $q(t) = at$. For what values of a are $p(t)$ and $q(t)$ orthogonal?

(b) Let $p(t) = 3t + 1$ and $q(t) = at + b$. For what values of a and b are $p(t)$ and $q(t)$ orthogonal?

24. Let V be the Euclidean space R^3 with the standard inner product.

(a) Let $\alpha = \begin{bmatrix} 1 \\ 1 \\ -2 \end{bmatrix}$ and $\beta = \begin{bmatrix} a \\ -1 \\ 2 \end{bmatrix}$. For what values of a are α and β orthogonal?

(b) Let $\alpha = \begin{bmatrix} 1/\sqrt{2} \\ 0 \\ 1/\sqrt{2} \end{bmatrix}$ and $\beta = \begin{bmatrix} a \\ -1 \\ -b \end{bmatrix}$. For what values of a and b is $\{\alpha, \beta\}$ an orthonormal set?

25. Let $C = [c_{ij}]$ be an $n \times n$ positive definite symmetric matrix and let V be an n-dimensional vector space with ordered basis $S = \{\alpha_1, \alpha_2, \ldots, \alpha_n\}$. For $\alpha = a_1\alpha_1 + a_2\alpha_2 + \cdots + a_n\alpha_n$ and $\beta = b_1\alpha_1 + b_2\alpha_2 + \cdots + b_n\alpha_n$ in V define $(\alpha, \beta) = \sum_{j=1}^{n} \sum_{i=1}^{n} b_j c_{ij} a_i$. Prove that this defines an inner product on V.

26. If A and B are $n \times n$ matrices, show that $(A\alpha, B\beta) = (\alpha, A'B\beta)$ for any vectors α and β in Euclidean space R^n with the standard inner product.

27. In the Euclidean space R^n with the standard inner product, prove that $(\alpha, \beta) = \alpha'\beta$.

3.3. The Gram–Schmidt Process

In this section we shall prove that for every Euclidean space V we can obtain a basis S for V such that S is an orthonormal set; such a basis is called an **orthonormal basis**, and the method we use to obtain it is the **Gram–Schmidt process**.

Theorem 3.4 (Gram–Schmidt Process). *Let V be an inner product space and $W \neq \{\theta\}$ a subspace of V with basis $S = \{\alpha_1, \alpha_2, \ldots, \alpha_n\}$. Then there exists an orthonormal basis $T = \{\gamma_1, \gamma_2, \ldots, \gamma_n\}$ for W.*

Proof: The proof is constructive; that is, we exhibit the desired basis T. However, we first find an orthogonal basis $T^* = \{\beta_1, \beta_2, \ldots, \beta_n\}$ for W.

First, we pick any one of the vectors in S, say α_1, and call it β_1. Thus $\beta_1 = \alpha_1$. We now look for a vector β_2 in the subspace W_1 of W spanned by $\{\alpha_1, \alpha_2\}$, which is orthogonal to β_1. Since $\beta_1 = \alpha_1$, W_1 is also the subspace spanned by $\{\beta_1, \alpha_2\}$. Thus $\beta_2 = a_1\beta_1 + a_2\alpha_2$. We determine a_1 and a_2, so that $(\beta_1, \beta_2) = 0$. Now $0 = (\beta_2, \beta_1) = (a_1\beta_1 + a_2\alpha_2, \beta_1) = a_1(\beta_1, \beta_1) + a_2(\alpha_2, \beta_1)$. Note that $\beta_1 \neq \theta$ (why?), so $(\beta_1, \beta_1) \neq 0$. Thus

$$a_1 = -a_2 \frac{(\alpha_2, \beta_1)}{(\beta_1, \beta_1)}.$$

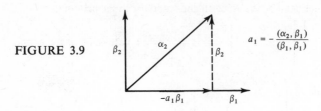

FIGURE 3.9

We may assign an arbitrary nonzero value to a_2. Thus, letting $a_2 = 1$, we obtain

$$a_1 = -\frac{(\alpha_2, \beta_1)}{(\beta_1, \beta_1)}.$$

Hence

$$\beta_2 = a_1\beta_1 + \alpha_2 = \alpha_2 - \frac{(\alpha_2, \beta_1)}{(\beta_1, \beta_1)} \beta_1.$$

At this point we have an orthogonal subset $\{\beta_1, \beta_2\}$ of W (Figure 3.9).

Next, we look for a vector β_3 in the subspace W_2 of W spanned by $\{\alpha_1, \alpha_2, \alpha_3\}$ which is orthogonal to both β_1 and β_2. Of course, W_2 is also the subspace spanned by $\{\beta_1, \beta_2, \alpha_3\}$ (why?). Thus $\beta_3 = b_1\beta_1 + b_2\beta_2 + b_3\alpha_3$. We let $b_3 = 1$ and try to find b_1 and b_2 so that $(\beta_3, \beta_1) = 0$ and $(\beta_3, \beta_2) = 0$. Now

$$0 = (\beta_3, \beta_1) = (b_1\beta_1 + b_2\beta_2 + \alpha_3, \beta_1) = b_1(\beta_1, \beta_1) + (\alpha_3, \beta_1)$$
$$0 = (\beta_3, \beta_2) = (b_1\beta_1 + b_2\beta_2 + \alpha_3, \beta_2) = b_2(\beta_2, \beta_2) + (\alpha_3, \beta_2).$$

Observe that $\beta_2 \neq \theta$ (why?). Solving for b_1 and b_2, we have

$$b_1 = -\frac{(\alpha_3, \beta_1)}{(\beta_1, \beta_1)} \quad \text{and} \quad b_2 = -\frac{(\alpha_3, \beta_2)}{(\beta_2, \beta_2)}.$$

Thus

$$\beta_3 = \alpha_3 - \frac{(\alpha_3, \beta_1)}{(\beta_1, \beta_1)} \beta_1 - \frac{(\alpha_3, \beta_2)}{(\beta_2, \beta_2)} \beta_2.$$

At this point we have an orthogonal subset $\{\beta_1, \beta_2, \beta_3\}$ of W (Figure 3.10).

We next seek a vector β_4 in the subspace W_3 spanned by $\{\alpha_1, \alpha_2, \alpha_3, \alpha_4\}$, and also by $\{\beta_1, \beta_2, \beta_3, \alpha_4\}$, which is orthogonal to $\beta_1, \beta_2, \beta_3$. We can then write

$$\beta_4 = \alpha_4 - \frac{(\alpha_4, \beta_1)}{(\beta_1, \beta_1)} \beta_1 - \frac{(\alpha_4, \beta_2)}{(\beta_2, \beta_2)} \beta_2 - \frac{(\alpha_4, \beta_3)}{(\beta_3, \beta_3)} \beta_3.$$

Continue in this manner until we have an orthogonal set

$$T^* = \{\beta_1, \beta_2, \ldots, \beta_n\}$$

of n vectors. By Theorem 3.3 we conclude that T^* is a basis for W. If we now let $\gamma_i = \beta_i/|\beta_i|$ for $i = 1, 2, \ldots, n$, then $T = \{\gamma_1, \gamma_2, \ldots, \gamma_n\}$ is an orthonormal basis for W.

EXAMPLE 1. Consider the basis $S = \{\alpha_1, \alpha_2, \alpha_3\}$ for the Euclidean space R^3 with the standard inner product where

$$\alpha_1 = \begin{bmatrix} 1 \\ 1 \\ 1 \end{bmatrix}, \; \alpha_2 = \begin{bmatrix} -1 \\ 0 \\ -1 \end{bmatrix}, \; \text{and } \alpha_3 = \begin{bmatrix} -1 \\ 2 \\ 3 \end{bmatrix}.$$

We transform S to an orthonormal basis $T = \{\gamma_1, \gamma_2, \gamma_3\}$ as follows.
First, let $\beta_1 = \alpha_1$. Then we find that

$$\beta_2 = \alpha_2 - \frac{(\alpha_2, \beta_1)}{(\beta_1, \beta_1)} \beta_1 = \begin{bmatrix} -1 \\ 0 \\ -1 \end{bmatrix} + \frac{2}{3} \begin{bmatrix} 1 \\ 1 \\ 1 \end{bmatrix} = \begin{bmatrix} -\frac{1}{3} \\ \frac{2}{3} \\ -\frac{1}{3} \end{bmatrix},$$

and

$$\beta_3 = \alpha_3 - \frac{(\alpha_3, \beta_1)}{(\beta_1, \beta_1)} \beta_1 - \frac{(\alpha_3, \beta_2)}{(\beta_2, \beta_2)} \beta_2 = \begin{bmatrix} -1 \\ 2 \\ 3 \end{bmatrix} - \left(\frac{4}{3}\right) \begin{bmatrix} 1 \\ 1 \\ 1 \end{bmatrix} - \left(\frac{\frac{2}{3}}{\frac{6}{9}}\right) \begin{bmatrix} -\frac{1}{3} \\ \frac{2}{3} \\ -\frac{1}{3} \end{bmatrix}$$

$$= \begin{bmatrix} -2 \\ 0 \\ 2 \end{bmatrix}.$$

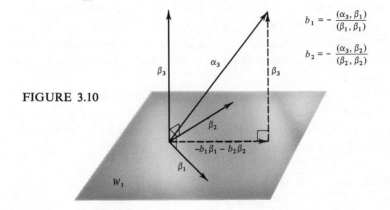

$$b_1 = -\frac{(\alpha_3, \beta_1)}{(\beta_1, \beta_1)}$$

$$b_2 = -\frac{(\alpha_3, \beta_2)}{(\beta_2, \beta_2)}$$

FIGURE 3.10

Then $\{\beta_1, \beta_2, \beta_3\}$ is an orthogonal basis. If we clear the fractions in each β_i by multiplying by a scalar, the resulting set is also an orthogonal basis. Hence

$$\left\{ \begin{bmatrix} 1 \\ 1 \\ 1 \end{bmatrix}, \begin{bmatrix} -1 \\ 2 \\ -1 \end{bmatrix}, \begin{bmatrix} -2 \\ 0 \\ 2 \end{bmatrix} \right\} \text{ is an orthogonal basis for } R^3, \text{ and an orthonormal}$$

basis is

$$\left\{ \begin{bmatrix} 1/\sqrt{3} \\ 1/\sqrt{3} \\ 1/\sqrt{3} \end{bmatrix}, \begin{bmatrix} -1/\sqrt{6} \\ 2/\sqrt{6} \\ -1/\sqrt{6} \end{bmatrix}, \begin{bmatrix} -1/\sqrt{2} \\ 0 \\ 1/\sqrt{2} \end{bmatrix} \right\}.$$

EXAMPLE 2. Let V be the Euclidean space P_3 with the inner product defined in Example 6 of Section 3.2. Let W be the subspace of P_3 having $S = \{t^2, t\}$ as a basis. We now find an orthonormal basis for W.

First, let $\alpha_1 = t^2$ and $\alpha_2 = t$. Now let $\beta_1 = \alpha_1 = t^2$. Then

$$\beta_2 = \alpha_2 - \frac{(\alpha_2, \beta_1)}{(\beta_1, \beta_1)} \beta_1 = t - \frac{\frac{1}{4}}{\frac{1}{5}} t^2 = t - \frac{5}{4} t^2,$$

where $(\beta_1, \beta_1) = \int_0^1 t^2 t^2 \, dt = \int_0^1 t^4 \, dt = \frac{1}{5}$, and $(\alpha_2, \beta_1) = \int_0^1 t t^2 \, dt = \int_0^1 t^3 \, dt = \frac{1}{4}$. Since $(\beta_2, \beta_2) = \int_0^1 (t - \frac{5}{4}t^2)^2 \, dt = 2$, $\{\sqrt{5}t^2, \sqrt{48}(t - \frac{5}{4}t^2)\}$ is an orthonormal basis for W.

In the proof of Theorem 3.4 we have also established the following result. At each stage of the Gram–Schmidt process, the ordered set $\{\gamma_1, \gamma_2, \ldots, \gamma_k\}$ is an orthonormal basis for the subspace spanned by $\{\alpha_1, \alpha_2, \ldots, \alpha_k\}$, $1 \le k \le m$. Also, the final orthonormal basis T depends upon the order of the vectors in the given basis S. Thus, if we change the order of the vectors in S, we might obtain a different orthonormal basis T for W.

We make one final observation with regard to the Gram–Schmidt process. In our proof of Theorem 3.4, we first obtained an orthogonal basis T^* and then normalized all the vectors in T^* to obtain the orthonormal basis T. Of course, an alternative course of action is to normalize each vector as soon as we produce it.

One of the useful consequences of having an orthonormal basis in a Euclidean space V is that an arbitrary inner product on V, when it is expressed in terms of coordinates with respect to the orthonormal basis, behaves like the standard inner product on R^n.

Theorem 3.5. *Let V be an n-dimensional Euclidean space, and let $S =$*

$\{\alpha_1, \alpha_2, \ldots, \alpha_n\}$ *be an orthonormal basis for V. If $\alpha = a_1\alpha_1 + a_2\alpha_2 + \cdots$*
$+ a_n\alpha_n$ and $\beta = b_1\alpha_1 + b_2\alpha_2 + \cdots + b_n\alpha_n$, then

$$(\alpha, \beta) = a_1 b_1 + a_2 b_2 + \cdots + a_n b_n.$$

Proof: We first compute the matrix $C = [c_{ij}]$ of the given inner product with respect to the ordered basis S. We have

$$\begin{aligned} c_{ij} = (\alpha_i, \alpha_j) &= 1 \quad \text{if } i = j \\ &= 0 \quad \text{if } i \neq j. \end{aligned}$$

Hence $C = I_n$, the identity matrix. Now we also know from Section 3.2 that

$$(\alpha, \beta) = [\alpha]'_S C [\beta]_S = [\alpha_S]' I_n [\beta]_S = [\alpha]'_S [\beta]_S$$

$$= [a_1 \quad a_2 \quad \cdots \quad a_n] \begin{bmatrix} b_1 \\ b_2 \\ \vdots \\ b_n \end{bmatrix} = a_1 b_1 + a_2 b_2 + \cdots + a_n b_n,$$

which establishes the result.

The theorem that we just proved has some additional implications. Consider the Euclidean space R_3 with the standard inner product and let W be the subspace with ordered basis $S = \{[2 \quad 1 \quad 1], [1 \quad 1 \quad 2]\}$. Let $\alpha = [5 \quad 3 \quad 4]$ be a vector in W. Then $[5 \quad 3 \quad 4] = 2[2 \quad 1 \quad 1] + 1[1 \quad 1 \quad 2]$, so $[5 \quad 3 \quad 4]_S = \begin{bmatrix} 2 \\ 1 \end{bmatrix}$. Now the length of α is $|\alpha| = \sqrt{5^2 + 3^2 + 4^2} = \sqrt{25 + 9 + 16} = \sqrt{50}$. We might expect to compute the length of α by using the coordinate vector with respect to S, that is, $|\alpha| = \sqrt{2^2 + 1^2} = \sqrt{5}$. Obviously, we have the wrong answer. However, let us transform the given basis S for W into an orthonormal basis T for W. Using the Gram–Schmidt process, we find that

$$\{[2 \quad 1 \quad 1], [-\tfrac{4}{6} \quad \tfrac{1}{6} \quad \tfrac{7}{6}]\}$$

is an orthogonal basis. Hence $\{[2 \quad 1 \quad 1], [-4 \quad 1 \quad 7]\}$ is also an orthogonal basis, and

$$T = \left\{ \left[\frac{2}{\sqrt{6}} \quad \frac{1}{\sqrt{6}} \quad \frac{1}{\sqrt{6}} \right], \left[-\frac{4}{\sqrt{66}} \quad \frac{1}{\sqrt{66}} \quad \frac{7}{\sqrt{66}} \right] \right\}$$

is an orthonormal basis. Then the coordinate vector of α with respect to T is

$$[\alpha]_T = \begin{bmatrix} \dfrac{17}{6}\sqrt{6} \\[2mm] \dfrac{\sqrt{66}}{6} \end{bmatrix}.$$

Computing the length of α using these coordinates, we find that

$$|\alpha|_T = \sqrt{\left(\frac{17}{6}\sqrt{6}\right)^2 + \left(\frac{\sqrt{66}}{6}\right)^2} = \sqrt{\frac{1800}{36}} = \sqrt{50}.$$

It is not difficult to show (Exercise 16) that if T is an orthonormal basis for an inner product space and $[\alpha]_T = \begin{bmatrix} a_1 \\ a_2 \\ \vdots \\ a_n \end{bmatrix}$, then $|\alpha| = \sqrt{a_1^2 + a_2^2 + \cdots + a_n^2}$.

3.3. Exercises

In this set of exercises, the Euclidean spaces R_n and R^n have the standard inner products on them. Euclidean space P_n has on it the inner product defined in Example 6 of Section 3.2.

1. Consider the Euclidean space R_4 and let W be the subspace that has

$$S = \{[1 \quad 1 \quad -1 \quad 0], [0 \quad 2 \quad 0 \quad 1]\}$$

   as a basis. Use the Gram–Schmidt process to obtain an orthonormal basis for W.

2. Let $S = \{t, 1\}$ be a basis for a subspace W of the Euclidean space P_2. Find an orthonormal basis for W.

3. Let $S = \{\alpha_1, \alpha_2, \ldots, \alpha_n\}$ be an orthonormal basis for a Euclidean space V. If α is any vector in V, then $\alpha = a_1\alpha_1 + a_2\alpha_2 + \cdots + a_n\alpha_n$, where the a_i are uniquely determined scalars. Prove that $a_i = (\alpha, \alpha_i)$ for $i = 1, 2, \ldots, n$.

4. Find an orthonormal basis for the Euclidean space R^3 containing the vectors

$$\begin{bmatrix} \frac{2}{3} \\ -\frac{2}{3} \\ \frac{1}{3} \end{bmatrix} \text{ and } \begin{bmatrix} \frac{2}{3} \\ \frac{1}{3} \\ -\frac{2}{3} \end{bmatrix}.$$

5. Repeat Exercise 2 with $S = \{t + 1, t - 1\}$.

6. Use the Gram–Schmidt process to transform the basis

$$\left\{ \begin{bmatrix} 1 \\ 1 \\ 1 \end{bmatrix}, \begin{bmatrix} 0 \\ 1 \\ 1 \end{bmatrix}, \begin{bmatrix} 1 \\ 2 \\ 3 \end{bmatrix} \right\}$$

for the Euclidean space R^3 into an orthonormal basis for R^3.

7. Construct an orthonormal basis for the subspace W of the Euclidean space R^3 *spanned* by $\left\{ \begin{bmatrix} 1 \\ 1 \\ 1 \end{bmatrix}, \begin{bmatrix} 2 \\ 2 \\ 2 \end{bmatrix}, \begin{bmatrix} 0 \\ 0 \\ 1 \end{bmatrix}, \begin{bmatrix} 1 \\ 2 \\ 3 \end{bmatrix} \right\}.$

8. Let $S = \{[1 \quad -1 \quad 0], [1 \quad 0 \quad -1]\}$ be a basis for a subspace W of the Euclidean space R_3.
   (a) Use the Gram–Schmidt process to obtain an orthonormal basis for W.
   (b) Using Exercise 3, write $\alpha = [5 \quad -2 \quad -3]$ as a linear combination of the vectors obtained in (a).

9. Construct an orthonormal basis for the subspace W of the Euclidean space R_3 *spanned* by
   $\{[1 \quad -1 \quad 1], [-2 \quad 2 \quad -2], [2 \quad -1 \quad 2], [0 \quad 0 \quad 0]\}.$

10. Use the Gram–Schmidt process to transform the basis $\left\{ \begin{bmatrix} 1 \\ 2 \end{bmatrix}, \begin{bmatrix} -3 \\ 4 \end{bmatrix} \right\}$ for the Euclidean space R^2 into:
    (a) An orthogonal basis.
    (b) An orthonormal basis.

11. Use the Gram–Schmidt process to transform the basis $\left\{ \begin{bmatrix} 1 \\ 0 \\ 1 \end{bmatrix}, \begin{bmatrix} -2 \\ 1 \\ 3 \end{bmatrix} \right\}$ for the subspace W of the Euclidean space R^3 into:
    (a) An orthogonal basis.
    (b) An orthonormal basis.

12. (a) Verify that $S = \left\{ \begin{bmatrix} \frac{1}{3} \\ \frac{2}{3} \\ \frac{2}{3} \end{bmatrix}, \begin{bmatrix} \frac{2}{3} \\ \frac{1}{3} \\ -\frac{2}{3} \end{bmatrix}, \begin{bmatrix} \frac{2}{3} \\ -\frac{2}{3} \\ \frac{1}{3} \end{bmatrix} \right\}$ is an orthonormal basis for the Euclidean space R^3.

    (b) Use Exercise 3 to find the coordinate vector of $\alpha = \begin{bmatrix} 15 \\ 3 \\ 3 \end{bmatrix}$ with respect to S.

    (c) Find the length of α directly and also using the coordinate vector found in (c).

13. Let V be the euclidean space of all 2×2 matrices with inner product defined by $(A, B) = \text{Tr}(B'A)$. Prove that
    $S = \left\{ \begin{bmatrix} 1 & 0 \\ 0 & 0 \end{bmatrix}, \begin{bmatrix} 0 & 1 \\ 0 & 0 \end{bmatrix}, \begin{bmatrix} 0 & 0 \\ 1 & 0 \end{bmatrix}, \begin{bmatrix} 0 & 0 \\ 0 & 1 \end{bmatrix} \right\}$ is an orthonormal basis for V.

14. Apply the Gram–Schmidt process to the basis $\{1, t, t^2\}$ for the Euclidean space P_2 and obtain an orthonormal basis for P_2.

15. Let W be the subspace of the Euclidean space R^3 with basis

$$S = \left\{ \begin{bmatrix} 1 \\ 0 \\ -2 \end{bmatrix}, \begin{bmatrix} -3 \\ 2 \\ 1 \end{bmatrix} \right\}. \text{ Let } \alpha = \begin{bmatrix} -1 \\ 2 \\ -3 \end{bmatrix} \text{ be in } W.$$

(a) Find the length of α directly.

(b) Using the Gram–Schmidt process, transform S into an orthonormal basis T for W.

(c) Find the length of α using the coordinate vector of α with respect to T.

16. Prove that if T is an orthonormal basis for a Euclidean space and

$$[\alpha]_T = \begin{bmatrix} a_1 \\ a_2 \\ \vdots \\ a_n \end{bmatrix}, \text{ then } |\alpha| = \sqrt{a_1^2 + a_2^2 + \cdots + a_n^2}.$$

CHAPTER 4

Linear Transformations and Matrices

4.1. Definition and Examples

As we have noted earlier, much of the calculus deals with the study of properties of functions. Indeed, properties of functions are of great importance in every branch of mathematics, and linear algebra is no exception. We have already encountered functions mapping one vector space into another vector space; these were the isomorphisms. If we drop some of the conditions that need to be satisfied by a function on a vector space to be an isomorphism, we get another very useful type of function, called a linear transformation. Linear transformations play an important role in many areas of mathematics, the physical and social sciences, and economics. All vector spaces considered henceforth are finite dimensional.

Definition 4.1. Let V and W be vector spaces. A function $L: V \to W$ is called
a **linear transformation** of V into W if:
   (a) $L(\alpha + \beta) = L(\alpha) + L(\beta)$ for α and β in V.
   (b) $L(c\alpha) = cL(\alpha)$ for α in V, and c a real number.
If $V = W$, the linear transformation $L: V \to W$ is also called a **linear operator** on V.

We now see that an isomorphism is a linear transformation that is one-one and onto. Linear transformations occur very frequently, and we now look at some examples. (At this point it might be profitable to review the material of Section 0.2.) It can be shown that $L: V \to W$ is a linear transformation if and only if $L(a\alpha + b\beta) = aL(\alpha) + bL(\beta)$ for any real numbers a, b and any vectors α, β in V (see Exercise 1).

EXAMPLE 1. Let $L: R^3 \to R^2$ be defined by

$$L\left(\begin{bmatrix} a_1 \\ a_2 \\ a_3 \end{bmatrix}\right) = \begin{bmatrix} a_1 \\ a_2 \end{bmatrix}.$$

To show that L is a linear transformation, we let

$$\alpha = \begin{bmatrix} a_1 \\ a_2 \\ a_3 \end{bmatrix} \quad \text{and} \quad \beta = \begin{bmatrix} b_1 \\ b_2 \\ b_3 \end{bmatrix}.$$

Then

$$L(\alpha + \beta) = L\left(\begin{bmatrix} a_1 + b_1 \\ a_2 + b_2 \\ a_3 + b_3 \end{bmatrix}\right) = \begin{bmatrix} a_1 + b_1 \\ a_2 + b_2 \end{bmatrix} = \begin{bmatrix} a_1 \\ a_2 \end{bmatrix} + \begin{bmatrix} b_1 \\ b_2 \end{bmatrix} = L(\alpha) + L(\beta).$$

Also, if c is a real number, then

$$L(c\alpha) = L\left(\begin{bmatrix} ca_1 \\ ca_2 \\ ca_3 \end{bmatrix}\right) = \begin{bmatrix} ca_1 \\ ca_2 \end{bmatrix} = c\begin{bmatrix} a_1 \\ a_2 \end{bmatrix} = cL(\alpha).$$

Hence L is a linear transformation, which is called **projection**. It is simple and helpful to describe geometrically the effect of L. The image under L of a vector in R^3 with end point $P(a_1, a_2, a_3)$ is found by drawing a line through P perpendicular to R^2, the (x, y)-plane. We obtain the point $Q(a_1, a_2)$ of intersection of this line with the (x, y)-plane. The vector in R^2 with end point Q is the image of α under L (Figure 4.1).

EXAMPLE 2. We consider the mapping L from the vector space P_2 into the vector space P_1 defined by $L(at^2 + bt + c) = 2at + b$. Then L is ordinary

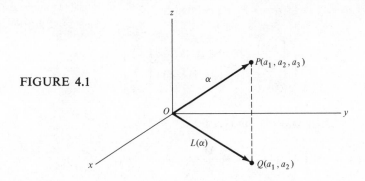

FIGURE 4.1

differentiation and it is not difficult to show that L is a linear transformation (verify).

EXAMPLE 3. Let $L: R^3 \to R^3$ be defined by

$$L\left(\begin{bmatrix} a_1 \\ a_2 \\ a_3 \end{bmatrix}\right) = r \begin{bmatrix} a_1 \\ a_2 \\ a_3 \end{bmatrix},$$

where r is a real number. Then L is a linear operator on R^2 (verify). If $r > 1$, L is called **dilation**; if $0 < r < 1$, L is called **contraction**. Thus dilation stretches a vector, while contraction shrinks it (Figure 4.2).

EXAMPLE 4. Let $L: P_2 \to R$ (R = the vector space of real numbers under the usual operations of addition and scalar multiplication) be defined by $L(at^2 + bt + c) = \int_0^1 (at^2 + bt + c)\, dt$. Then L is a linear transformation (verify).

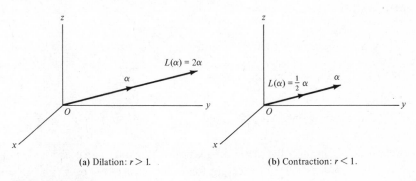

(a) Dilation: $r > 1$.   (b) Contraction: $r < 1$.

FIGURE 4.2

EXAMPLE 5. Let $L: R^3 \to R^2$ be defined by

$$L\left(\begin{bmatrix} a_1 \\ a_2 \\ a_3 \end{bmatrix}\right) = \begin{bmatrix} 1 & 0 & 1 \\ 0 & 1 & -1 \end{bmatrix} \begin{bmatrix} a_1 \\ a_2 \\ a_3 \end{bmatrix}.$$

Then L is a linear transformation (verify).

EXAMPLE 6. Let $L: R^2 \to R^2$ be defined by

$$L\left(\begin{bmatrix} a_1 \\ a_2 \end{bmatrix}\right) = \begin{bmatrix} a_1 \\ -a_2 \end{bmatrix}.$$

Then L is a linear operator on R^2 (verify). Geometrically, L does the following (Figure 4.3). Thus L is **reflection** with respect to the x-axis. The student may also consider reflection with respect to the y-axis.

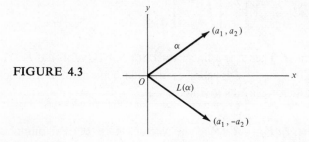

FIGURE 4.3

EXAMPLE 7. In the calculus we rotated every point in the (x, y)-plane counterclockwise through an angle ϕ about the origin of a Cartesian coordinate system. Thus, if P has coordinates (x, y), then after rotating, we get P' with coordinates (x', y'), where

$$x' = x \cos \phi - y \sin \phi \quad \text{and} \quad y' = x \sin \phi + y \cos \phi,$$

or

$$x = x' \cos \phi + y' \sin \phi, \quad \text{and} \quad y = -x' \sin \phi + y' \cos \phi.$$

We did this to simplify the general equation of second degree $ax^2 + bxy + cy^2 + dx + ey + f = 0$. Substituting for x and y in terms of x' and y', we obtain $a'x'^2 + b'x'y' + c'y'^2 + d'x' + e'y' + f' = 0$. The key point is to choose ϕ so that $b' = 0$. Once this is done (we might now have to perform a translation of coordinates), we identify the general equation of second degree as a circle, ellipse, hyperbola, parabola, or a degenerate form of these. Note that we may perform this change of coordinates by considering the map $L: R^2 \to R^2$ defined by

$$L\left(\begin{bmatrix} x \\ y \end{bmatrix}\right) = \begin{bmatrix} \cos \phi & -\sin \phi \\ \sin \phi & \cos \phi \end{bmatrix}\begin{bmatrix} x \\ y \end{bmatrix}.$$

If $\alpha = \begin{bmatrix} x \\ y \end{bmatrix}$ is the vector from the origin to $P(x, y)$, then $L(\alpha)$, the vector obtained by rotating α counterclockwise through the angle ϕ (Figure 4.4), is the vector from the origin to $P'(x', y')$. Then L is a linear transformation (verify) called **rotation**.

FIGURE 4.4

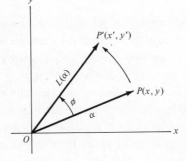

EXAMPLE 8. Let $L: R^3 \to R^3$ be defined by

$$L\left(\begin{bmatrix} a_1 \\ a_2 \\ a_3 \end{bmatrix}\right) = \begin{bmatrix} a_1 + 1 \\ 2a_2 \\ a_3 \end{bmatrix}.$$

To determine whether L is a linear transformation, let

$$\alpha = \begin{bmatrix} a_1 \\ a_2 \\ a_3 \end{bmatrix} \quad \text{and} \quad \beta = \begin{bmatrix} b_1 \\ b_2 \\ b_3 \end{bmatrix}.$$

Then

$$L(\alpha + \beta) = L\left(\begin{bmatrix} a_1 \\ a_2 \\ a_3 \end{bmatrix} + \begin{bmatrix} b_1 \\ b_2 \\ b_3 \end{bmatrix}\right) = L\left(\begin{bmatrix} a_1 + b_1 \\ a_2 + b_2 \\ a_3 + b_3 \end{bmatrix}\right)$$

$$= \begin{bmatrix} (a_1 + b_1) + 1 \\ 2(a_2 + b_2) \\ a_3 + b_3 \end{bmatrix}.$$

On the other hand,

$$L(\alpha) + L(\beta) = \begin{bmatrix} a_1 + 1 \\ 2a_2 \\ a_3 \end{bmatrix} + \begin{bmatrix} b_1 + 1 \\ 2b_2 \\ b_3 \end{bmatrix} = \begin{bmatrix} (a_1 + b_1) + 2 \\ 2(a_2 + b_2) \\ a_3 + b_3 \end{bmatrix}.$$

Since $L(\alpha + \beta) \neq L(\alpha) + L(\beta)$, we conclude that the function L is not a linear transformation.

We know, from the calculus, that a function can be specified by a formula that assigns to every member of the domain a unique element of the range. On the other hand, we can also specify a function by listing next to each member of the domain its assigned element of the range. An example of this would be provided by listing the names of all charge account customers of a department store along with their charge account number. At first glance it appears impossible to describe a linear transformation $L: V \to W$ of a vector space $V \neq \{\theta\}$ into a vector space W in this latter manner, for V has infinitely many members in it. However, the following very useful theorem tells us that once we say what a linear transformation L does to a basis of V, then we have completely specified L. Thus, since in this book we only deal with finite-dimensional vector spaces, it is possible to describe L by giving only the images of a finite number of vectors in the domain V.

Theorem 4.1. *Let* $L: V \to W$ *be a linear transformation of an n-dimensional vector space* V *into a vector space* W. *Let* $S = \{\alpha_1, \alpha_2, \ldots, \alpha_n\}$ *be a basis for* V. *If* α *is any vector in* V, *then* $L(\alpha)$ *is completely determined by* $\{L(\alpha_1), L(\alpha_2), \ldots, L(\alpha_n)\}$.

Proof: Since α is in V, we can write $\alpha = a_1\alpha_1 + a_2\alpha_2 + \cdots + a_n\alpha_n$, where $a_1, a_2, \ldots, a_n$ are uniquely determined real numbers. Then

$$L(\alpha) = L(a_1\alpha_1 + a_2\alpha_2 + \cdots + a_n\alpha_n) = L(a_1\alpha_1) + L(a_2\alpha_2) + \cdots + L(a_n\alpha_n)$$
$$= a_1 L(\alpha_1) + a_2 L(\alpha_2) + \cdots + a_n L(\alpha_n).$$

Thus $L(\alpha)$ has been completely determined by the elements $L(\alpha_1), L(\alpha_2), \ldots,$ $L(\alpha_n)$.

Theorem 4.2. *Let $L: V \to W$ be a linear transformation. Then*
(a) $L(\theta_V) = \theta_W$.
(b) $L(\alpha - \beta) = L(\alpha) - L(\beta)$, *for α, β in V.*

Proof:
(a) We have $\theta_V + \theta_V = \theta_V$, so $L(\theta_V + \theta_V) = L(\theta_V) + L(\theta_V) = L(\theta_V)$. If we subtract $L(\theta_V)$ from both sides, we obtain $L(\theta_V) = \theta_W$.
(b) $L(\alpha - \beta) = L(\alpha + (-1)\beta) = L(\alpha) + L((-1)\beta) = L(\alpha) + (-1)L(\beta) = L(\alpha) - L(\beta)$.

4.1. Exercises

1. Let $L: V \to W$ be a mapping of a vector space V into a vector space W. Prove that L is a linear transformation if and only if $L(a\alpha + b\beta) = aL(\alpha) + bL(\beta)$ for any real numbers a, b and any vectors α, β in V.
2. Prove that the functions in Examples 2, 3, 4, 5, 6, and 7 are linear transformations.
3. Which of the following functions are linear transformations?
   (a) $L: R_2 \to R_3$ defined by $L([a_1 \quad a_2]) = [a_1 + 1 \quad a_2 \quad a_1 + a_2]$.
   (b) $L: R_2 \to R_3$ defined by $L([a_1 \quad a_2]) = [a_1 + a_2 \quad a_2 \quad a_1 - a_2]$.
4. Which of the following functions are linear transformations? [Here $p'(t)$ denotes the derivative of $p(t)$ with respect to t.]
   (a) $L: P_2 \to P_3$ defined by $L(p(t)) = t^3 p'(0) + t^2 p(0)$.
   (b) $L: P_1 \to P_2$ defined by $L(p(t)) = tp(t) + p(0)$.
   (c) $L: P_1 \to P_2$ defined by $L(p(t)) = tp(t) + 1$.
5. Let $L: R^3 \to R^2$ be a linear transformation for which we know that

$$L\left(\begin{bmatrix} 1 \\ 0 \\ 0 \end{bmatrix}\right) = \begin{bmatrix} 2 \\ -4 \end{bmatrix}, \quad L\left(\begin{bmatrix} 0 \\ 1 \\ 0 \end{bmatrix}\right) = \begin{bmatrix} 3 \\ -5 \end{bmatrix}, \quad L\left(\begin{bmatrix} 0 \\ 0 \\ 1 \end{bmatrix}\right) = \begin{bmatrix} 2 \\ 3 \end{bmatrix}.$$

   (a) What is $L\left(\begin{bmatrix} 1 \\ -2 \\ 3 \end{bmatrix}\right)$?   (b) What is $L\left(\begin{bmatrix} a_1 \\ a_2 \\ a_3 \end{bmatrix}\right)$?
6. Which of the following functions are linear transformations?
   (a) $L: R_3 \to R_3$ defined by $L([a_1 \quad a_2 \quad a_3]) = [a_1 \quad a_2^2 + a_3^2 \quad a_3^2]$.
   (b) $L: R_3 \to R_3$ defined by $L([a_1 \quad a_2 \quad a_3]) = [1 \quad a_3 \quad a_2]$.
   (c) $L: R_3 \to R_3$ defined by $L([a_1 \quad a_2 \quad a_3]) = [0 \quad a_3 \quad a_2]$.

7. Consider the function $L: {_3}R_4 \rightarrow {_2}R_4$ defined by $L(A) = \begin{bmatrix} 2 & 3 & 1 \\ 1 & 2 & -3 \end{bmatrix} A$ for A in ${_3}R_4$.

   (a) Find $L\left(\begin{bmatrix} 1 & 2 & 0 & -1 \\ 3 & 0 & 2 & 3 \\ 4 & 1 & -2 & 1 \end{bmatrix}\right)$.

   (b) Show that L is a linear transformation.

8. Let $L: P_2 \rightarrow P_3$ be a linear transformation for which we know that $L(1) = 1$, $L(t) = t^2$, $L(t^2) = t^3 + t$.

   (a) Find $L(2t^2 - 5t + 3)$.

   (b) Find $L(at^2 + bt + c)$.

9. Let A be a fixed 3×3 matrix, and let $L: {_3}R_3 \rightarrow {_3}R_3$ be defined by $L(X) = AX - XA$, for X in ${_3}R_3$. Show that L is a linear transformation.

10. Let $L: R \rightarrow R$ be defined by $L(\alpha) = a\alpha + b$, where a and b are real numbers (of course, α is a vector in R, which in this case means that α is also a real number). Find all values of a and b such that L is a linear transformation.

11. Let V be an inner product space, and let β be a fixed vector in V. Let $L: V \rightarrow R$ be defined by $L(\alpha) = (\alpha, \beta)$ for α in V. Show that L is a linear transformation.

12. Describe the following linear transformations geometrically.

   (a) $L\left(\begin{bmatrix} a_1 \\ a_2 \end{bmatrix}\right) = \begin{bmatrix} -a_1 \\ a_2 \end{bmatrix}$.   (b) $L\left(\begin{bmatrix} a_1 \\ a_2 \end{bmatrix}\right) = \begin{bmatrix} -a_1 \\ -a_2 \end{bmatrix}$.

   (c) $L\left(\begin{bmatrix} a_1 \\ a_2 \end{bmatrix}\right) = \begin{bmatrix} -a_1 \\ a_2 \end{bmatrix}$.

13. Let V be a vector space and r a fixed scalar. Prove that the function $L: V \rightarrow V$ defined by $L(\alpha) = r\alpha$ is a linear operator on V.

14. Let V and W be vector spaces. Prove that the function $O: V \rightarrow W$ defined by $O(\alpha) = \theta_W$ is a linear transformation, which is called the **zero linear transformation**.

15. Let $I: V \rightarrow V$ be defined by $I(\alpha) = \alpha$, for α in V. Show that I is a linear transformation, which is called the **identity operator** on V.

16. Let A be an $m \times n$ matrix, and suppose that $L: R^n \rightarrow R^m$ is defined by $L(\alpha) = A\alpha$, for α in R^n. Show that L is a linear transformation.

4.2. The Kernel and Range of a Linear Transformation

In this section we study special types of linear transformations; we formulate the notions of one-one linear transformations and onto linear transformations. We also develop methods for determining when a linear transformation is one-one or onto, and examine some applications of these notions.

Definition 4.2. A linear transformation $L: V \rightarrow W$ is called **one-one** if it is a one-one function, that is, if $\alpha_1 \neq \alpha_2$ implies that $L(\alpha_1) \neq L(\alpha_2)$. An equivalent statement is that L is one-one if $L(\alpha_1) = L(\alpha_2)$ implies that $\alpha_1 = \alpha_2$ (see Figure 0.2).

EXAMPLE 1. Let $L: R^2 \to R^2$ be defined by

$$L\left(\begin{bmatrix} a_1 \\ a_2 \end{bmatrix}\right) = \begin{bmatrix} a_1 + a_2 \\ a_1 - a_2 \end{bmatrix}.$$

To determine whether L is one-one, we let

$$\alpha_1 = \begin{bmatrix} a_1 \\ a_2 \end{bmatrix} \quad \text{and} \quad \alpha_2 = \begin{bmatrix} b_1 \\ b_2 \end{bmatrix}.$$

Then if $L(\alpha_1) = L(\alpha_2)$, we have

$$a_1 + a_2 = b_1 + b_2$$
$$a_1 - a_2 = b_1 - b_2.$$

Adding these equations, we obtain $2a_1 = 2b_1$, or $a_1 = b_1$, which implies that $a_2 = b_2$. Hence $\alpha_1 = \alpha_2$ and L is one-one.

EXAMPLE 2. Let $L: R^3 \to R^2$ be the linear transformation defined in Example 1 of Section 4.1 (the projection map) by

$$L\left(\begin{bmatrix} a_1 \\ a_2 \\ a_3 \end{bmatrix}\right) = \begin{bmatrix} a_1 \\ a_2 \end{bmatrix}.$$

Since $L\left(\begin{bmatrix} 1 \\ 3 \\ 3 \end{bmatrix}\right) = L\left(\begin{bmatrix} 1 \\ 3 \\ -2 \end{bmatrix}\right)$, we conclude that L is not one-one.

We shall now develop some more efficient ways of determining whether or not a linear transformation is one-one.

Definition 4.3. Let $L: V \to W$ be a linear transformation of a vector space V into a vector space W. The **kernel of** L, ker L, is the subset of V consisting of all elements α of V such that $L(\alpha) = \theta_W$.

We observe that Theorem 4.2 assures us that ker L is never an empty set, because if $L: V \to W$ is a linear transformation, then θ_V is in ker L. An examination of the elements in ker L allows us to decide whether L is or is not one-one.

Theorem 4.3. *Let $L: V \to W$ be a linear transformation of a vector space V into a vector space W. Then*
   (a) *ker L is a subspace of V.*
   (b) *L is one-one if and only if ker $L = \{\theta_V\}$.*

Proof: (a) We show that if α and β are in ker L, then so are $\alpha + \beta$ and $c\alpha$ for any real number c. If α and β are in ker L, then $L(\alpha) = \theta_W$, and $L(\beta) = \theta_W$. Then

$$L(\alpha + \beta) = L(\alpha) + L(\beta) = \theta_W + \theta_W = \theta_W.$$

Thus $\alpha + \beta$ is in ker L. Also,

$$L(c\alpha) = cL(\alpha) = c\theta_W = \theta_W,$$

so $c\alpha$ is in ker L. Hence ker L is a subspace of V.

(b) Let L be one-one. We show that ker $L = \{\theta_V\}$. Let α be in ker L. Then $L(\alpha) = \theta_W$. Also, we already know that $L(\theta_V) = \theta_W$. Then $L(\alpha) = L(\theta_V)$. Since L is one-one, we conclude that $\alpha = \theta_V$. Hence ker $L = \{\theta_V\}$.

Conversely, suppose that ker $L = \{\theta_V\}$. We wish to show that L is one-one. Let $L(\alpha_1) = L(\alpha_2)$ for α_1 and α_2 in V. Then $L(\alpha_1) - L(\alpha_2) = \theta_W$, so $L(\alpha_1 - \alpha_2) = \theta_W$. This means that $\alpha_1 - \alpha_2$ is in ker L, so $\alpha_1 - \alpha_2 = \theta_V$. Hence $\alpha_1 = \alpha_2$, and L is one-one.

Note that we can also state Theorem 4.3(b) as: L is one-one if and only if dim ker $L = 0$.

EXAMPLE 3. Let $L: R^3 \to R^2$ be as defined in Example 2. To find ker L, we must determine all α in V so that $L(\alpha) = \theta_W$. That is, we seek $\alpha = \begin{bmatrix} a_1 \\ a_2 \\ a_3 \end{bmatrix}$

so that $L(\alpha) = L\left(\begin{bmatrix} a_1 \\ a_2 \\ a_3 \end{bmatrix} \right) = \begin{bmatrix} 0 \\ 0 \end{bmatrix} = \theta_W$. However, $L(\alpha) = \begin{bmatrix} a_1 \\ a_2 \end{bmatrix}$. Thus

$\begin{bmatrix} a_1 \\ a_2 \end{bmatrix} = \begin{bmatrix} 0 \\ 0 \end{bmatrix}$, so $a_1 = 0$, $a_2 = 0$, and a_3 can be any real number. Hence ker L

consists of all vectors in R^3 of the form $\begin{bmatrix} 0 \\ 0 \\ a \end{bmatrix}$, where a is any real number.

It is clear that ker L consists of the z-axis in (x, y, z) three-dimensional space R^3. Moreover, the dimension of the subspace ker L is 1. Finally, since ker $L \neq \{\theta_V\}$, L is not one-one, as we saw in Example 2.

EXAMPLE 4. Let $L: P_2 \to R$ be as defined in Example 4 of Section 4.1: $L(at^2 + bt + c) = \int_0^1 (at^2 + bt + c)\, dt$. To find ker L, we seek an element $\alpha = at^2 + bt + c$ in P_2 such that $L(\alpha) = L(at^2 + bt + c) = \theta_R = 0$. Now

$$L(\alpha) = \frac{at^3}{3} + \frac{bt^2}{2} + ct \Big|_0^1 = \frac{a}{3} + \frac{b}{2} + c.$$

Thus $c = -a/3 - b/2$. Then ker L consists of all polynomials in P_2 of the form $at^2 + bt + (-a/3 - b/2)$, for a and b any real numbers. Of course, L is not one-one. To find the dimension of ker L, we obtain a basis for ker L. Any vector in ker L can be written as

$$at^2 + bt + \left(-\frac{a}{3} - \frac{b}{2}\right) = a\left(t^2 - \frac{1}{3}\right) + b\left(t - \frac{1}{2}\right).$$

Thus the elements $(t^2 - \frac{1}{3})$ and $(t - \frac{1}{2})$ in P_2 span ker L. Now, these elements are also linearly independent, for if

$$a_1(t^2 - \tfrac{1}{3}) + a_2(t - \tfrac{1}{2}) = 0,$$

then $a_1 t^2 + a_2 t + (-a_1/3 - a_2/2) = 0$, which implies that $a_1 = a_2 = 0$. Thus $\{t^2 - \frac{1}{3}, t - \frac{1}{2}\}$ is a basis for ker L, and dimension ker $L = 2$.

Definition 4.4. If $L: V \to W$ is a linear transformation of a vector space V into a vector space W, then the **range** of L or **image** of V under L, denoted by range L, consists of all those vectors in W that are images under L of vectors in V. Thus β is in range L if we can find some vector α in V such that $L(\alpha) = \beta$; L is called **onto** if range $L = W$.

Theorem 4.4. *If $L: V \to W$ is a linear transformation of a vector space V into a vector space W, then range L is a subspace of W.*

Proof: Let β_1 and β_2 be in range L. Then $\beta_1 = L(\alpha_1)$ and $\beta_2 = L(\alpha_2)$ for some α_1 and α_2 in V. Now $\beta_1 + \beta_2 = L(\alpha_1) + L(\alpha_2) = L(\alpha_1 + \alpha_2)$, which implies that $\beta_1 + \beta_2$ is in range L. Also, if β is in range L, then $\beta = L(\alpha)$ for some α in V. Then $c\beta = cL(\alpha) = L(c\alpha)$, so $c\beta$ is in range L. Hence range L is a subspace of W.

EXAMPLE 5. Consider Example 2 of this section again. To find out whether the projection L is onto, we choose any vector $\beta = \begin{bmatrix} c \\ d \end{bmatrix}$ in R^2 and we seek a vector $\alpha = \begin{bmatrix} a_1 \\ a_2 \\ a_3 \end{bmatrix}$ in V such that $L(\alpha) = \beta$. Now $L(\alpha) = \begin{bmatrix} a_1 \\ a_2 \end{bmatrix}$, so if $a_1 = c$ and $a_2 = d$, then $L(\alpha) = \beta$. Therefore, L is onto and dim range $L = 2$.

EXAMPLE 6. Consider Example 4 of this section; is L onto? Given a vector β in R, $\beta = r$, a real number, can we find a vector $\alpha = at^2 + bt + c$ in P_2 so that $L(\alpha) = \beta = r$? Now $L(\alpha) = \int_0^1 (at^2 + bt + c)\,dt = a/3 + b/2 + c$. We can let $a = b = 0$ and $c = r$. Hence L is onto, and dim range $L = 1$.

EXAMPLE 7. Let $L: R^3 \to R^3$ be defined by

$$L\left(\begin{bmatrix} a_1 \\ a_2 \\ a_3 \end{bmatrix}\right) = \begin{bmatrix} 1 & 0 & 1 \\ 1 & 1 & 2 \\ 2 & 1 & 3 \end{bmatrix} \begin{bmatrix} a_1 \\ a_2 \\ a_3 \end{bmatrix}.$$

Is L onto? Given any $\beta = \begin{bmatrix} a \\ b \\ c \end{bmatrix}$ in R^3, where a, b, and c are any real numbers, can we find $\alpha = \begin{bmatrix} a_1 \\ a_2 \\ a_3 \end{bmatrix}$ so that $L(\alpha) = \beta$? We seek a solution to the linear system

$$\begin{bmatrix} 1 & 0 & 1 \\ 1 & 1 & 2 \\ 2 & 1 & 3 \end{bmatrix} \begin{bmatrix} a_1 \\ a_2 \\ a_3 \end{bmatrix} = \begin{bmatrix} a \\ b \\ c \end{bmatrix}$$

and we find the reduced row echelon form of the augmented matrix to be

$$\begin{bmatrix} 1 & 0 & 1 & \vdots & a \\ 0 & 1 & 1 & \vdots & b - a \\ 0 & 0 & 0 & \vdots & c - b - a \end{bmatrix}.$$

Thus a solution exists only for $c - b - a = 0$, so L is not onto.

To find a basis for range L, we note that

$$L\left(\begin{bmatrix} a_1 \\ a_2 \\ a_3 \end{bmatrix}\right) = \begin{bmatrix} 1 & 0 & 1 \\ 1 & 1 & 2 \\ 2 & 1 & 3 \end{bmatrix} \begin{bmatrix} a_1 \\ a_2 \\ a_3 \end{bmatrix} = \begin{bmatrix} a_1 + & & a_3 \\ a_1 + a_2 + 2a_3 \\ 2a_1 + a_2 + 3a_3 \end{bmatrix}$$

$$= a_1 \begin{bmatrix} 1 \\ 1 \\ 2 \end{bmatrix} + a_2 \begin{bmatrix} 0 \\ 1 \\ 1 \end{bmatrix} + a_3 \begin{bmatrix} 1 \\ 2 \\ 3 \end{bmatrix}.$$

This means that $\left\{ \begin{bmatrix} 1 \\ 1 \\ 2 \end{bmatrix}, \begin{bmatrix} 0 \\ 1 \\ 1 \end{bmatrix}, \begin{bmatrix} 1 \\ 2 \\ 3 \end{bmatrix} \right\}$ spans range L. Now the first two

vectors in this set are linearly independent and the third vector is the sum of the first two. Thus the first two vectors form a basis for range L, and dim range $L = 2$.

It is also interesting to find dim ker L for this example; that is, we wish to find all α in R^3 so that $L(\alpha) = \theta_{R^3}$. Solving the resulting homogeneous system, we find (verify) that $a_1 = -a_3$ and $a_2 = -a_3$. Thus ker L consists of all

vectors of the form $\begin{bmatrix} -a \\ -a \\ a \end{bmatrix}$, where a is any real number; therefore, dim ker L

$= 1$ and L is not one-one.

Observe that we can find a basis for ker L and range L by the method discussed in Section 2.4. Also, to find out if a linear transformation is one-one or onto, we must solve a linear system. This is one further demonstration of the frequency with which linear systems must be solved to answer many questions in linear algebra. Finally, in the last example we saw that dim ker L + dim range L = dim domain L. This very important result is always true and we now prove it in the following theorem.

Theorem 4.5. *If $L: V \to W$ is a linear transformation of an n-dimensional vector space V into a vector space W, then dim ker L + dim range L = dim V.*

Proof: If $L = O$, the zero linear transformation defined by $O(\alpha) = \theta_W$ for all α in V (see Exercise 14 in Section 4.1), then ker $L = V$ and range $L = \{\theta_W\}$. In this case dim ker L = dim V, and dim range $L = 0$, so the conclusion of the theorem holds. Now suppose that $L \neq O$ and let k = dim ker L. If

$n = k$, then ker $L = V$ (Exercise 29, Section 2.2), which implies that $L(\alpha) = \theta_W$ for every α in V. Hence range $L = \{\theta_W\}$, dim range $L = 0$, and again the conclusion holds. Next, suppose that $1 \le k < n$. We shall prove that dim range $L = n - k$. Let $\{\alpha_1, \alpha_2, \ldots, \alpha_k\}$ be a basis for ker L. By Theorem 2.8 we can extend this basis to a basis $S = \{\alpha_1, \alpha_2, \ldots, \alpha_k, \alpha_{k+1}, \ldots, \alpha_n\}$ for V. We prove that the set $T = \{L(\alpha_{k+1}), L(\alpha_{k+2}), \ldots, L(\alpha_n)\}$ is a basis for range L.

First, we show that T spans range L. Let β be any vector in range L. Then $\beta = L(\alpha)$ for some α in V. Since S is a basis for V, we can find a unique set of real numbers $a_1, a_2, \ldots, a_n$ such that $\alpha = a_1\alpha_1 + a_2\alpha_2 + \cdots + a_n\alpha_n$. Then

$$
\begin{aligned}
\beta = L(\alpha) &= L(a_1\alpha_1 + a_2\alpha_2 + \cdots + a_k\alpha_k + a_{k+1}\alpha_{k+1} + \cdots + a_n\alpha_n) \\
&= a_1L(\alpha_1) + a_2L(\alpha_2) + \cdots + a_kL(\alpha_k) + a_{k+1}L(\alpha_{k+1}) + \cdots + a_nL(\alpha_n) \\
&= a_{k+1}L(\alpha_{k+1}) + \cdots + a_nL(\alpha_n),
\end{aligned}
$$

because $\alpha_1, \alpha_2, \ldots, \alpha_k$ are in ker L. Hence T spans range L.

Now we show that T is linearly independent. Suppose that

$$a_{k+1}L(\alpha_{k+1}) + a_{k+2}L(\alpha_{k+2}) + \cdots + a_nL(\alpha_n) = \theta_W.$$

Then

$$L(a_{k+1}\alpha_{k+1} + a_{k+2}\alpha_{k+2} + \cdots + a_n\alpha_n) = \theta_W.$$

Hence the vector $a_{k+1}\alpha_{k+1} + a_{k+2}\alpha_{k+2} + \cdots + a_n\alpha_n$ is in ker L, and we can write

$$a_{k+1}\alpha_{k+1} + a_{k+2}\alpha_{k+2} + \cdots + a_n\alpha_n = b_1\alpha_1 + b_2\alpha_2 + \cdots + b_k\alpha_k,$$

where $b_1, b_2, \ldots, b_k$ are uniquely determined real numbers. We then write

$$b_1\alpha_1 + b_2\alpha_2 + \cdots + b_k\alpha_k - a_{k+1}\alpha_{k+1} - a_{k+2}\alpha_{k+2} \cdots - a_n\alpha_n = \theta_W.$$

Since S is linearly independent, we find that $b_1 = b_2 = \cdots = b_k = a_{k+1} = \cdots = a_n = 0$. Hence T is linearly independent and forms a basis for range L.

The dimension of ker L is also called the **nullity of L.**

We have seen that a linear transformation may be one-one and onto or onto and not one-one. However, the following corollary shows that each of these properties implies the other if the vector spaces V and W have the same dimensions.

Corollary 4.1. *If $L: V \to W$ is a linear transformation of a vector space V into a vector space W and dim $V = $ dim W, then:*
  (a) *If L is one-one, then it is onto.*
  (b) *If L is onto, then it is one-one.*

Proof: Exercise.

If $A = [a_{ij}]$ is an $m \times n$ matrix, then we can define a function $L: R^n \to R^m$ by $L(\alpha) = A\alpha$ for α a vector in R^n. It is easy to verify that L is a linear transformation. (Exercise 16 of Section 4.1.) Now if $AX = B$ is a system of m linear equations in n unknowns, then finding a solution to $AX = B$ is equivalent to obtaining, for a given vector B in R^m, a vector X in R^n such that $L(X) = B$. Then ker L consists of all vectors X in R^n such that $L(X) = AX = O$, that is, of all solutions to the homogeneous system $AX = O$. Moreover, it is easy to see that range L is the subspace of R^m spanned by the columns of A; that is, range $L = $ the column space of A (Exercise 8). This means that dim range $L = $ rank A. It then follows from Theorem 4.5 that dim ker $L = n - $ rank A. Thus we have established the result that the solution space of a homogeneous system $AX = O$, where A is an $m \times n$ matrix, has dimension $n - k$, where $k = $ rank A.

A linear transformation $L: V \to W$ of a vector space V into a vector space W is called **invertible** if it is an invertible function, that is, if there exists a unique function $L^{-1}: W \to V$ such that $L \circ L^{-1} = I_W$ and $L^{-1} \circ L = I_V$, where $I_V = $ *identity linear transformation* on V and $I_W = $ *identity linear transformation* on W. We now prove the following theorem.

Theorem 4.6. *A linear transformation $L: V \to W$ is invertible if and only if L is one-one and onto. Moreover, L^{-1} is a linear transformation and $(L^{-1})^{-1} = L$.*

Proof: Let L be one-one and onto. We define a function $H: W \to V$ as follows. If β is in W, then since L is onto, $\beta = L(\alpha)$ for some α in V, and since L is one-one, α is unique. Let $H(\beta) = \alpha$; H is a function and $L(H(\beta)) = L(\alpha) = \beta$, so that $L \circ H = I_W$. Also, $H(L(\alpha)) = H(\beta) = \alpha$, so $H \circ L = I_V$. Thus H is an inverse of L. Now H is unique, for if $H_1: W \to V$ is a function such that $L \circ H_1 = I_W$ and $H_1 \circ L = I_V$, then $L(H(\beta)) = \beta = L(H_1(\beta))$ for any β in W. Since L is one-one, we conclude that $H(\beta) = H_1(\beta)$. Hence $H = H_1$. Thus $H = L^{-1}$ and L is invertible.

Conversely, let L be invertible; that is, $L \circ L^{-1} = I_W$ and $L^{-1} \circ L = I_V$. We show that L is one-one and onto. Suppose that $L(\alpha_1) = L(\alpha_2)$ for α_1, α_2 in V. Then $L^{-1}(L(\alpha_1)) = L^{-1}(L(\alpha_2))$, so $\alpha_1 = \alpha_2$, which means that L is one-one. Also, if β is a vector in W, then $L(L^{-1}(\beta)) = \beta$, so if we let $L^{-1}(\beta) = \alpha$, then $L(\alpha) = \beta$. Thus L is onto.

We now show that L^{-1} is a linear transformation. Let β_1, β_2 be in W, where $L(\alpha_1) = \beta_1$ and $L(\alpha_2) = \beta_2$ for α_1, α_2 in V. Then since

$$L(a\alpha_1 + b\alpha_2) = aL(\alpha_1) + bL(\alpha_2) = a\beta_1 + b\beta_2 \qquad \text{for } a, b \text{ real numbers,}$$

we have

$$L^{-1}(a\beta_1 + b\beta_2) = a\alpha_1 + b\alpha_2 = aL^{-1}(\beta_1) + bL^{-1}(\beta_2),$$

which implies that L^{-1} is a linear transformation.

Finally, since $L \circ L^{-1} = I_W$, $L^{-1} \circ L = I_V$ and inverses are unique, we conclude that $(L^{-1})^{-1} = L$.

EXAMPLE 8. Consider the linear transformation $L: R^3 \to R^3$ defined by

$$L\left(\begin{bmatrix} a_1 \\ a_2 \\ a_3 \end{bmatrix}\right) = \begin{bmatrix} 1 & 1 & 1 \\ 2 & 2 & 1 \\ 0 & 1 & 1 \end{bmatrix}\begin{bmatrix} a_1 \\ a_2 \\ a_3 \end{bmatrix}.$$

Since $\ker L = \{\theta\}$ (verify), L is one-one and so is invertible; to obtain L^{-1}, we let

$$L(\alpha) = L\left(\begin{bmatrix} a_1 \\ a_2 \\ a_3 \end{bmatrix}\right) = \beta = \begin{bmatrix} b_1 \\ b_2 \\ b_3 \end{bmatrix}.$$

We are then solving the linear system

$$\begin{aligned} a_1 + a_2 + a_3 &= b_1 \\ 2a_1 + 2a_2 + a_3 &= b_2 \\ a_2 + a_3 &= b_3. \end{aligned}$$

for a_1, a_2, and a_3. We find that

$$L^{-1}\left(\begin{bmatrix} b_1 \\ b_2 \\ b_3 \end{bmatrix}\right) = \begin{bmatrix} 1 & 1 & 1 \\ 2 & 2 & 1 \\ 0 & 1 & 1 \end{bmatrix}^{-1}\begin{bmatrix} b_1 \\ b_2 \\ b_3 \end{bmatrix} = \begin{bmatrix} 1 & 0 & -1 \\ -2 & 1 & 1 \\ 2 & -1 & 0 \end{bmatrix}\begin{bmatrix} b_1 \\ b_2 \\ b_3 \end{bmatrix}$$

$$= \begin{bmatrix} b_1 - b_3 \\ -2b_1 + b_2 + b_3 \\ 2b_1 - b_2 \end{bmatrix}.$$

The following useful theorem shows that one-one linear transformations preserve linear independence of a set of vectors. Moreover, if this property holds, then L is one-one.

Theorem 4.7. *A linear transformation $L: V \to W$ is one-one if and only if the image of every linearly independent set of vectors in V is a linearly independent set of vectors in W.*

Proof: Let $S = \{\alpha_1, \alpha_2, \ldots, \alpha_k\}$ be a linearly independent set of vectors in V; let $T = \{L(\alpha_1), L(\alpha_2), \ldots, L(\alpha_k)\}$. Suppose that L is one-one; we show that T is linearly independent. Let

$$a_1 L(\alpha_1) + a_2 L(\alpha_2) + \cdots + a_k L(\alpha_k) = \theta_W,$$

where $a_1, a_2, \ldots, a_k$ are real numbers. Then

$$L(a_1\alpha_1 + a_2\alpha_2 + \cdots + a_k\alpha_k) = \theta_W = L(\theta_V).$$

Since L is one-one, we conclude that

$$a_1\alpha_1 + a_2\alpha_2 + \cdots + a_k\alpha_k = \theta_V.$$

Now S is linearly independent, so $a_1 = a_2 = \cdots = a_k = 0$. Hence T is linearly independent.

Conversely, let the image of any linearly independent set of vectors in V be a linearly independent set of vectors in W. Now $\{\alpha\}$, where $\alpha \neq \theta_V$, is a linearly independent set in V. Since the set $\{L(\alpha)\}$ is linearly independent, $L(\alpha) \neq \theta_W$, so $\ker L = \{\theta_V\}$, which means that L is one-one.

It follows from this theorem that if $L: V \to W$ is a linear transformation and dim $V = $ dim W, then L is one-one, and thus invertible, if and only if the image of a basis for V under L is a basis for W (see Exercise 11).

We now make one final remark for a linear system $AX = B$, where A is $n \times n$. We again consider the linear transformation $L: R^n \to R^n$ defined by $L(\alpha) = A\alpha$, for α in R^n. If A is a nonsingular matrix, then dim range $L = $ rank $A = n$, so dim ker $L = 0$. Thus L is one-one and hence onto. This means that the given linear system has a unique solution (of course, we already knew this result from other considerations). However, if A is singular, then rank $A < n$. This means that dim ker $L = n - $ rank $A > 0$, so L is not one-one and not onto. This means that there exists a vector B in R^n, for which the system $AX = B$ has no solution. Moreover, since A is singular, $AX = O$ has a nontrivial solution X_0. If $AX = B$ has a solution Y, then $X_0 + Y$ is a solution to $AX = B$ (verify). Thus, for A singular, if a solution to $AX = B$ exists, then it is not unique.

4.2. Exercises

1. Let $L: R_2 \to R_3$ be the linear transformation defined by $L([a_1 \quad a_2]) = [a_1 \quad a_1 + a_2 \quad a_2]$.
   (a) Find ker L.   (b) Is L one-one?   (c) Is L onto?
2. Let $L: R_4 \to R_3$ be the linear transformation defined by $L([a_1 \quad a_2 \quad a_3 \quad a_4]) = [a_1 + a_2 \quad a_3 + a_4 \quad a_1 + a_3]$.
   (a) Find a basis for ker L.   (b) What is dim ker L?
   (c) Find a basis for range L.   (d) What is dim range L?
3. Let $L: P_2 \to P_3$ be the linear transformation defined by $L(p(t)) = t^2 p'(t)$.
   (a) Find a basis for and the dimension of ker L.
   (b) Find a basis for and the dimension of range L.
4. Let $L: {}_2R_3 \to {}_3R_3$ be the linear transformation defined by

$$L(A) = \begin{bmatrix} 2 & -1 \\ 1 & 2 \\ 3 & 1 \end{bmatrix} A \qquad \text{for } A \text{ in } {}_2R_3.$$

   (a) Find the dimension of ker L.
   (b) Find the dimension of range L.
5. Let $L: V \to W$ be a linear transformation.
   (a) Show that dim range $L \le$ dim V.
   (b) Prove that if L is onto, then dim $W \le$ dim V.
6. Verify Theorem 4.5 for the following linear transformations.
   (a) $L: P_2 \to P_2$ defined by $L(p(t)) = tp'(t)$.
   (b) $L: R_3 \to R_2$ defined by $L([a_1 \quad a_2 \quad a_3]) = [a_1 + a_2 \quad a_1 + a_3]$.
   (c) $L: {}_2R_2 \to {}_2R_3$ defined by

$$L(A) = A \begin{bmatrix} 1 & 2 & 3 \\ 2 & 1 & 3 \end{bmatrix} \qquad \text{for } A \text{ in } {}_2R_2.$$

7. Prove Corollary 4.1.
8. Let A be an $m \times n$ matrix, and consider the linear transformation $L: R^n \to R^m$ defined by $L(\alpha) = A\alpha$, for α in R^n. Show that range $L =$ column space of A.
9. Let $L: R^5 \to R^4$ be the linear transformation defined by

$$L\left(\begin{bmatrix} a_1 \\ a_2 \\ a_3 \\ a_4 \\ a_5 \end{bmatrix}\right) = \begin{bmatrix} 1 & 0 & -1 & 3 & -1 \\ 1 & 0 & 0 & 2 & -1 \\ 2 & 0 & -1 & 5 & -1 \\ 0 & 0 & -1 & 1 & 0 \end{bmatrix} \begin{bmatrix} a_1 \\ a_2 \\ a_3 \\ a_4 \\ a_5 \end{bmatrix}.$$

   (a) Find a basis for and the dimension of ker L.
   (b) Find a basis for and the dimension of range L.

10. Let $L: R_3 \rightarrow R_3$ be the linear transformation defined by $L(\varepsilon_1') = L([1 \quad 0 \quad 0])$ $= [3 \quad 0 \quad 0], L(\varepsilon_2') = L([0 \quad 1 \quad 0]) = [1 \quad 1 \quad 1]$, and $L(\varepsilon_3') = L([0 \quad 0 \quad 1]) =$ $[2 \quad 1 \quad 1]$. Is the set $\{L(\varepsilon_1'), L(\varepsilon_2'), L(\varepsilon_3')\} = \{[3 \quad 0 \quad 0], [1 \quad 1 \quad 1], [2 \quad 1 \quad 1]\}$ a basis for R_3?

11. Let $L: V \rightarrow W$ be a linear transformation, and let dim $V =$ dim W. Prove that L is invertible if and only if the image of a basis for V under L is a basis for W.

12. Let $L: R^3 \rightarrow R^3$ be defined by

$$L\left(\begin{bmatrix} 1 \\ 0 \\ 0 \end{bmatrix}\right) = \begin{bmatrix} 1 \\ 2 \\ 3 \end{bmatrix}, \quad L\left(\begin{bmatrix} 0 \\ 1 \\ 0 \end{bmatrix}\right) = \begin{bmatrix} 0 \\ 1 \\ 1 \end{bmatrix}, \quad L\left(\begin{bmatrix} 0 \\ 0 \\ 1 \end{bmatrix}\right) = \begin{bmatrix} 1 \\ 1 \\ 0 \end{bmatrix}.$$

(a) Prove that L is invertible.

(b) Find $L^{-1}\left(\begin{bmatrix} 2 \\ 3 \\ 4 \end{bmatrix}\right)$.

13. Let $L: V \rightarrow W$ be a linear transformation, and let $S = \{\alpha_1, \alpha_2, \ldots, \alpha_n\}$ be a set of vectors in V. Prove that if $T = \{L(\alpha_1), L(\alpha_2), \ldots, L(\alpha_n)\}$ is linearly independent, then so is S. What can we say about the converse?

14. Find the dimension of the solution space for the following homogeneous system

$$\begin{bmatrix} 1 & 2 & 1 & 3 \\ 2 & 1 & -1 & 2 \\ 1 & 0 & 0 & -1 \\ 4 & 1 & -1 & 0 \end{bmatrix} \begin{bmatrix} x_1 \\ x_2 \\ x_3 \\ x_4 \end{bmatrix} = \begin{bmatrix} 0 \\ 0 \\ 0 \\ 0 \end{bmatrix}.$$

15. Find a linear transformation $L: R_2 \rightarrow R_3$ such that $S = \{[1 \ -1 \ 2], [3 \ 1 \ -1]\}$ is a basis for range L.

16. Let $L: R^3 \rightarrow R^3$ be the linear transformation defined by

$$L\left(\begin{bmatrix} a_1 \\ a_2 \\ a_3 \end{bmatrix}\right) = \begin{bmatrix} 1 & 1 & 1 \\ 0 & 1 & 2 \\ 1 & 2 & 2 \end{bmatrix} \begin{bmatrix} a_1 \\ a_2 \\ a_3 \end{bmatrix}.$$

(a) Prove that L is invertible.

(b) Find $L^{-1}\left(\begin{bmatrix} a_1 \\ a_2 \\ a_3 \end{bmatrix}\right)$.

17. Let $L: V \rightarrow W$ be a linear transformation. Prove that L is one-one if and only if dim range $L =$ dim V.

18. Let $L: R^4 \rightarrow R^6$ be a linear transformation.
    (a) If dim ker $L = 2$, what is dim range L?
    (b) If dim range $L = 3$, what is dim ker L?

19. Let $L: V \rightarrow R^5$ be a linear transformation.
    (a) If L is onto and dim ker $L = 2$, what is dim V?
    (b) If L is one-one and onto, what is dim V?

4.3. The Matrix of a Linear Transformation

In Section 4.2 we saw that if A is an $m \times n$ matrix, we can define a linear transformation $L: R^n \to R^m$ by $L(\alpha) = A\alpha$ for α in R^n. In Section 4.4 we shall prove that if A is an $m \times n$ matrix, V is an n-dimensional vector space with ordered basis S and W is an m-dimensional vector space with ordered basis T, then we can define a linear transformation $L: V \to W$ associated with A and the two bases. We shall now develop the converse notion; if $L: V \to W$ is a linear transformation of an n-dimensional vector space V into an m-dimensional vector space W, and if we choose ordered bases for V and W, then we can associate a unique $m \times n$ matrix A with L that will enable us to find $L(\alpha)$ for α in V by merely performing matrix multiplication.

Theorem 4.8. *Let $L: V \to W$ be a linear transformation of an n-dimensional vector space V into an m-dimensional vector space W ($n \neq 0$, $m \neq 0$) and let $S = \{\alpha_1, \alpha_2, \ldots, \alpha_n\}$ and $T = \{\beta_1, \beta_2, \ldots, \beta_m\}$ be ordered bases for V and W, respectively. Then the $m \times n$ matrix A whose jth column is the coordinate vector $[L(\alpha_j)]_T$ of $L(\alpha_j)$ with respect to T has the following property: If $\beta = L(\alpha)$ for some α in V, then $[\beta]_T = A[\alpha]_S$, where $[\alpha]_S$ and $[\beta]_T$ are the coordinate vectors of α and β with respect to S and T, respectively. Moreover, A is the only matrix with this property.*

Proof: We show how to construct the matrix A. Consider the vector α_j in V for $j = 1, 2, \ldots, n$. Then $L(\alpha_j)$ is a vector in W, and since T is an ordered basis for W, we can then express this vector as a linear combination of the vectors in T in a unique manner. Thus $L(\alpha_j) = c_{1j}\beta_1 + c_{2j}\beta_2 + \cdots + c_{mj}\beta_m$. This means that the coordinate vector of $L(\alpha_j)$ with respect to T is

$$[L(\alpha_j)]_T = \begin{bmatrix} c_{1j} \\ c_{2j} \\ \vdots \\ c_{mj} \end{bmatrix}.$$ We now define an $m \times n$ matrix A by choosing $[L(\alpha_j)]_T$

as the jth column of A. We show that this matrix satisfies the properties stated in the theorem.

Let α be any vector in V, and let $L(\alpha) = \beta$, where β is in W. Now let

$$[\alpha]_S = \begin{bmatrix} a_1 \\ a_2 \\ \vdots \\ a_n \end{bmatrix} \quad \text{and} \quad [\beta]_T = \begin{bmatrix} b_1 \\ b_2 \\ \vdots \\ b_m \end{bmatrix}.$$

This means that $\alpha = a_1\alpha_1 + a_2\alpha_2 + \cdots + a_n\alpha_n$. Then

$$
\begin{aligned}
L(\alpha) &= a_1 L(\alpha_1) + a_2 L(\alpha_2) + \cdots + a_n L(\alpha_n) \\
&= a_1(c_{11}\beta_1 + c_{21}\beta_2 + \cdots + c_{m1}\beta_m) + a_2(c_{12}\beta_1 + c_{22}\beta_2 + \cdots + c_{m2}\beta_m) \\
&\quad + \cdots + a_n(c_{1n}\beta_1 + c_{2n}\beta_2 + \cdots + c_{mn}\beta_m) \\
&= (c_{11}a_1 + c_{12}a_2 + \cdots + c_{1n}a_n)\beta_1 + (c_{21}a_1 + c_{22}a_2 + \cdots + c_{2n}a_n)\beta_2 \\
&\quad + \cdots + (c_{m1}a_1 + c_{m2}a_2 + \cdots + c_{mn}a_n)\beta_m.
\end{aligned}
$$

Now $L(\alpha) = \beta = b_1\beta_1 + b_2\beta_2 + \cdots + b_m\beta_m$. Hence

$$
b_i = c_{i1}a_1 + c_{i2}a_2 + \cdots + c_{in}a_n \qquad \text{for } i = 1, 2, \ldots, m.
$$

Next, we verify that $[\beta]_T = A[\alpha]_S$. We have

$$
\begin{bmatrix}
c_{11} & c_{12} & \cdots & c_{1n} \\
c_{21} & c_{22} & \cdots & c_{2n} \\
\vdots & & & \vdots \\
c_{m1} & c_{m2} & \cdots & c_{mn}
\end{bmatrix}
\begin{bmatrix}
a_1 \\ a_2 \\ \vdots \\ a_n
\end{bmatrix}
=
\begin{bmatrix}
c_{11}a_1 + c_{12}a_2 + \cdots + c_{1n}a_n \\
c_{21}a_1 + c_{22}a_2 + \cdots + c_{2n}a_n \\
\vdots \\
c_{m1}a_1 + c_{m2}a_2 + \cdots + c_{mn}a_n
\end{bmatrix}
=
\begin{bmatrix}
b_1 \\ b_2 \\ \vdots \\ b_m
\end{bmatrix}.
$$

Finally, we show that $A = [c_{ij}]$ is the only matrix with this property. Suppose that we have another matrix $A^* = [c_{ij}^*]$ with the same properties as A, and that $A^* \neq A$. All the elements of A and A^* cannot be equal, so say that the kth columns of these matrices are unequal. Now the coordinate vector of α_k with respect to the basis S is

$$
[\alpha_k]_S =
\begin{bmatrix}
0 \\ 0 \\ \vdots \\ 1 \\ 0 \\ \vdots \\ 0
\end{bmatrix}
\leftarrow k\text{th row.}
$$

Then

$$
[L(\alpha_k)]_T = A
\begin{bmatrix}
0 \\ 0 \\ \vdots \\ 1 \\ 0 \\ \vdots \\ 0
\end{bmatrix}
=
\begin{bmatrix}
a_{1k} \\ a_{2k} \\ \vdots \\ a_{mk}
\end{bmatrix}
= \text{the } k\text{th column of } A
$$

and

$$[L(\alpha_k)]_T = A^* \begin{bmatrix} 0 \\ 0 \\ \vdots \\ 1 \\ 0 \\ \vdots \\ 0 \end{bmatrix} = \begin{bmatrix} a^*_{1k} \\ a^*_{2k} \\ \vdots \\ a^*_{mk} \end{bmatrix} = \text{the } k\text{th column of } A^*.$$

This means that $L(\alpha_k)$ has two different coordinate vectors with respect to the same basis, which is impossible. Hence the matrix A is unique.

EXAMPLE 1. Let $L: P_2 \to P_1$ be defined by $L(p(t)) = p'(t)$ and consider the ordered bases $S = \{t^2, t, 1\}$ and $T = \{t, 1\}$ for P_2 and P_1, respectively. We now find the matrix A associated with L. We have

$$L(t^2) = 2t = 2t + 0(1), \qquad \text{so } [L(t^2)]_T = \begin{bmatrix} 2 \\ 0 \end{bmatrix}.$$

$$L(t) = 1 = 0(t) + 1(1), \qquad \text{so } [L(t)]_T = \begin{bmatrix} 0 \\ 1 \end{bmatrix}.$$

$$L(1) = 0 = 0(t) + 0(1), \qquad \text{so } [L(1)]_T = \begin{bmatrix} 0 \\ 0 \end{bmatrix}.$$

Thus $A = \begin{bmatrix} 2 & 0 & 0 \\ 0 & 1 & 0 \end{bmatrix}$. Now let $p(t) = 5t^2 - 3t + 2$. Then $L(p(t)) = 10t - 3$.
However, we can find $L(p(t))$ using the matrix A as follows. Since

$$[p(t)]_S = \begin{bmatrix} 5 \\ -3 \\ 2 \end{bmatrix},$$

then

$$[L(p(t))]_T = \begin{bmatrix} 2 & 0 & 0 \\ 0 & 1 & 0 \end{bmatrix} \begin{bmatrix} 5 \\ -3 \\ 2 \end{bmatrix} = \begin{bmatrix} 10 \\ -3 \end{bmatrix}.$$

This means that $L(p(t)) = 10t - 3$.

EXAMPLE 2. Let $L: P_2 \to P_1$ be defined as in Example 1 and consider the ordered bases $S = \{1, t, t^2\}$ and $T = \{t, 1\}$ for P_2 and P_1, respectively.

We then find that the matrix A associated with L is $\begin{bmatrix} 0 & 0 & 2 \\ 0 & 1 & 0 \end{bmatrix}$. Notice that if we change the order of the vectors in S or T, the matrix changes.

EXAMPLE 3. Let $L: P_2 \to P_1$ be defined as in Example 1 and consider the ordered bases $S = \{t^2, t, 1\}$ and $T = \{t + 1, t - 1\}$ for P_2 and P_1, respectively. To find the matrix A associated with L, we proceed as follows:

$$L(t^2) = 2t = 1(t + 1) + 1(t - 1), \qquad \text{so } [L(t^2)]_T = \begin{bmatrix} 1 \\ 1 \end{bmatrix}.$$

$$L(t) = 1 = \tfrac{1}{2}(t + 1) - \tfrac{1}{2}(t - 1), \qquad \text{so } [L(t)]_T = \begin{bmatrix} \tfrac{1}{2} \\ -\tfrac{1}{2} \end{bmatrix}.$$

$$L(1) = 0 = 0(t + 1) + 0(t - 1), \qquad \text{so } [L(1)]_T = \begin{bmatrix} 0 \\ 0 \end{bmatrix}.$$

Hence $A = \begin{bmatrix} 1 & \tfrac{1}{2} & 0 \\ 1 & -\tfrac{1}{2} & 0 \end{bmatrix}$. Again, if $p(t) = 5t^2 - 3t + 2$, then $L(p(t))$ $= 10t - 3$. On the other hand,

$$[L(p(t))]_T = \begin{bmatrix} 1 & \tfrac{1}{2} & 0 \\ 1 & -\tfrac{1}{2} & 0 \end{bmatrix} \begin{bmatrix} 5 \\ -3 \\ 2 \end{bmatrix} = \begin{bmatrix} \tfrac{7}{2} \\ \tfrac{13}{2} \end{bmatrix},$$

so $L(p(t)) = \tfrac{7}{2}(t + 1) + \tfrac{13}{2}(t - 1) = 10t - 3$.

The matrix A is called the **representation of L with respect to the ordered bases S and T.** We also say that A **represents L with respect to S and T.** Having A enables us to replace L by A, α by $[\alpha]_S$, β by $[\beta]_T$ in the expression $L(\alpha) = \beta$ to get $A[\alpha]_S = [\beta]_T$. We can thus work with matrices rather than with linear transformations. Physicists and others who deal at great length with linear transformations perform most of their computations with the matrix representations of the linear transformations. We might also mention that if $L: R^n \to R^m$ is a linear transformation, then one sometimes uses the natural bases for R^n and R^m, which simplifies the task of obtaining a representation of L.

EXAMPLE 4. Let $L: R^3 \to R^2$ be defined by

$$L\left(\begin{bmatrix} a_1 \\ a_2 \\ a_3 \end{bmatrix} \right) = \begin{bmatrix} 1 & 1 & 1 \\ 1 & 2 & 3 \end{bmatrix} \begin{bmatrix} a_1 \\ a_2 \\ a_3 \end{bmatrix}.$$

Let

$$\varepsilon_1 = \begin{bmatrix} 1 \\ 0 \\ 0 \end{bmatrix}, \qquad \varepsilon_2 = \begin{bmatrix} 0 \\ 1 \\ 0 \end{bmatrix}, \qquad \varepsilon_3 = \begin{bmatrix} 0 \\ 0 \\ 1 \end{bmatrix},$$

$$\bar{\varepsilon}_1 = \begin{bmatrix} 1 \\ 0 \end{bmatrix}, \qquad \text{and} \qquad \bar{\varepsilon}_2 = \begin{bmatrix} 0 \\ 1 \end{bmatrix}.$$

Then $S = \{\varepsilon_1, \varepsilon_2, \varepsilon_3\}$ and $T = \{\bar{\varepsilon}_1, \bar{\varepsilon}_2\}$ are the natural bases for R^3 and R^2, respectively.

Now

$$L(\varepsilon_1) = \begin{bmatrix} 1 & 1 & 1 \\ 1 & 2 & 3 \end{bmatrix} \begin{bmatrix} 1 \\ 0 \\ 0 \end{bmatrix} = \begin{bmatrix} 1 \\ 1 \end{bmatrix} = 1\bar{\varepsilon}_1 + 1\bar{\varepsilon}_2, \qquad \text{so } [L(\varepsilon_1)]_T = \begin{bmatrix} 1 \\ 1 \end{bmatrix}$$

$$L(\varepsilon_2) = \begin{bmatrix} 1 & 1 & 1 \\ 1 & 2 & 3 \end{bmatrix} \begin{bmatrix} 0 \\ 1 \\ 0 \end{bmatrix} = \begin{bmatrix} 1 \\ 2 \end{bmatrix} = 1\bar{\varepsilon}_1 + 2\bar{\varepsilon}_2, \qquad \text{so } [L(\varepsilon_2)]_T = \begin{bmatrix} 1 \\ 2 \end{bmatrix}$$

$$L(\varepsilon_3) = \begin{bmatrix} 1 & 1 & 1 \\ 1 & 2 & 3 \end{bmatrix} \begin{bmatrix} 0 \\ 0 \\ 1 \end{bmatrix} = \begin{bmatrix} 1 \\ 3 \end{bmatrix} = 1\bar{\varepsilon}_1 + 3\bar{\varepsilon}_2, \qquad \text{so } [L(\varepsilon_3)]_T = \begin{bmatrix} 1 \\ 3 \end{bmatrix}.$$

Then

$$A = \begin{bmatrix} 1 & 1 & 1 \\ 1 & 2 & 3 \end{bmatrix}.$$

The reason that A is the same matrix as the one involved in the definition of L is that the natural bases are being used for R^3 and R^2.

EXAMPLE 5. Let $L: R^3 \to R^2$ be defined as in Example 4 and consider the ordered bases

$$S = \left\{ \begin{bmatrix} 1 \\ 1 \\ 0 \end{bmatrix}, \begin{bmatrix} 0 \\ 1 \\ 1 \end{bmatrix}, \begin{bmatrix} 0 \\ 0 \\ 1 \end{bmatrix} \right\} \qquad \text{and} \qquad T = \left\{ \begin{bmatrix} 1 \\ 2 \end{bmatrix}, \begin{bmatrix} 1 \\ 3 \end{bmatrix} \right\}$$

for R^3 and R^2, respectively. Then

$$L\left(\begin{bmatrix} 1 \\ 1 \\ 0 \end{bmatrix}\right) = \begin{bmatrix} 1 & 1 & 1 \\ 1 & 2 & 3 \end{bmatrix}\begin{bmatrix} 1 \\ 1 \\ 0 \end{bmatrix} = \begin{bmatrix} 2 \\ 3 \end{bmatrix} = 3\begin{bmatrix} 1 \\ 2 \end{bmatrix} - 1\begin{bmatrix} 1 \\ 3 \end{bmatrix},$$

$$\text{so} \quad \left[L\left(\begin{bmatrix} 1 \\ 1 \\ 0 \end{bmatrix}\right)\right]_T = \begin{bmatrix} 3 \\ -1 \end{bmatrix}$$

$$L\left(\begin{bmatrix} 0 \\ 1 \\ 1 \end{bmatrix}\right) = \begin{bmatrix} 2 \\ 5 \end{bmatrix} = 1\begin{bmatrix} 1 \\ 2 \end{bmatrix} + 1\begin{bmatrix} 1 \\ 3 \end{bmatrix}, \qquad \text{so} \quad \left[L\left(\begin{bmatrix} 0 \\ 1 \\ 1 \end{bmatrix}\right)\right]_T = \begin{bmatrix} 1 \\ 1 \end{bmatrix}$$

$$L\left(\begin{bmatrix} 0 \\ 0 \\ 1 \end{bmatrix}\right) = \begin{bmatrix} 1 \\ 3 \end{bmatrix} = 0\begin{bmatrix} 1 \\ 2 \end{bmatrix} + 1\begin{bmatrix} 1 \\ 3 \end{bmatrix}, \qquad \text{so} \quad \left[L\left(\begin{bmatrix} 0 \\ 0 \\ 1 \end{bmatrix}\right)\right]_T = \begin{bmatrix} 0 \\ 1 \end{bmatrix}.$$

Hence $A = \begin{bmatrix} 3 & 1 & 0 \\ -1 & 1 & 1 \end{bmatrix}$. This matrix, of course, differs from the one that

defined L. Thus, although a matrix A may be involved in the definition of a linear transformation L, we cannot conclude that it is necessarily the representation of L that we seek.

If $L: V \to V$ is a linear operator on an n-dimensional space V, then to obtain a representation of L we fix ordered bases S and T for V, and obtain a matrix A representing L with respect to S and T. However, it is often convenient in this case to choose $S = T$. To avoid verbosity in this case, we refer to A as the **representation of L with respect to S.**

It is, of course, clear that the matrix of the identity operator (see Exercise 15 in Section 4.1) on an n-dimensional space, with respect to any basis, is I_n.

If $L: V \to V$ is an invertible linear operator and if A is the representation of L with respect to an ordered basis S for V, then A^{-1} is the representation of L^{-1} with respect to S. This fact, which can be proved directly at this point, will follow almost trivially in Section 4.4.

4.3. Exercises

1. Let $L: R^2 \to R^2$ be defined by

$$L\left(\begin{bmatrix} a_1 \\ a_2 \end{bmatrix}\right) = \begin{bmatrix} a_1 + 2a_2 \\ 2a_1 - a_2 \end{bmatrix}.$$

Let S be the natural basis for R^2 and let $T = \left\{ \begin{bmatrix} -1 \\ 2 \end{bmatrix}, \begin{bmatrix} 2 \\ 0 \end{bmatrix} \right\}$.

Find the representation of L with respect to

(a) S.   (b) S and T.   (c) T and S.   (d) T.

(e) Find $L\left(\begin{bmatrix} 1 \\ 2 \end{bmatrix} \right)$ using the definition of L and also using the matrices obtained in (a), (b), (c), and (d).

2. Let $L: R_4 \to R_3$ be defined by $L([a_1 \ a_2 \ a_3 \ a_4]) = [a_1 \quad a_2 + a_3 \quad a_3 + a_4]$.

Let S and T be the natural bases for R_4 and R_3, respectively, Let $S' = \{[1 \ 0 \ 0 \ 1], [0 \ 0 \ 0 \ 1], [1 \ 1 \ 0 \ 0], [0 \ 1 \ 1 \ 0]\}$ and $T' = \{[1 \ 1 \ 0], [0 \ 1 \ 0], [1 \ 0 \ 1]\}$.

(a) Find the representation of L with respect to S and T.

(b) Find the representation of L with respect to S' and T'.

(c) Find $L([2 \ 1 \ -1 \ 3])$ using the matrices obtained in (a) and (b) and compare this answer with that obtained from the definition for L.

3. Let $L: R^4 \to R^3$ be defined by

$$L\left(\begin{bmatrix} a_1 \\ a_2 \\ a_3 \\ a_4 \end{bmatrix} \right) = \begin{bmatrix} 1 & 0 & 1 & 1 \\ 0 & 1 & 2 & 1 \\ -1 & -2 & 1 & 0 \end{bmatrix} \begin{bmatrix} a_1 \\ a_2 \\ a_3 \\ a_4 \end{bmatrix}.$$

Let S and T be the natural bases for R^4 and R^3, respectively, and consider the ordered bases

$$S' = \left\{ \begin{bmatrix} 1 \\ 1 \\ 0 \\ 0 \end{bmatrix}, \begin{bmatrix} 0 \\ 1 \\ 0 \\ 0 \end{bmatrix}, \begin{bmatrix} 0 \\ 0 \\ 1 \\ 1 \end{bmatrix}, \begin{bmatrix} 0 \\ 1 \\ 1 \\ 0 \end{bmatrix} \right\} \quad \text{and} \quad T' = \left\{ \begin{bmatrix} 1 \\ 0 \\ 1 \end{bmatrix}, \begin{bmatrix} 0 \\ 1 \\ 1 \end{bmatrix}, \begin{bmatrix} 0 \\ 0 \\ 1 \end{bmatrix} \right\}$$

for R^4 and R^3, respectively. Find the representation of L with respect to

(a) S and T.   (b) S' and T'.

4. Let $L: R^2 \to R^2$ be the linear transformation rotating R^2 counterclockwise through an angle ϕ. Find the representation of L with respect to the natural basis for R^2.

5. Let $L: R^3 \to R^3$ be defined by

$$L\left(\begin{bmatrix} 1 \\ 0 \\ 0 \end{bmatrix} \right) = \begin{bmatrix} 1 \\ 1 \\ 0 \end{bmatrix}, \quad L\left(\begin{bmatrix} 0 \\ 1 \\ 0 \end{bmatrix} \right) = \begin{bmatrix} 2 \\ 0 \\ 1 \end{bmatrix}, \quad L\left(\begin{bmatrix} 0 \\ 0 \\ 1 \end{bmatrix} \right) = \begin{bmatrix} 1 \\ 0 \\ 1 \end{bmatrix}.$$

(a) Find the representation of L with respect to the natural basis S for R^3.

(b) Find $L\left(\begin{bmatrix} 1 \\ 2 \\ 3 \end{bmatrix} \right)$ using the definition of L and also using the matrix obtained in (a).

6. Let $L: R^3 \rightarrow R^3$ be defined as in Exercise 5. Let $T = \{L(\varepsilon_1), L(\varepsilon_2), L(\varepsilon_3)\}$ be an ordered basis for R^3, and let S be the natural basis for R^3.
   (a) Find the representation of L with respect to S and T.
   (b) Find $L\left(\begin{bmatrix} 1 \\ 2 \\ 3 \end{bmatrix}\right)$ using the matrix obtained in (a).

7. Let $L: R^3 \rightarrow R^3$ be the linear transformation represented by the matrix $\begin{bmatrix} 1 & 3 & 1 \\ 1 & 2 & 0 \\ 0 & 1 & 1 \end{bmatrix}$ with respect to the natural basis for R^3. Find:
   (a) $L\left(\begin{bmatrix} 1 \\ 2 \\ 3 \end{bmatrix}\right)$. (b) $L\left(\begin{bmatrix} 0 \\ 1 \\ 1 \end{bmatrix}\right)$.

8. Let $L: {}_2R_2 \rightarrow {}_2R_2$ be defined by $L(A) = \begin{bmatrix} 1 & 2 \\ 3 & 4 \end{bmatrix} A$ for A in ${}_2R_2$. Consider the ordered bases

$$S = \left\{ \begin{bmatrix} 1 & 0 \\ 0 & 0 \end{bmatrix}, \begin{bmatrix} 0 & 1 \\ 0 & 0 \end{bmatrix}, \begin{bmatrix} 0 & 0 \\ 1 & 0 \end{bmatrix}, \begin{bmatrix} 0 & 0 \\ 0 & 1 \end{bmatrix} \right\}$$

and

$$T = \left\{ \begin{bmatrix} 1 & 0 \\ 0 & 1 \end{bmatrix}, \begin{bmatrix} 1 & 1 \\ 0 & 0 \end{bmatrix}, \begin{bmatrix} 1 & 0 \\ 1 & 0 \end{bmatrix}, \begin{bmatrix} 0 & 1 \\ 0 & 0 \end{bmatrix} \right\}$$

   for ${}_2R_2$. Find the representation of L with respect to
   (a) S. (b) T. (c) S and T. (d) T and S.

9. Let V be the vector space with basis $S = \{1, t, e^t, te^t\}$, and let $L: V \rightarrow V$ be a linear operator defined by $L(f) = f' = df/dt$. Find the representation of L with respect to S.

10. Let $L: P_1 \rightarrow P_2$ be defined by $L(p(t)) = tp(t) + p(0)$. Consider the ordered bases $S = \{t, 1\}$ and $S' = \{t + 1, t - 1\}$ for P_1, and $T = \{t^2, t, 1\}$ and $T' = \{t^2 + 1, t - 1, t + 1\}$ for P_2. Find the representation of L with respect to
    (a) S and T. (b) S' and T'.
    (c) Find $L(-3t + 3)$ using the definition of L and the matrices obtained in (a) and (b).

11. Let $A = \begin{bmatrix} 1 & 2 \\ 3 & 4 \end{bmatrix}$, and let $L: {}_2R_2 \rightarrow {}_2R_2$ be the linear transformation defined by $L(X) = AX - XA$ for X in ${}_2R_2$. Let S and T be the ordered bases for ${}_2R_2$ defined in Exercise 8. Find the representation of L with respect to
    (a) S. (b) T. (c) S and T. (d) T and S.

12. Let $L: V \rightarrow V$ be a linear operator. A nontrivial subspace U of V is called **invariant** under L if $L(U)$ is contained in U. Show that if dim $U = m$, and

dim $V = n$, then L has a representation with respect to a basis S for V of the form $\begin{bmatrix} A & B \\ O & C \end{bmatrix}$, where A is $m \times m$, B is $m \times (n - m)$, O is the zero $(n - m) \times m$ matrix, and C is $(n - m) \times (n - m)$.

13. Let $L: R^2 \rightarrow R^2$ be defined by

$$L\left(\begin{bmatrix} a_1 \\ a_2 \end{bmatrix}\right) = \begin{bmatrix} a_1 \\ -a_2 \end{bmatrix},$$

reflection about the x-axis.

Consider the natural basis S and the ordered basis $T = \left\{ \begin{bmatrix} 1 \\ 1 \end{bmatrix}, \begin{bmatrix} -1 \\ 1 \end{bmatrix} \right\}$, for R^2. Find the representation of L with respect to

(a) S.   (b) T.   (c) S and T.   (d) T and S.

14. If $L: R_3 \rightarrow R_2$ is the linear transformation whose representation with respect to the natural bases for R_3 and R_2 is $\begin{bmatrix} 1 & -1 & 2 \\ 2 & 1 & 3 \end{bmatrix}$, find

(a) $L([1 \quad 2 \quad 3])$.   (b) $L([-1 \quad 2 \quad -1])$.   (c) $L([0 \quad 1 \quad 2])$.

(d) $L([0 \quad 1 \quad 0])$.   (e) $L([0 \quad 0 \quad 1])$.

15. If $O: V \rightarrow W$ is the zero linear transformation, show that the representation of L with respect to any ordered bases for V and W is the $m \times n$ zero matrix, where $n = \dim V$, $m = \dim W$.

16. If $I: V \rightarrow V$ is the identity linear operator on V defined by $I(\alpha) = \alpha$ for α in V prove that the matrix of I with respect to any ordered basis S for V is I_n, where $\dim V = n$.

17. Let $I: R_2 \rightarrow R_2$ be the identity linear operator on R_2. Let $S = \{[1 \quad 0], [0 \quad 1]\}$ and $T = \{[1 \quad -1], [2 \quad 3]\}$.
Find the representation of I with respect to

(a) S.   (b) T.   (c) S and T.   (d) T and S.

18. Let V be the vector space of continuous functions with basis $S = \{e^t, e^{-t}\}$. Find the representation of the linear operator $L: V \rightarrow V$ defined by $L(f) = f'$ with respect to S.

19. Let V be the vector space of continuous functions with ordered basis $S = \{\sin t, \cos t\}$. Find the representation of the linear operator $L: V \rightarrow V$ defined by $L(f) = f'$ with respect to S.

20. Let V be the vector space of continuous functions with ordered basis $S = \{\sin t, \cos t\}$ and consider $T = \{\sin t - \cos t, \sin t + \cos t\}$, another ordered basis for V. Find the representation of the linear operator $L: V \rightarrow V$ defined by $L(f) = f'$ with respect to

(a) S.   (b) T.   (c) S and T.   (d) T and S.

21. Let $L: V \rightarrow V$ be a linear operator defined by $L(\alpha) = c\alpha$, where c is a fixed constant. Prove that the representation of L with respect to any ordered basis for V is a scalar matrix. (See Section 1.4.)

22. Let the representation of $L: R^3 \to R^2$ with respect to the ordered bases $S = \{\alpha_1, \alpha_2, \alpha_3\}$ and $T = \{\beta_1, \beta_2\}$ be

$$A = \begin{bmatrix} 1 & 2 & 1 \\ -1 & 1 & 0 \end{bmatrix},$$

where

$$\alpha_1 = \begin{bmatrix} -1 \\ 1 \\ 0 \end{bmatrix}, \qquad \alpha_2 = \begin{bmatrix} 0 \\ 1 \\ 1 \end{bmatrix}, \qquad \alpha_3 = \begin{bmatrix} 1 \\ 0 \\ 0 \end{bmatrix},$$

$$\beta_1 = \begin{bmatrix} 1 \\ 2 \end{bmatrix}, \qquad \text{and} \qquad \beta_2 = \begin{bmatrix} 1 \\ -1 \end{bmatrix}.$$

(a) Compute $[L(\alpha_1)]_T$, $[L(\alpha_2)]_T$, and $[L(\alpha_3)]_T$.
(b) Compute $L(\alpha_1)$, $L(\alpha_2)$, and $L(\alpha_3)$.
(c) Compute $L\left(\begin{bmatrix} 2 \\ 1 \\ -1 \end{bmatrix}\right)$.

4.4. The Vector Space of Matrices and the Vector Space of Linear Transformations

We have already seen in Section 2.1 that the set $_mR_n$ of all $m \times n$ matrices is a vector space under the operations of matrix addition and scalar multiplication. We now show in this section that the set of all linear transformations of an n-dimensional vector space V into an m-dimensional vector space W is also a vector space U under two suitably defined operations, and we shall examine the relation between U and $_mR_n$.

Definition 4.5. Let V and W be two vector spaces of dimensions n and m, respectively. Let $L_1: V \to W$ and $L_2: V \to W$ be linear transformations. We define a mapping $L: V \to W$ by $L(\alpha) = L_1(\alpha) + L_2(\alpha)$, for α in V. Of course, the $+$ here is vector addition in W. We shall denote L by $L_1 \boxplus L_2$ and call it the **sum** of L_1 and L_2. Also, if $L: V \to W$ is a linear transformation and c is a real number, we define a mapping $H: V \to W$ by $H(\alpha) = cL(\alpha)$ for α in V. Of course, the operation on the right-hand side is scalar multiplication in W. We denote H by $c \boxdot L$ and call it the **scalar multiple** of L by c.

EXAMPLE 1. Let $V = R_3$ and $W = R_2$. Let $L_1: R_3 \to R_2$ and $L_2: R_3 \to R_2$ be defined by

$$L_1(\alpha) = L_1([a_1 \quad a_2 \quad a_3]) = [a_1 + a_2 \quad a_2 + a_3]$$

and

$$L_2(\alpha) = L_2([a_1 \quad a_2 \quad a_3]) = [a_1 + a_3 \quad a_2].$$

Then

$$(L_1 \boxplus L_2)(\alpha) = [2a_1 + a_2 + a_3 \quad 2a_2 + a_3],$$

and

$$(3 \boxdot L_1)(\alpha) = [3a_1 + 3a_2 \quad 3a_2 + 3a_3].$$

We leave it to the reader (see the exercises in this section) to verify that if $L, L_1,$ and L_2 are linear transformations of V into W and if c is a real number, then $L_1 \boxplus L_2$ and $c \boxdot L$ are linear transformations. We also let the reader show that the set U of all linear transformations of V into W is a vector space under the operations $\boxplus$ and $\boxdot$. The linear transformation $O: V \to W$ defined by $O(\alpha) = \theta_W$ for α in V is the zero vector in U. That is, $L \boxplus O = O \boxplus L = L$ for any L in U. Also, if L is in U, then $L \boxplus (-1 \boxdot L) = O$, so we may write $(-1) \boxdot L$ as $-L$. Of course, to say that $S = \{L_1, L_2, \ldots, L_k\}$ is a linearly dependent set in U merely means that there exist k scalars $a_1, a_2, \ldots, a_k$ not all zero such that $a_1 L_1 + a_2 L_2 + \cdots + a_k L_k = O$, where O is the zero linear transformation.

EXAMPLE 2. Let $L_1: R_2 \to R_3, L_2: R_2 \to R_3,$ and $L_3: R_2 \to R_3$ be defined by

$$L_1([a_1 \quad a_2]) = [a_1 + a_2 \quad 2a_1 \quad a_2],$$
$$L_2([a_1 \quad a_2]) = [a_2 - a_1 \quad 2a_1 + a_2 \quad a_1],$$

and

$$L_3([a_1 \quad a_2]) = [3a_1 \quad -2a_2 \quad a_1 + 2a_2].$$

To determine whether $S = \{L_1, L_2, L_3\}$ is linearly independent, we proceed as follows. Suppose that

$$a_1 L_1 + a_2 L_2 + a_3 L_3 = O,$$

where $a_1, a_2,$ and a_3 are real numbers. Then for $\varepsilon_1' = [1 \quad 0]$, we have

$$(a_1 L_1 + a_2 L_2 + a_3 L_3)(\varepsilon_1') = O(\varepsilon_1') = [0 \quad 0 \quad 0].$$

We then have

$$a_1 L_1(\varepsilon_1') + a_2 L_2(\varepsilon_1') + a_3 L_3(\varepsilon_1') = a_1[1 \quad 2 \quad 0] + a_2[-1 \quad 2 \quad 1]$$
$$+ a_3[3 \quad 0 \quad 1] = [0 \quad 0 \quad 0].$$

Solving the resulting linear system, we find that $a_1 = a_2 = a_3 = 0$. Hence S is linearly independent.

Theorem 4.9. *Let U be the vector space of all linear transformations of an n-dimensional vector space V into an m-dimensional vector space W, $n \neq 0$ and $m \neq 0$, under the operations $\boxplus$ and $\boxdot$. Then U is isomorphic to the vector space $_mR_n$ of all $m \times n$ matrices.*

Proof: Let $S = \{\alpha_1, \alpha_2, \ldots, \alpha_n\}$ and $T = \{\beta_1, \beta_2, \ldots, \beta_m\}$ be ordered bases for V and W, respectively. We define a function $M: U \rightarrow {}_mR_n$ by letting $M(L) =$ the matrix representing L with respect to the bases S and T. We now show that M is an isomorphism.

First, M is one-one, for if L_1 and L_2 are two different elements in U, then $L_1(\alpha_j) \neq L_2(\alpha_j)$ for some $j = 1, 2, \ldots, n$. This means that the jth columns of $M(L_1)$ and $M(L_2)$, which are the coordinate vectors of $L_1(\alpha_j)$ and $L_2(\alpha_j)$ with respect to T, are different, so $M(L_1) \neq M(L_2)$. Hence M is one-one.

Next, M is onto. Let $A = [a_{ij}]$ be a given $m \times n$ matrix; that is, A is an element of $_mR_n$. Then we define a function $L: V \rightarrow W$ by

$$L(\alpha_i) = \sum_{k=1}^{m} a_{ki}\beta_k, \qquad i = 1, 2, \ldots, n,$$

and if $\alpha = c_1\alpha_1 + c_2\alpha_2 + \cdots + c_n\alpha_n$, we define $L(\alpha)$ by

$$L(\alpha) = \sum_{i=1}^{n} c_i L(\alpha_i).$$

It is easy to show that L is a linear transformation; moreover, the matrix representing L with respect to S and T is $A = [a_{ij}]$ (verify). Thus $M(L) = A$, and so M is onto.

Now let $M(L_1) = A = [a_{ij}]$ and $M(L_2) = B = [b_{ij}]$. We show that $M(L_1 \boxplus L_2) = A + B$. First, note that the jth column of $M(L_1 \boxplus L_2)$ is $[(L_1 \boxplus L_2)(\alpha_j)]_T = [L_1(\alpha_j) + L_2(\alpha_j)]_T = [L_1(\alpha_j)]_T + [L_2(\alpha_j)]_T$. Thus the jth column of $M(L_1 \boxplus L_2)$ is the sum of the jth columns of $M(L_1) = A$ and $M(L_2) = B$. Hence $M(L_1 \boxplus L_2) = A + B$.

Finally, let $M(L) = A$, and $c = $ a real number. Following the idea in the above paragraph, we can show that $M(c \boxdot L) = cA$ (verify). Hence U and $_nR_m$ are isomorphic.

This theorem implies that the dimension of U is mn, for dim $_mR_n = mn$. Also, it means that when dealing with finite-dimensional vector spaces, we can always replace all linear transformations by their matrix representations and work only with the matrices. Moreover, it should be noted that matrices lend themselves much more readily than linear transformations to computer implementations.

EXAMPLE 3. Let $A = \begin{bmatrix} 1 & 2 & -1 \\ 2 & -1 & 3 \end{bmatrix}$. Let $S = \{\varepsilon_1, \varepsilon_2, \varepsilon_3\}$ and $T = \{\bar{\varepsilon}_1, \bar{\varepsilon}_2\}$ be the natural bases for R^3 and R^2, respectively. We now find the unique linear transformation $L: R^3 \to R^2$ whose representation with respect to S and T is A. Let

$$L(\varepsilon_1) = 1\bar{\varepsilon}_1 + 2\bar{\varepsilon}_2 = \begin{bmatrix} 1 \\ 2 \end{bmatrix},$$

$$L(\varepsilon_2) = 2\bar{\varepsilon}_1 - 1\bar{\varepsilon}_2 = \begin{bmatrix} 2 \\ -1 \end{bmatrix},$$

$$L(\varepsilon_3) = -\bar{\varepsilon}_1 + 3\bar{\varepsilon}_2 = \begin{bmatrix} -1 \\ 3 \end{bmatrix}.$$

Now if $\alpha = \begin{bmatrix} a_1 \\ a_2 \\ a_3 \end{bmatrix}$ is in R^3, we define $L(\alpha)$ by

$$L(\alpha) = L(a_1\varepsilon_1 + a_2\varepsilon_2 + a_3\varepsilon_3)$$

$$= a_1L(\varepsilon_1) + a_2L(\varepsilon_2) + a_3L(\varepsilon_3) = \begin{bmatrix} a_1 + 2a_2 - a_3 \\ 2a_1 - a_2 + 3a_3 \end{bmatrix}.$$

Note that $L(\alpha) = \begin{bmatrix} 1 & 2 & -1 \\ 2 & -1 & 3 \end{bmatrix} \begin{bmatrix} a_1 \\ a_2 \\ a_3 \end{bmatrix}$, so we could have defined L by $L(\alpha) = A\alpha$ for α in R^3. We can do this when the bases S and T are the natural bases.

Next, consider the ordered bases

$$S' = \left\{ \begin{bmatrix} 1 \\ 0 \\ 1 \end{bmatrix}, \begin{bmatrix} 1 \\ 1 \\ 0 \end{bmatrix}, \begin{bmatrix} 0 \\ 1 \\ 1 \end{bmatrix} \right\} \quad \text{and} \quad T' = \left\{ \begin{bmatrix} 1 \\ 3 \end{bmatrix}, \begin{bmatrix} 2 \\ -1 \end{bmatrix} \right\}$$

for R^3 and R^2, respectively. To determine $L: R^3 \to R^2$, we let

$$L\left(\begin{bmatrix} 1 \\ 0 \\ 1 \end{bmatrix}\right) = 1\begin{bmatrix} 1 \\ 3 \end{bmatrix} + 2\begin{bmatrix} 2 \\ -1 \end{bmatrix} = \begin{bmatrix} 5 \\ 1 \end{bmatrix},$$

$$L\left(\begin{bmatrix} 1 \\ 1 \\ 0 \end{bmatrix}\right) = 2\begin{bmatrix} 1 \\ 3 \end{bmatrix} - 1\begin{bmatrix} 2 \\ -1 \end{bmatrix} = \begin{bmatrix} 0 \\ 7 \end{bmatrix},$$

and

$$L\left(\begin{bmatrix} 0 \\ 1 \\ 1 \end{bmatrix}\right) = -1\begin{bmatrix} 1 \\ 3 \end{bmatrix} + 3\begin{bmatrix} 2 \\ -1 \end{bmatrix} = \begin{bmatrix} 5 \\ -6 \end{bmatrix}.$$

Then if $\alpha = \begin{bmatrix} a_1 \\ a_2 \\ a_3 \end{bmatrix}$, we express α in terms of the basis S as

$$\alpha = b_1 \begin{bmatrix} 1 \\ 0 \\ 1 \end{bmatrix} + b_2 \begin{bmatrix} 1 \\ 1 \\ 0 \end{bmatrix} + b_3 \begin{bmatrix} 0 \\ 1 \\ 1 \end{bmatrix},$$

and we define $L(\alpha)$ by

$$L(\alpha) = b_1 L\left(\begin{bmatrix} 1 \\ 0 \\ 1 \end{bmatrix}\right) + b_2 L\left(\begin{bmatrix} 1 \\ 1 \\ 0 \end{bmatrix}\right) + b_3 L\left(\begin{bmatrix} 0 \\ 1 \\ 1 \end{bmatrix}\right) = \begin{bmatrix} 5b_1 + 5b_3 \\ b_1 + 7b_2 - 6b_3 \end{bmatrix}.$$

Thus to find $L\left(\begin{bmatrix} 1 \\ 2 \\ 3 \end{bmatrix}\right)$ we first have (verify)

$$\begin{bmatrix} 1 \\ 2 \\ 3 \end{bmatrix} = 1\begin{bmatrix} 1 \\ 0 \\ 1 \end{bmatrix} + 0\begin{bmatrix} 1 \\ 1 \\ 0 \end{bmatrix} + 2\begin{bmatrix} 0 \\ 1 \\ 1 \end{bmatrix}.$$

Then

$$L\left(\begin{bmatrix} 1 \\ 2 \\ 3 \end{bmatrix}\right) = \begin{bmatrix} 15 \\ -11 \end{bmatrix}.$$

Now let V_1 be an n-dimensional vector space, V_2 an m-dimensional vector space, and V_3 a p-dimensional vector space. Let $L_1: V_1 \to V_2$ and $L_2: V_2 \to V_3$ be linear transformations. We can define the **composite** function $L_2 \circ L_1: V_1 \to V_3$ by $(L_2 \circ L_1)(\alpha) = L_2(L_1(\alpha))$ for α in V_1. Then it is easy to see that $L_2 \circ L_1$ is a linear transformation. If $L: V \to V$, then $L \circ L$ is written as L^2.

EXAMPLE 4. Let $L_1: R^2 \to R^2$ and $L_2: R^2 \to R^2$ be defined by

$$L_1\left(\begin{bmatrix} a_1 \\ a_2 \end{bmatrix}\right) = \begin{bmatrix} a_1 \\ -a_2 \end{bmatrix} \quad \text{and} \quad L_2\left(\begin{bmatrix} a_1 \\ a_2 \end{bmatrix}\right) = \begin{bmatrix} a_2 \\ a_1 \end{bmatrix}.$$

Then

$$(L_1 \circ L_2)\left(\begin{bmatrix} a_1 \\ a_2 \end{bmatrix}\right) = L_1\left(\begin{bmatrix} a_2 \\ a_1 \end{bmatrix}\right) = \begin{bmatrix} a_2 \\ -a_1 \end{bmatrix},$$

while

$$(L_2 \circ L_1)\left(\begin{bmatrix} a_1 \\ a_2 \end{bmatrix}\right) = L_2\left(\begin{bmatrix} a_1 \\ -a_2 \end{bmatrix}\right) = \begin{bmatrix} -a_2 \\ a_1 \end{bmatrix}.$$

Thus $L_1 \circ L_2 \neq L_2 \circ L_1$.

It is not difficult to show that if ordered bases P, S, and T are chosen for V_1, V_2, and V_3, respectively, then $M(L_2 \circ L_1) = M(L_2)M(L_1)$. Since AB need not equal BA for matrices A and B, it is thus not surprising that $L_1 \circ L_2$ may not be the same linear transformation as $L_2 \circ L_1$.

If $L: V \to V$ is an invertible linear operator and if A is a representation of L with respect to an ordered basis S for V, then the representation of the identity operator $L \circ L^{-1}$ is the identity matrix I_n. Thus $M(L)M(L^{-1}) = I_n$, which means that A^{-1} is the representation of L^{-1} with respect to S.

4.4. Exercises

1. Let L_1, L_2, and L_3 be linear transformations of V into W. Prove the following:
   (a) $L_1 \boxplus L_2$ is a linear transformation of V into W.
   (b) If c is a real number, then $c \boxdot L$ is a linear transformation of V into W.
   (c) If A represents L with respect to the ordered bases S and T for V and W,

respectively, then cA represents $c \boxdot L$ with respect to S and T, where c = a real number.

2. Let U be the set of all linear transformations of V into W, and let $O: V \to W$ be the zero linear transformation defined by $O(\alpha) = \theta_W$ for α in V.

   (a) Prove that $O \boxplus L = L \boxplus O = L$ for any L in U.

   (b) Show that if L is in U, then $L \boxplus ((-1) \boxdot L) = O$.

3. Verify that the set U of all linear transformations of V into W is a vector space under the operations $\boxplus$ and $\boxdot$.

4. Let $L_1: R_3 \to R_3$ and $L_2: R_3 \to R_3$ be linear transformations such that $L_1(\varepsilon_1') = [-1 \quad 2 \quad 1]$, $L_1(\varepsilon_2') = [0 \quad 1 \quad 2]$, $L_1(\varepsilon_3') = [-1 \quad 1 \quad 3]$, and $L_2(\varepsilon_1') = [0 \quad 1 \quad 3]$, $L_2(\varepsilon_2') = [4 \quad -2 \quad 1]$, $L_2(\varepsilon_3') = [0 \quad 2 \quad 2]$, where $S = \{\varepsilon_1', \varepsilon_2', \varepsilon_3'\}$ is the natural basis for R_3. Find:

   (a) $(L_1 \boxplus L_2)([a_1 \quad a_2 \quad a_3])$.   (b) $(L_1 \boxplus L_2)([2 \quad 1 \quad -3])$.

   (c) The representation of $L_1 \boxplus L_2$ with respect to S.

   (d) $(-2 \boxdot L_1)([a_1 \quad a_2 \quad a_3])$.

   (e) $(-2 \boxdot L_1 \boxplus 4 \boxdot L_2)([2 \quad 1 \quad -3])$.

   (f) The representation of $(-2 \boxdot L_1 \boxplus 4 \boxdot L_2)$ with respect to S.

5. Let V_1, V_2, and V_3 be vector spaces of dimensions n, m, and p, respectively. Let $L_1: V_1 \to V_2$ and $L_2: V_2 \to V_3$ be linear transformations. Prove that $L_2 \circ L_1: V_1 \to V_3$ is a linear transformation.

6. Let V_1, V_2, and V_3 be vector spaces of dimensions n, m, and p, respectively. Let $L_1: V_1 \to V_2$ and $L_2: V_2 \to V_3$ be linear transformations. Let P, S, and T be ordered bases for V_1, V_2, and V_3, respectively. If A is the representation of L_1 with respect to P and S, and B is the representation of L_2 with respect to S and T, prove that BA is the representation of $L_2 \circ L_1$ with respect to P and T.

7. Let L_1, L_2, and S be as in Exercise 4. Find

   (a) $(L_1 \circ L_2)([a_1 \quad a_2 \quad a_3])$.   (b) $(L_2 \circ L_1)([a_1 \quad a_2 \quad a_3])$.

   (c) The representation of $L_1 \circ L_2$ with respect to S.

   (d) The representation of $L_2 \circ L_1$ with respect to S.

8. If $\begin{bmatrix} 1 & 4 & -1 \\ 2 & 1 & 3 \\ 1 & -1 & 2 \end{bmatrix}$ is the representation of a linear operator $L: R^3 \to R^3$ with respect to ordered bases S and T for R^3, find the representation with respect to S and T of

   (a) $2 \boxdot L$.

   (b) $2 \boxdot L \boxplus 5 \boxdot I$, where $I: R^3 \to R^3$ is the identity operator on R^3.

9. Let L_1, L_2, and L_3 be linear transformations of R_3 into R_2 defined by $L_1([a_1 \quad a_2 \quad a_3]) = [a_1 + a_2 \quad a_1 - a_3]$, $L_2([a_1 \quad a_2 \quad a_3]) = [a_1 - a_2 \quad a_3]$, and $L_3([a_1 \quad a_2 \quad a_3]) = [a_1 \quad a_2 + a_3]$. Prove that $S = \{L_1, L_2, L_3\}$ is a linearly independent set in the vector space U of all linear transformations of R_3 into R_2.

10. Find the dimension of the vector space U of all linear transformations of V into W for each of the following:

    (a) $V = R^2$, $W = R^3$.   (b) $V = P_2$, $W = P_1$.

    (c) $V = {}_2R_1$, $W = {}_3R_2$.   (d) $V = R_3$, $W = R_4$.

11. Repeat Exercise 10 for each of the following:
    (a) $V = W$ is the vector space with basis $\{\sin t, \cos t\}$.
    (b) $V = W$ is the vector space with basis $\{1, t, e^t, te^t\}$.
    (c) V is the vector space *spanned* by $\{1, t, 2t\}$, and W is the vector space with basis $\{t^2, t, 1\}$.

12. Let $A = [a_{ij}]$ be a given $m \times n$ matrix, and let V and W be given vector spaces of dimensions n and m, respectively. Let $S = \{\alpha_1, \alpha_2, \ldots, \alpha_n\}$ be an ordered basis for V and $T = \{\beta_1, \beta_2, \ldots, \beta_m\}$ be an ordered basis for W. Define a function $L: V \to W$ by $L(\alpha_i) = \sum_{k=1}^{m} a_{ki}\beta_k$, $i = 1, 2, \ldots, n$, and if $\alpha = c_1\alpha_1 + c_2\alpha_2 + \cdots + c_n\alpha_n$, we define $L(\alpha)$ by $L(\alpha) = \sum_{i=1}^{n} c_i L(\alpha_i)$.
    (a) Show that L is a linear transformation.
    (b) Show that A represents L with respect to S and T.

13. Let $A = \begin{bmatrix} 1 & 2 & -2 \\ 3 & 4 & -1 \end{bmatrix}$. Let S be the natural basis for R^3 and T the natural basis for R^2.
    (a) Find the linear transformation $L: R^3 \to R^2$ determined by A.
    (b) Find $L\left(\begin{bmatrix} a_1 \\ a_2 \\ a_3 \end{bmatrix}\right)$.   (c) Find $L\left(\begin{bmatrix} 1 \\ 2 \\ 3 \end{bmatrix}\right)$.

14. Let A be as in Exercise 13. Consider the ordered bases $S = \{t^2, t, 1\}$ and $T = \{t, 1\}$ for P_2 and P_1, respectively.
    (a) Find the linear transformation $L: P_2 \to P_1$ determined by A.
    (b) Find $L(at^2 + bt + c)$.   (c) Find $L(2t^2 - 5t + 4)$.

15. Find two linear transformations $L_1: R^2 \to R^2$ and $L_2: R^2 \to R^2$ such that $L_2 \circ L_1 \neq L_1 \circ L_2$.

16. Find a linear transformation $L: R^2 \to R^2$, $L \neq I$, the identity operator, such that $L^2 = L \circ L = I$.

17. Find a linear transformation $L: R^2 \to R^2$, $L \neq O$, the zero transformation, such that $L^2 = O$.

18. Find a linear transformation $L: R^2 \to R^2$, $L \neq I$, $L \neq O$, such that $L^2 = L$.

19. Let $L: R^3 \to R^3$ be the linear transformation defined in Exercise 12 of Section 4.2. Find the matrix representing L^{-1} with respect to the natural basis for R^3.

20. Let $L: R^3 \to R^3$ be the linear transformation defined in Exercise 16 of Section 4.2. Find the matrix representing L^{-1} with respect to the natural basis for R^3.

21. Let $L: R^3 \to R^3$ be the invertible linear transformation represented by
$$A = \begin{bmatrix} 2 & 0 & 4 \\ -1 & 1 & -2 \\ 2 & 3 & 3 \end{bmatrix}$$
with respect to an ordered basis S for R^3. Find the representation of L^{-1} with respect to S.

22. Let $L: V \to V$ be a linear transformation represented by a matrix A with respect to an ordered basis S for V. Show that A^2 represents $L^2 = L \circ L$ with

respect to S. Moreover, show that if k is a positive integer, then A^k represents $L^k = L \circ L \circ \cdots \circ L$ (k times) with respect to S.

23. Let $L: P_1 \to P_2$ be the invertible linear transformation represented by $A = \begin{bmatrix} 4 & 2 \\ 3 & -1 \end{bmatrix}$ with respect to an ordered basis S for P_1. Find the representation of L^{-1} with respect to S.

4.5. Similarity

In Section 4.3 we saw how the matrix representing a linear transformation of an n-dimensional vector space V into an m-dimensional vector space W depends upon the ordered bases we choose for V and W. We now see how this matrix changes when the bases for V and W are changed.

Theorem 4.10. *Let $L: V \to W$ be a linear transformation of an n-dimensional vector space V into an m-dimensional vector space W. Let $S = \{\alpha_1, \alpha_2, \ldots, \alpha_n\}$ and $S' = \{\alpha'_1, \alpha'_2, \ldots, \alpha'_m\}$ be ordered bases for V, with transition matrix P from S' to S; let $T = \{\beta_1, \beta_2, \ldots, \beta_m\}$ and $T' = \{\beta'_1, \beta'_2, \ldots, \beta'_m\}$ be ordered bases for W with transition matrix Q from T' to T. If A is the representation of L with respect to S and T, then $Q^{-1}AP$ is the representation of L with respect to S' and T'.*

Proof: Recalling Section 2.3, where the transition matrix was first introduced, if P is the transition matrix from S' to S, then

$$[\alpha]_S = P[\alpha]_{S'}. \tag{1}$$

The jth column of P is the coordinate vector $[\alpha_j]_S$ of α'_j with respect to S. Similarly, if Q is the transition matrix from T' to T, then

$$[\beta]_T = Q[\beta]_{T'} \tag{2}$$

and the jth column of Q is the coordinate vector $[\beta'_j]_T$ of β'_j with respect to T. Now if A is representation of L with respect to S and T, then

$$[L(\alpha)]_T = A[\alpha]_S, \tag{3}$$

for α in V. Substituting $\beta = L(\alpha)$ in (2), we have $[L(\alpha)]_T = Q[L(\alpha)]_{T'}$. Now using first (3) and then (1) in this last equation, we obtain $Q[L(\alpha)]_{T'} = AP[\alpha]_{S'}$ so $[L(\alpha)]_{T'} = Q^{-1}AP[\alpha]_{S'}$. This means that $Q^{-1}AP$ is the representation of L with respect to S' and T'.

From Section 1.7 we see that two representations of a linear transformation with respect to different pairs of bases are equivalent.

EXAMPLE 1. Let $L: R^3 \to R^2$ be defined by $L\left(\begin{bmatrix} a_1 \\ a_2 \\ a_3 \end{bmatrix}\right) = \begin{bmatrix} a_1 + a_3 \\ a_2 - a_3 \end{bmatrix}$.

Consider the ordered bases

$$S = \left\{ \begin{bmatrix} 1 \\ 0 \\ 0 \end{bmatrix}, \begin{bmatrix} 0 \\ 1 \\ 0 \end{bmatrix}, \begin{bmatrix} 0 \\ 0 \\ 1 \end{bmatrix} \right\} \quad \text{and} \quad S' = \left\{ \begin{bmatrix} 1 \\ 1 \\ 0 \end{bmatrix}, \begin{bmatrix} 0 \\ 1 \\ 1 \end{bmatrix}, \begin{bmatrix} 0 \\ 0 \\ 1 \end{bmatrix} \right\}$$

for R^3, and

$$T = \left\{ \begin{bmatrix} 1 \\ 0 \end{bmatrix}, \begin{bmatrix} 0 \\ 1 \end{bmatrix} \right\} \quad \text{and} \quad T' = \left\{ \begin{bmatrix} 1 \\ 1 \end{bmatrix}, \begin{bmatrix} 1 \\ 3 \end{bmatrix} \right\}$$

for R^2. We can easily establish that $A = \begin{bmatrix} 1 & 0 & 1 \\ 0 & 1 & -1 \end{bmatrix}$ is the representation of L with respect to S and T.

The transition matrix P from S' to S is the matrix whose jth column is the coordinate vector of the jth vector in the basis S' with respect to S. Thus

$$P = \begin{bmatrix} 1 & 0 & 0 \\ 1 & 1 & 0 \\ 0 & 1 & 1 \end{bmatrix} \text{ and the transition matrix } Q \text{ from } T' \text{ to } T \text{ is } Q = \begin{bmatrix} 1 & 1 \\ 1 & 3 \end{bmatrix}.$$

Now $Q^{-1} = \begin{bmatrix} \frac{3}{2} & -\frac{1}{2} \\ -\frac{1}{2} & \frac{1}{2} \end{bmatrix}$. (We could also obtain Q^{-1} as the transition matrix from T to T'.) Then the representation of L with respect to S' and T' is

$$B = Q^{-1}AP = \begin{bmatrix} 1 & \frac{3}{2} & 2 \\ 0 & -\frac{1}{2} & -1 \end{bmatrix}.$$

On the other hand, we can compute the representation of L with respect to S' and T' directly. We have

$$L\left(\begin{bmatrix} 1 \\ 1 \\ 0 \end{bmatrix}\right) = \begin{bmatrix} 1 \\ 1 \end{bmatrix} = 1\begin{bmatrix} 1 \\ 1 \end{bmatrix} + 0\begin{bmatrix} 1 \\ 3 \end{bmatrix}, \qquad \text{so} \qquad \left[L\left(\begin{bmatrix} 1 \\ 1 \\ 0 \end{bmatrix}\right) \right]_{T'} = \begin{bmatrix} 1 \\ 0 \end{bmatrix}$$

$$L\left(\begin{bmatrix} 0 \\ 1 \\ 1 \end{bmatrix}\right) = \begin{bmatrix} 1 \\ 0 \end{bmatrix} = \frac{3}{2}\begin{bmatrix} 1 \\ 1 \end{bmatrix} - \frac{1}{2}\begin{bmatrix} 1 \\ 3 \end{bmatrix}, \qquad \text{so} \quad \left[L\left(\begin{bmatrix} 0 \\ 1 \\ 1 \end{bmatrix}\right)\right]_{T'} = \begin{bmatrix} \frac{3}{3} \\ -\frac{1}{2} \end{bmatrix}$$

$$L\left(\begin{bmatrix} 0 \\ 0 \\ 1 \end{bmatrix}\right) = \begin{bmatrix} 1 \\ -1 \end{bmatrix} = 2\begin{bmatrix} 1 \\ 1 \end{bmatrix} - 1\begin{bmatrix} 1 \\ 3 \end{bmatrix}, \qquad \text{so} \quad \left[L\left(\begin{bmatrix} 0 \\ 0 \\ 1 \end{bmatrix}\right)\right]_{T'} = \begin{bmatrix} 2 \\ -1 \end{bmatrix}.$$

Then the representation of L with respect to S' and T' is

$$\begin{bmatrix} 1 & \frac{3}{2} & 2 \\ 0 & -\frac{1}{2} & -1 \end{bmatrix}.$$

We may define the rank of a linear transformation $L: V \to W$, rank L, as the rank of any matrix representing L. This definition makes sense, since if A and B represent L, then A and B are equivalent; by Section 1.7 we know that equivalent matrices have the same rank. We can now restate Theorem 4.5 as follows. If $L: V \to W$ is a linear transformation, then nullity L + rank L = dim V. We also speak of the **nullity** of an $m \times n$ matrix A. This is the nullity of the linear transformation $L: R^n \to R^m$ defined by $L(\alpha) = A\alpha$, for α in R^n (see Section 4.2). That is, nullity A = dim ker L.

Taking $V = W$ in Theorem 4.10 we obtain an important result, which we state as Corollary 4.2.

Corollary 4.2. *Let $L: V \to V$ be a linear operator on an n-dimensional vector space V. Let $S = \{\alpha_1, \alpha_2, \ldots, \alpha_n\}$ and $S' = \{\alpha_1', \alpha_2', \ldots, \alpha_n'\}$ be ordered bases for V with transition matrix P from S' to S. If A is the representation of L with respect to S, then $P^{-1}AP$ is the representation of L with respect to S'.*

Definition 4.6. If A and B are $n \times n$ matrices, we say that B is **similar** to A if there is a nonsingular matrix P such that $B = P^{-1}AP$.

It is easy to show (Exercise 1) that B is similar to A if and only if A is similar to B. Thus we replace the statements "A is similar to B" and "B is similar to A" by "A and B are similar."

We thus see that any two representations of a linear transformation $L: V \to V$ are similar. Conversely, let $A = [a_{ij}]$ and $B = [b_{ij}]$ be similar $n \times n$ matrices and let V be an n-dimensional vector space (we may take V as R^n). We wish to show that A and B represent the same linear transformation $L: V \to V$ with respect to different bases. Since A and B are similar, $B =$

$P^{-1}AP$ for some nonsingular matrix $P = [p_{ij}]$. Let $S = \{\alpha_1, \alpha_2, \ldots, \alpha_n\}$ be an ordered basis for V; from the proof of Theorem 4.9, we know that there exists a linear transformation $L: V \to V$ which is represented by A with respect to S. Then

$$[L(\alpha)]_S = A[\alpha]_S. \tag{4}$$

We wish to prove that B also represents L with respect to some basis for V. Let

$$\beta_i = \sum_{k=1}^{n} p_{ki}\alpha_k. \tag{5}$$

We first show that $T = \{\beta_1, \beta_2, \ldots, \beta_n\}$ is also a basis for V. Suppose that

$$a_1\beta_1 + a_2\beta_2 + \cdots + a_n\beta_n = \theta.$$

Then, from (5), we have

$$a_1\left(\sum_{k=1}^{n} p_{k1}\alpha_k\right) + a_2\left(\sum_{k=1}^{n} p_{k2}\alpha_k\right) + \cdots + a_n\left(\sum_{k=1}^{n} p_{kn}\alpha_k\right) = \theta,$$

which can be rewritten as

$$\left(\sum_{s=1}^{n} p_{1s}a_s\right)\alpha_1 + \left(\sum_{s=1}^{n} p_{2s}a_s\right)\alpha_2 + \cdots + \left(\sum_{s=1}^{n} p_{ns}a_s\right)\alpha_n = \theta. \tag{6}$$

Since S is linearly independent, each of the coefficients in (6) is zero. Thus

$$\sum_{s=1}^{n} p_{rs}a_s = 0 \qquad r = 1, 2, \ldots, n.$$

This is a homogeneous system of n equations in the n unknowns $a_1, a_2, \ldots, a_n$, whose coefficient matrix is P. Since P is nonsingular, the only solution is the trivial one. Hence, $a_1 = a_2 = \cdots = a_n = 0$ and T is linearly independent. Moreover, Equation (5) implies that P is the transition matrix from T to S (see Section 2.3). Thus

$$[\alpha]_S = P[\alpha]_T. \tag{7}$$

Using (7), Equation (4), the matrix equation of L, becomes

$$P[L(\alpha)]_T = AP[\alpha]_T \quad \text{or} \quad [L(\alpha)]_T = P^{-1}AP[\alpha]_T,$$

which means that the representation of L with respect to T is $B = P^{-1}AP$. We can summarize these results in the following theorem.

Theorem 4.11. *Let V be any n-dimensional vector space and let A and B be any $n \times n$ matrices. Then A and B are similar if and only if A and B represent the same linear transformation $L: V \to V$ with respect to different ordered bases.*

EXAMPLE 2. Let $L: R_3 \to R_3$ be defined by

$$L([a_1 \quad a_2 \quad a_3]) = [2a_1 - a_3 \quad a_1 + a_2 - a_3 \quad a_3].$$

Let $S = \{[1 \quad 0 \quad 0], [0 \quad 1 \quad 0], [0 \quad 0 \quad 1]\}$ be the natural basis for R_3. The representation of L with respect to S is $A = \begin{bmatrix} 2 & 0 & -1 \\ 1 & 1 & -1 \\ 0 & 0 & 1 \end{bmatrix}$. Now consider the ordered basis

$$S' = \{[1 \quad 0 \quad 1], [0 \quad 1 \quad 0], [1 \quad 1 \quad 0]\}$$

for R_3. The transition matrix P from S' to S is $P = \begin{bmatrix} 1 & 0 & 1 \\ 0 & 1 & 1 \\ 1 & 0 & 0 \end{bmatrix}$; moreover,

$$P^{-1} = \begin{bmatrix} 0 & 0 & 1 \\ -1 & 1 & 1 \\ 1 & 0 & -1 \end{bmatrix}.$$

Then the representation of L with respect to S' is

$$B = P^{-1}AP = \begin{bmatrix} 1 & 0 & 0 \\ 0 & 1 & 0 \\ 0 & 0 & 2 \end{bmatrix}.$$

The same result can be obtained directly (verify). The matrices A and B are similar.

Observe that the matrix B obtained in Example 2 is diagonal. We can now ask a number of related questions. First, given $L: V \to V$, when can we choose a basis S for V such that the representation of L with respect to S is diagonal? How do we choose such a basis? In Example 2 we apparently pulled our basis S' "out of the air." If we cannot choose a basis giving a representation of L which is diagonal, can we choose a basis giving a matrix that is close in appearance to a diagonal matrix? What do we gain from having such simple representations? First, we already know from Section 4.4 that if A represents $L: V \to V$ with respect to some ordered basis S for V, then A^k represents $L \circ L \circ \cdots \circ L = L^k$ with respect to S; now if A is similar to B, then B^k also represents L^k. Of course, if B is diagonal, then it is a trivial matter to compute B^k: the diagonal elements of B^k are those of B raised to the k power. We shall also find that if A is similar to a diagonal matrix, then we can easily solve a homogeneous linear system of differential equations with constant coefficients. The answers to these questions will be taken up in detail in Chapter 6.

Similar matrices enjoy some other nice properties. For example, if A and B are similar, then Tr (A) = Tr (B) (see Exercise 12 in Section 1.2 for a definition of trace). Also, if A and B are similar, then A^k and B^k are similar, if k is a positive integer. Proofs of these results are not difficult to supply and are left as exercises.

We obtain one final result on similar matrices.

Theorem 4.12. *If A and B are similar $n \times n$ matrices, then rank A = rank B.*

Proof: We know from Theorem 4.11 that A and B represent the same linear transformation $L: R^n \to R^n$ with respect to different bases. Now we have range L = column space of A (Exercise 8 in Section 4.2) and range L = column space of B. Thus dim range L = rank A and dim range L = rank B, so that rank A = rank B.

We might also note that Theorem 4.12 could be established by merely noting that if A and B are similar, then they are equivalent; we have already proved in Section 2.4 that A and B are equivalent if and only if they have equal ranks. However, the proof given above illustrates the notions of similarity discussed in the present section.

4.5. Exercises

1. Let A, B, and C be square matrices. Show that
   (a) A is similar to A.
   (b) If A is similar to B, then B is similar to A.
   (c) If A is similar to B and B is similar to C, then A is similar to C.
2. Let L be the linear transformation defined in Exercise 2, Section 4.3.
   (a) Find the transition matrix P from S' to S.
   (b) Find the transition matrix from S to S' and verify that it is P^{-1}.
   (c) Find the transition matrix Q from T' to T.
   (d) Find the representation of L with respect to S' and T'.
   (e) What is the dimension of range L?

3. Do Exercise 1(d) of Section 4.3 using transition matrices.
4. Do Exercise 8(b) of Section 4.3 using transition matrices.
5. Let $L: R^2 \rightarrow R^2$ be defined by

$$L\left(\begin{bmatrix} a_1 \\ a_2 \end{bmatrix}\right) = \begin{bmatrix} a_1 \\ -a_2 \end{bmatrix}.$$

  (a) Find the representation of L with respect to the natural basis S for R^2.
  (b) Find the representation of L with respect to the ordered basis

$$T = \left\{ \begin{bmatrix} 1 \\ -1 \end{bmatrix}, \begin{bmatrix} 1 \\ 2 \end{bmatrix} \right\}.$$

  (c) Verify that the matrices obtained in (a) and (b) are similar.
  (d) Verify that the ranks of the matrices obtained in (a) and (b) are equal.
6. Show that if A and B are similar matrices, then A^k and B^k are similar for any positive integer k. (*Hint:* If $B = P^{-1}AP$, find $B^2 = BB$, and so on.)
7. Show that if A and B are similar, then A' and B' are similar.
8. Prove that if A and B are similar, then Tr $(A) =$ Tr (B). (*Hint:* See Exercise 12 in Section 1.2 for a definition of trace.)
9. Let $L: R_3 \rightarrow R_2$ be the linear transformation whose representation is

$A = \begin{bmatrix} 2 & -1 & 3 \\ 3 & 1 & 0 \end{bmatrix}$ with respect to the ordered bases

$S = \{[1 \quad 0 \quad -1], [0 \quad 2 \quad 0], [1 \quad 2 \quad 3]\}$ and $T = \{[1 \quad -1], [2 \quad 0]\}$.

Find the representation of L with respect to the natural bases for R_3 and R_2.
10. Let $L: R^3 \rightarrow R^3$ be the linear transformation whose representation with respect to the natural basis for R^3 is $A = [a_{ij}]$. Let $P = \begin{bmatrix} 0 & 1 & 1 \\ 1 & 0 & 1 \\ 1 & 1 & 0 \end{bmatrix}$ be a nonsingular matrix. Find a basis T for R^3 with respect to which $B = P^{-1}AP$ represents L. (*Hint:* See the discussion following Definition 4.6.)
11. Let A and B be similar. Prove:
  (a) If A is nonsingular, then B is nonsingular.
  (b) If A is nonsingular, then A^{-1} and B^{-1} are similar.
12. Do Exercise 13(b) of Section 4.3 using transition matrices.
13. Do Exercise 17(b) of Section 4.3 using transition matrices.
14. Do Exercise 10(b) of Section 4.3 using transition matrices.
15. Do Exercise 20(b) of Section 4.3 using transition matrices.
16. Prove that A and O_n are similar if and only if $A = O_n$.

CHAPTER 5

Determinants

5.1. Definition

In Exercise 12, Section 1.2, we defined the trace of a square ($n \times n$) matrix $A = [a_{ij}]$ by $\text{Tr}(A) = \sum_{i=1}^{n} a_{ii}$. Another very important number associated with a square matrix A is the determinant of A, which we now define. Determinants first arose in the solution of linear systems. Although the methods given in Chapter 1 for solving such systems are more efficient than those involving determinants, determinants will be vital for our further study of a linear transformation $L: V \rightarrow V$. First, we deal briefly with permutations, which are used in our definition of determinant. Throughout this chapter, when we use the term "matrix" we mean "square matrix."

Definition 5.1. Let $S = \{1, 2, \ldots, n\}$ be the set of integers from 1 to n, arranged in ascending order. A rearrangement $j_1 j_2 \cdots j_n$ of the elements of S is called a **permutation** of S. Thus a permutation of S is a one-one mapping of S onto itself.

We can put any one of the n elements of S in first position, any one of the remaining $n - 1$ elements in second position, any one of the remaining $n - 2$

elements in third position, and so on until the nth position can only be filled by the last remaining element. Thus there are $n(n - 1)(n - 2)\cdots2\cdot1 = n!$ permutations of S; we denote the set of all permutations of S by S_n.

EXAMPLE 1. Let $S = \{1, 2, 3\}$. The set S_3 of all permutations of S consists of the $3! = 6$ permutations 123, 132, 213, 231, 312, and 321.

A permutation $j_1 j_2 \cdots j_n$ of S is said to have an **inversion** if a larger integer, j_s, precedes a smaller one, j_r. A permutation is called **even** if the total number of inversions in it is even, or **odd** if the total number of inversions in it is odd. If $n \geq 2$, there are $n!/2$ even and $n!/2$ odd permutations in S_n.

EXAMPLE 2. S_1 has only $1! = 1$ permutation: 1, which is even because there are no inversions.

EXAMPLE 3. S_2 has $2! = 2$ permutations: 12, which is even (no inversions), and 21, which is odd (one inversion).

EXAMPLE 4. In the permutation 4312 in S_4, 4 precedes 3, 4 precedes 1, 4 precedes 2, 3 precedes 1, and 3 precedes 2. Thus the total number of inversions in this permutation is 5, and 4312 is odd.

EXAMPLE 5. S_3 has $3! = 3\cdot2\cdot1 = 6$ permutations: 123, 231, and 312, which are even, and 132, 213, and 321, which are odd.

Definition 5.2. Let $A = [a_{ij}]$ be an $n \times n$ matrix. We define the **determinant** $|A|$ of A by

$$|A| = \sum (\pm)a_{1j_1}a_{2j_2}\cdots a_{nj_n},$$

where the summation is over all permutations $j_1 j_2 \cdots j_n$ of the set $S = \{1, 2, \ldots, n\}$. The sign is taken as $+$ or $-$ according as the permutation $j_1 j_2 \cdots j_n$ is even or odd.

In each term $(\pm)a_{1j_1}a_{2j_2}\cdots a_{nj_n}$ of $|A|$, the row subscripts are in natural order while the column subscripts are in the order $j_1 j_2 \cdots j_n$. Thus each term in $|A|$ is a product of n entries of A, each with its appropriate sign, with exactly one entry from each row and exactly one entry from each column. Since we sum over all permutations of S, $|A|$ has $n!$ terms in the sum.

EXAMPLE 6. If $A = [a_{11}]$ is a 1×1 matrix, then $|A| = a_{11}$.

EXAMPLE 7. If $A = \begin{bmatrix} a_{11} & a_{12} \\ a_{21} & a_{22} \end{bmatrix}$, then to obtain $|A|$ we write down the terms $a_{1-}a_{2-}$ and $a_{1-}a_{2-}$, and replace the dashes with all possible elements of S_2: These are 12 and 21. Now 12 is an even permutation and 21 is an odd permutation. Thus

$$|A| = a_{11}a_{22} - a_{12}a_{21}.$$

We can also obtain $|A|$ by forming the product of the entries on the line from left to right and subtracting from this number the product of the entries on the line from right to left:

$$\begin{matrix} a_{11} & a_{12} \\ a_{21} & a_{22}. \end{matrix}$$

Thus, if $A = \begin{bmatrix} 2 & -3 \\ 4 & 5 \end{bmatrix}$, then $|A| = (2)(5) - (-3)(4) = 22$.

EXAMPLE 8. If $A = \begin{bmatrix} a_{11} & a_{12} & a_{13} \\ a_{21} & a_{22} & a_{23} \\ a_{31} & a_{32} & a_{33} \end{bmatrix}$, then to compute $|A|$ we write down the six terms $a_{1-}a_{2-}a_{3-}, a_{1-}a_{2-}a_{3-}, a_{1-}a_{2-}a_{3-}, a_{1-}a_{2-}a_{3-}, a_{1-}a_{2-}a_{3-}, a_{1-}a_{2-}a_{3-}$. All the elements of S_3 are used to replace the dashes, and if we prefix each term by $+$ or $-$ according as the permutation is even or odd, we find that

$$|A| = a_{11}a_{22}a_{33} + a_{12}a_{23}a_{31} + a_{13}a_{21}a_{32} - a_{11}a_{23}a_{32} - a_{12}a_{21}a_{33} - a_{13}a_{22}a_{31}.$$

$$(1)$$

We can also obtain $|A|$ as follows. Repeat the first and second columns of A as shown below. Form the sum of the products of the entries on the lines from left to right, and subtract from this number the products of the entries on the lines from right to left (verify):

$$\begin{matrix} a_{11} & a_{12} & a_{13} & a_{11} & a_{12} \\ a_{21} & a_{22} & a_{23} & a_{21} & a_{22} \\ a_{31} & a_{32} & a_{33} & a_{31} & a_{32} \end{matrix}$$

It should be emphasized that for $n \geq 4$, there is no "easy" method for evaluating $|A|$ as in Examples 7 and 8.

EXAMPLE 9. Let

$$A = \begin{bmatrix} 1 & 2 & 3 \\ 2 & 1 & 3 \\ 3 & 1 & 2 \end{bmatrix}.$$

Substituting in (1), we find that

$$|A| = (1)(1)(2) + (2)(3)(3) + (3)(2)(1) - (1)(3)(1)$$
$$- (2)(2)(2) - (3)(1)(3) = 6.$$

We could obtain the same result by using the method of lines illustrated above, as follows:

$$|A| = (1)(1)(2) + (2)(3)(3) + (3)(2)(1) - (3)(1)(3) - (1)(3)(1)$$
$$- (2)(2)(2) = 6.$$

It may already have struck the reader than this is an extremely tedious way of computing determinants. In fact, $10! = 3.6288 \times 10^6$ and $20! = 2.4329 \times 10^{18}$, each an enormous number. In Section 5.2 we shall develop properties of determinants that will greatly reduce the computational effort.

Permutations are studied to some depth in abstract algebra courses or in courses dealing with group theory. As we just noted, we shall develop methods for evaluating determinants other than those involving permutations. However, we do require the following important property of permutations. If we interchange two numbers in the permutation $j_1 j_2 \cdots j_n$, then the number of inversions is either increased or decreased by an odd number.

A proof of this fact can be given by first noting that if two adjacent numbers in the permutation $j_1 j_2 \cdots j_n$ are interchanged, then the number of inversions is either increased or decreased by 1. Thus consider the permutations $j_1 j_2 \cdots j_e j_f \cdots j_n$ and $j_1 j_2 \cdots j_f j_e \cdots j_n$. If $j_e j_f$ is an inversion, then $j_f j_e$ is not an inversion and the second permutation has one fewer inversion than the first one; if $j_e j_f$ is not an inversion, then $j_f j_e$ is, and so the second permutation has one more inversion than the first. Now an interchange of any two numbers in a permutation $j_1 j_2 \cdots j_n$ can always be achieved by an odd number of successive interchanges of adjacent numbers. Thus, if we wish to interchange j_c and $j_k (j_c < j_k)$ and there are s numbers between j_c and j_k, we move j_c to the right, by interchanging adjacent numbers, until j_c follows j_k. This requires $s + 1$

steps. Next, we move j_k to the left, by interchanging adjacent numbers until it is where j_c was. This requires s steps. Thus the total number of adjacent interchanges required is $(s + 1) + s = 2s + 1$, which is always odd. Since each adjacent interchange changes the number of inversions by 1 or -1, and since a sum of an odd number of numbers each of which is 1 or -1 is always odd, we conclude that the number of inversions is changed by an odd number. Thus the number of inversions in 54132 is 8 and the number of inversions in 52134 (obtained by interchanging 2 and 4) is 5.

5.1. Exercises

1. Find the number of inversions in each of the following permutations of $S = \{1, 2, 3, 4, 5\}$.
   (a) 52134.   (b) 45213.   (c) 42135.
   (d) 13542.   (e) 35241.   (f) 12345.

2. Determine whether each of the following permutations of $S = \{1, 2, 3, 4\}$ is even or odd.
   (a) 4213.   (b) 1243.   (c) 1234.   (d) 3214.   (e) 1423.

3. Determine the sign associated with each of the following permutations of the column indices in the expansion of the determinant of a 5×5 matrix.
   (a) 25431.   (b) 31245.   (c) 21345.   (d) 52341.   (e) 34125.

4. (a) Find the number of inversions in the permutation 436215.
   (b) Verify that the number of inversions in the permutation 416235, obtained from that in (a) by interchanging two numbers, differs from the answer in (a) by an odd number.

5. Evaluate:
   (a) $\begin{vmatrix} 2 & -1 \\ 3 & 2 \end{vmatrix}$.   (b) $\begin{vmatrix} 2 & 1 \\ 4 & 3 \end{vmatrix}$.   (c) $\begin{vmatrix} 1 & 2 \\ 2 & 4 \end{vmatrix}$.

6. Let $A = [a_{ij}]$ be a 4×4 matrix. Develop the general expression for $|A|$.

7. Evaluate:
   (a) $\begin{vmatrix} 2 & 1 & 3 \\ 3 & 2 & 1 \\ 0 & 1 & 2 \end{vmatrix}$.   (b) $\begin{vmatrix} 2 & 1 & 3 \\ -3 & 2 & 1 \\ -1 & 3 & 4 \end{vmatrix}$.   (c) $\begin{vmatrix} 0 & 0 & 0 & 3 \\ 0 & 0 & 4 & 0 \\ 0 & 2 & 0 & 0 \\ 6 & 0 & 0 & 0 \end{vmatrix}$.

8. Evaluate:
   (a) $\begin{vmatrix} 2 & 0 & 0 \\ 0 & -3 & 0 \\ 0 & 0 & 4 \end{vmatrix}$.   (b) $\begin{vmatrix} 2 & 4 & 5 \\ 0 & -6 & 2 \\ 0 & 0 & 3 \end{vmatrix}$.   (c) $\begin{vmatrix} 0 & 0 & 2 & 0 \\ 0 & 3 & 0 & 0 \\ 6 & 0 & 0 & 0 \\ 0 & 0 & 0 & 5 \end{vmatrix}$.

9. Evaluate:
   (a) $\begin{vmatrix} t-1 & 2 \\ 3 & t-2 \end{vmatrix}$.   (b) $\begin{vmatrix} t & 4 \\ 5 & t-8 \end{vmatrix}$.

10. Evaluate:

(a) $\begin{vmatrix} t - 1 & -1 & -2 \\ 0 & t - 2 & 2 \\ 0 & 0 & t - 3 \end{vmatrix}.$ (b) $\begin{vmatrix} t - 1 & 0 & 1 \\ -2 & t & -1 \\ 0 & 0 & t + 1 \end{vmatrix}.$

11. For each of the matrices in Exercise 9, find values of t for which the determinant is 0.

12. For each of the matrices in Exercise 10, find values of t for which the determinant is 0.

5.2. Properties of Determinants

In this section we examine properties of determinants that simplify their computation.

Theorem 5.1. *If A is a matrix, then $|A| = |A'|$.*

Proof: Let $A = [a_{ij}]$ and $A' = [b_{ij}]$, where $b_{ij} = a_{ji}$. We have

$$|A'| = \sum (\pm) b_{1j_1} b_{2j_2} \cdots b_{nj_n} = \sum (\pm) a_{j_1 1} a_{j_2 2} \cdots a_{j_n n}.$$

We can then write $b_{1j_1} b_{2j_2} \cdots b_{nj_n} = a_{j_1 1} a_{j_2 2} \cdots a_{j_n n} = a_{1k_1} a_{2k_2} \cdots a_{nk_n}$, which is a term of $|A|$. Thus the terms in $|A'|$ and $|A|$ are identical. We must now check that the signs of corresponding terms are also identical. It can be shown that the number of inversions in the permutation $k_1 k_2 \cdots k_n$, which determines the sign associated with the term $a_{1k_1} a_{2k_2} \cdots a_{nk_n}$, is the same as the number of inversions in the permutation $j_1 j_2 \cdots j_n$, which determines the sign associated with the term $b_{1j_1} b_{2j_2} \cdots b_{nj_n}$. (As an example,

$$b_{13} b_{24} b_{35} b_{41} b_{52} = a_{31} a_{42} a_{53} a_{14} a_{25} = a_{14} a_{25} a_{31} a_{42} a_{53};$$

the number of inversions in the permutation 45123 is 6 and the number of inversions in the permutation 34512 is also 6.) Since the signs of corresponding terms are identical, we conclude that $|A'| = |A|$.

EXAMPLE 1. Let A be the matrix in Example 9 of Section 5.1. Then

$$A' = \begin{bmatrix} 1 & 2 & 3 \\ 2 & 1 & 1 \\ 3 & 3 & 2 \end{bmatrix}.$$

Substituting in (1) of Section 5.1 (or using the method of lines given in Example 8 of Section 5.1), we find that

$$|A'| = (1)(1)(2) + (2)(1)(3) + (3)(2)(3)$$
$$- (1)(1)(3) - (2)(2)(2) - (3)(1)(3) = 6 = |A|.$$

Theorem 5.1 will enable us to replace "row" by "column" in many of the additional properties of determinants; we see how to do this in the following theorem.

Theorem 5.2. *If matrix B results from matrix A by interchanging two rows (columns) of A, then $|B| = -|A|$.*

Proof: Suppose that B arises from A by interchanging rows r and s of A, say $r < s$. Then we have $b_{rj} = a_{sj}$, $b_{sj} = a_{rj}$, and $b_{ij} = a_{ij}$ for $i \neq r$, $i \neq s$. Now

$$|B| = \sum (\pm) b_{1j_1} b_{2j_2} \cdots b_{rj_r} \cdots b_{sj_s} \cdots b_{nj_n}$$
$$= \sum (\pm) a_{1j_1} a_{2j_2} \cdots a_{sj_r} \cdots a_{rj_s} \cdots a_{nj_n}$$
$$= \sum (\pm) a_{1j_1} a_{2j_2} \cdots a_{rj_s} \cdots a_{sj_r} \cdots a_{nj_n}.$$

The permutation $j_1 j_2 \cdots j_s \cdots j_r \cdots j_n$ results from the permutation $j_1 j_2 \cdots j_r \cdots j_s \cdots j_n$ by an interchange of two numbers and the number of inversions in the former differs by an odd number from the number of inversions in the latter. This means that the sign of each term in $|B|$ is the negative of the corresponding term in $|A|$. Hence $|B| = -|A|$.

Now let B arise from A by interchanging two columns of A. Then B' arises from A' by interchanging two rows of A'. So $|B'| = -|A'|$, but $|B'| = |B|$ and $|A'| = |A|$. Hence $|B| = -|A|$.

In the results to follow, proofs will be given only for the rows of A; the proofs for the corresponding column cases proceed as at the end of the proof of Theorem 5.2.

EXAMPLE 2. We have $\begin{vmatrix} 2 & -1 \\ 3 & 2 \end{vmatrix} = -\begin{vmatrix} 3 & 2 \\ 2 & -1 \end{vmatrix} = \begin{vmatrix} 2 & 3 \\ -1 & 2 \end{vmatrix} = 7.$

Theorem 5.3. *If two rows (columns) of A are equal, then $|A| = 0$.*

Proof: Suppose that rows r and s of A are equal. Interchange rows r and s of A to obtain a matrix B. Then $|B| = -|A|$. On the other hand, $B = A$, so $|B| = |A|$. Thus $|A| = -|A|$, and so $|A| = 0$.

EXAMPLE 3. We have $\begin{vmatrix} 1 & 2 & 3 \\ -1 & 0 & 7 \\ 1 & 2 & 3 \end{vmatrix} = 0$ (verify by the use of Definition 5.2).

Theorem 5.4. *If a row (column) of A consists entirely of zeros, then* $|A| = 0$.

Proof: Let the ith row of A consist entirely of zeros. Since each term in Definition 5.2 for the determinant of A contains a factor from the ith row, each term in $|A|$ is zero. Hence $|A| = 0$.

EXAMPLE 4. We have $\begin{vmatrix} 1 & 2 & 3 \\ 4 & 5 & 6 \\ 0 & 0 & 0 \end{vmatrix} = 0$ (verify by the use of Definition 5.2).

Theorem 5.5. *If B is obtained from A by multiplying a row (column) of A by a real number c, then* $|B| = c|A|$.

Proof: Suppose that the rth row of $A = [a_{ij}]$ is multiplied by c to obtain $B = [b_{ij}]$. Then $b_{ij} = a_{ij}$ if $i \neq r$ and $b_{rj} = ca_{rj}$. Using Definition 5.2, we obtain $|B|$ as

$$|B| = \sum (\pm) b_{1j_1} b_{2j_2} \cdots b_{rj_r} \cdots b_{nj_n}$$
$$= \sum (\pm) a_{1j_1} a_{2j_2} \cdots (ca_{rj_r}) \cdots a_{nj_n}$$
$$= c\left(\sum (\pm) a_{1j_1} a_{2j_2} \cdots a_{rj_r} \cdots a_{nj_n}\right) = c|A|.$$

EXAMPLE 5. We have $\begin{vmatrix} 2 & 6 \\ 1 & 12 \end{vmatrix} = 2\begin{vmatrix} 1 & 3 \\ 1 & 12 \end{vmatrix} = 6\begin{vmatrix} 2 & 1 \\ 1 & 2 \end{vmatrix} = 18$.

We can use Theorem 5.5 to simplify the computation of $|A|$ by factoring out common factors from rows and columns of A.

EXAMPLE 6. We have

$$\begin{vmatrix} 1 & 2 & 3 \\ 1 & 5 & 3 \\ 2 & 8 & 6 \end{vmatrix} = 2\begin{vmatrix} 1 & 2 & 3 \\ 1 & 5 & 3 \\ 1 & 4 & 3 \end{vmatrix} = (2)(3)\begin{vmatrix} 1 & 2 & 1 \\ 1 & 5 & 1 \\ 1 & 4 & 1 \end{vmatrix} = (2)(3)(0) = 0.$$

Here we have factored out 2 from the third row and 3 from the third column, and then used Theorem 5.3, since the first and third columns are equal.

Theorem 5.6. *If* $B = [b_{ij}]$ *is obtained from* $A = [a_{ij}]$ *by adding to each element of the rth row (column) c times the corresponding element of the sth row (column)* $(r \neq s)$, *then* $|B| = |A|$.

Proof: We prove the theorem for rows. We have $b_{ij} = a_{ij}$ for $i \neq r$, and $b_{rj} = a_{rj} + ca_{sj}$, $r \neq s$, say $r < s$. Then

$$
\begin{aligned}
|B| &= \sum (\pm) b_{1j_1} b_{2j_2} \cdots b_{rj_r} \cdots b_{nj_n} \\
&= \sum (\pm) a_{1j_1} a_{2j_2} \cdots (a_{rj_r} + ca_{sj_r}) \cdots a_{nj_n} \\
&= \sum (\pm) a_{1j_1} a_{2j_2} \cdots a_{rj_r} \cdots a_{sj_s} \cdots a_{nj_n} \\
&\quad + \sum (\pm) a_{1j_1} a_{2j_2} \cdots (ca_{sj_r}) \cdots a_{sj_s} \cdots a_{nj_n}.
\end{aligned}
$$

Now the first term in this last expression is $|A|$, while the second is $c[\sum (\pm) a_{1j_1} a_{2j_2} \cdots a_{sj_r} \cdots a_{sj_s} \cdots a_{nj_n}]$. Note that

$$
\sum (\pm) a_{1j_1} a_{2j_2} \cdots a_{sj_r} \cdots a_{sj_s} \cdots a_{nj_n} =
\begin{vmatrix}
a_{11} & a_{12} & \cdots & a_{1n} \\
a_{21} & a_{22} & \cdots & a_{2n} \\
& & \vdots & \\
a_{s1} & a_{s2} & \cdots & a_{sn} \\
& & \vdots & \\
a_{s1} & a_{s2} & \cdots & a_{sn} \\
& & \vdots & \\
a_{n1} & a_{n2} & \cdots & a_{nn}
\end{vmatrix} = 0,
$$

because this matrix has two equal rows. Hence $|B| = |A| + 0 = |A|$.

EXAMPLE 7. We have $\begin{vmatrix} 1 & 2 & 3 \\ 2 & -1 & 3 \\ 1 & 0 & 1 \end{vmatrix} = \begin{vmatrix} 5 & 0 & 9 \\ 2 & -1 & 3 \\ 1 & 0 & 1 \end{vmatrix} = 4$ (verify by the use of Definition 5.2), obtained by adding twice the second row to the first row.

Theorem 5.7. *If a matrix* $A = [a_{ij}]$ *is upper (lower) triangular, then* $|A| = a_{11}a_{22} \cdots a_{nn}$; *that is, the determinant of a triangular matrix is the product of the elements on the main diagonal.*

Proof: Let $A = [a_{ij}]$ be upper triangular (that is, $a_{ij} = 0$ for $i > j$). Then a term $a_{1j_1} a_{2j_2} \cdots a_{nj_n}$ in the expression for $|A|$ can be nonzero only for

$1 \le j_1, 2 \le j_2, \ldots, n \le j_n$. Now $j_1 j_2 \cdots j_n$ must be a permutation, or re-arrangement, of $\{1, 2, \ldots, n\}$. Hence we must have $j_1 = 1, j_2 = 2, \ldots, j_n = n$. Thus the only term of $|A|$ that can be nonzero is the product of the elements on the main diagonal of A. Hence $|A| = a_{11} a_{22} \cdots a_{nn}$.

We leave the proof of the lower triangular case to the reader.

Theorems 5.2, 5.5, and 5.6 are very useful in evaluating determinants. What we do is transform A by means of our elementary row or column operations to a triangular matrix. Of course, we must keep track of how the determinant of the resulting matrices changes as we perform the elementary row or column operations.

EXAMPLE 8. We have

$$\begin{vmatrix} 4 & 3 & 2 \\ 3 & -2 & 5 \\ 2 & 4 & 6 \end{vmatrix}_{\frac{1}{2}\mathbf{r}_3 \to \mathbf{r}_3} = 2\begin{vmatrix} 4 & 3 & 2 \\ 3 & -2 & 5 \\ 1 & 2 & 3 \end{vmatrix}_{\mathbf{r}_1 \leftrightarrow \mathbf{r}_3} =$$

$$-2\begin{vmatrix} 1 & 2 & 3 \\ 3 & -2 & 5 \\ 4 & 3 & 2 \end{vmatrix}_{\mathbf{r}_2 - 3\mathbf{r}_1 \to \mathbf{r}_2} = -2\begin{vmatrix} 1 & 2 & 3 \\ 0 & -8 & -4 \\ 4 & 3 & 2 \end{vmatrix}_{\mathbf{r}_3 - 4\mathbf{r}_1 \to \mathbf{r}_3} =$$

$$-2\begin{vmatrix} 1 & 2 & 3 \\ 0 & -8 & -4 \\ 0 & -5 & -10 \end{vmatrix}_{\frac{1}{4}\mathbf{r}_2 \to \mathbf{r}_2} = (-2)(4)\begin{vmatrix} 1 & 2 & 3 \\ 0 & -2 & -1 \\ 0 & -5 & -10 \end{vmatrix}_{\frac{1}{5}\mathbf{r}_3 \to \mathbf{r}_3} =$$

$$(-2)(4)(5)\begin{vmatrix} 1 & 2 & 3 \\ 0 & -2 & -1 \\ 0 & -1 & -2 \end{vmatrix}_{\mathbf{r}_3 - \frac{1}{2}\mathbf{r}_2 \to \mathbf{r}_3} = (-2)(4)(5)\begin{vmatrix} 1 & 2 & 3 \\ 0 & -2 & -1 \\ 0 & 0 & -\frac{3}{2} \end{vmatrix}$$

$$= (-2)(4)(5)(1)(-2)(-\tfrac{3}{2}) = -120.$$

Here $\frac{1}{2}\mathbf{r}_3 \to \mathbf{r}_3$ means that $\frac{1}{2}$ times the third row $\mathbf{r}_3$ replaces the third row; $\mathbf{r}_1 \leftrightarrow \mathbf{r}_3$ means that we interchange the first and third rows; $\mathbf{r}_2 - 3\mathbf{r}_1 \to \mathbf{r}_2$ means that the second row minus three times the first row replaces the second row.

We can now compute the determinant of the identity matrix $I_n : |I_n| = 1$. We can also compute the determinants of the elementary matrices discussed in Section 1.6, as follows.

Let E_1 be an elementary matrix of type I; that is, E_1 is obtained from I_n

by interchanging, say, the ith and jth rows of I_n. By Theorem 5.2 we have that $|E_1| = -|I_n| = -1$. Now let E_2 be an elementary matrix of type II; that is, E_2 is obtained from I_n by multiplying, say, the ith row of I_n by $c \neq 0$. By Theorem 5.5 we have that $|E_2| = c|I_n| = c$. Finally, let E_3 be an elementary matrix of type III; that is, E_3 is obtained from I_n by adding to the rth row of I_n c times the sth row ($r \neq s$) of I_n. By Theorem 5.6 we have that $|E_3| = |I_n| = 1$.

We shall now turn to prove that the determinant of a product of two matrices is the product of their determinants and that A is nonsingular if and only if $|A| \neq 0$.

Lemma 5.1. *If E is an elementary matrix, then $|EA| = |E|\,|A|$, and $|AE| = |A|\,|E|$.*

Proof: If E is an elementary matrix of type I, then EA is obtained from A by interchanging two rows of A, so $|EA| = -|A|$. Also, $|E| = -1$. Thus $|EA| = |E|\,|A|$.

If E is an elementary matrix of type II, then EA is obtained from A by multiplying a given row of A by $c \neq 0$. Then $|EA| = c|A|$ and $|E| = c$, so $|EA| = |E|\,|A|$.

Finally, if E is an elementary matrix of type III, then EA is obtained from A by adding to a given row of A a multiple of a different row of A. Then $|EA| = |A|$ and $|E| = 1$, so $|EA| = |E|\,|A|$.

Thus, in all cases, $|EA| = |E|\,|A|$. By a similar proof, we can show that $|AE| = |A|\,|E|$.

It also follows from Lemma 5.1 that if $B = E_r E_{r-1} \cdots E_2 E_1 A$, then

$$|B| = |E_r(E_{r-1} \cdots E_2 E_1 A)|$$
$$= |E_r|\,|E_{r-1} E_{r-2} \cdots E_2 E_1 A| = |E_r|\,|E_{r-1}| \cdots |E_2|\,|E_1|\,|A|.$$

Theorem 5.8. *If A is an $n \times n$ matrix, then A is nonsingular if and only if $|A| \neq 0$.*

Proof: We recall Theorem 1.29 in Section 1.7, from which it follows that every $n \times n$ matrix is equivalent to a diagonal matrix $D = \begin{bmatrix} I_k & O \\ O & O \end{bmatrix}$, where $k = \text{rank } A$, $1 \leq k \leq n$. If $k = n$, there are no zero rows or columns and $D = I_n$. This means that we can obtain A from D by a finite sequence of elementary row or elementary column operations. Thus there exist elementary matrices $E_1, E_2, \ldots, E_r$ and $F_1, F_2, \ldots, F_s$ such that

$$A = E_r E_{r-1} \cdots E_2 E_1 D F_1 F_2 \cdots F_{s-1} F_s.$$

We then have

$$|A| = |E_r| \, |E_{r-1}| \cdots |E_2| \, |E_1| \, |D| \, |F_1| \, |F_2| \cdots |F_{s-1}| \, |F_s|.$$

Since E_i and F_j are elementary matrices, $|E_i| \neq 0$ and $|F_j| \neq 0$ for $i = 1, 2, \ldots, r$ and $j = 1, 2, \ldots, s$. Moreover, $|D|$ is the product of the elements on its main diagonal, and D has k ones along its main diagonal. Hence $|D| \neq 0$ if and only if $k = n$. Since $|A| \neq 0$ if and only if $|D| \neq 0$, we see that $|A| \neq 0$ if and only if $k = n$. From Theorem 1.31 we conclude that $k = n$ if and only if A is nonsingular, so A is nonsingular if and only if $|A| \neq 0$.

Corollary 5.1. *If A is an $n \times n$ matrix, then rank $A = n$ if and only if $|A| \neq 0$.*

Proof: By Corollary 2.4 we know that A is nonsingular if and only if rank $A = n$.

Corollary 5.2. *If A is an $n \times n$ matrix, then $AX = O$ has a nontrivial solution if and only if $|A| = 0$.*

Proof: Exercise.

Theorem 5.9. *If A and B are $n \times n$ matrices, then $|AB| = |A| \, |B|$.*

Proof: If A is nonsingular, then A is row equivalent to I_n. Thus $A = E_k E_{k-1} \cdots E_2 E_1 I_n = E_k E_{k-1} \cdots E_2 E_1$, where $E_1, E_2, \ldots, E_k$ are elementary matrices. Then

$$|A| = |E_k E_{k-1} \cdots E_2 E_1| = |E_k| \, |E_{k-1}| \cdots |E_2| \, |E_1|.$$

Now

$$|AB| = |E_k E_{k-1} \cdots E_2 E_1 B| = |E_k| \, |E_{k-1}| \cdots |E_2| \, |E_1| \, |B| = |A| \, |B|.$$

If A is singular, then $|A| = 0$ by Theorem 5.8. Moreover, if A is singular, then A is row equivalent to a matrix C that has a row consisting entirely of zeros (Theorem 1.27). Thus $C = E_k E_{k-1} \cdots E_2 E_1 A$, so $CB = E_k E_{k-1} \cdots E_2 E_1 AB$.

This means that AB is row equivalent to CB, and since CB has a row consisting entirely of zeros, it follows that AB is singular. Hence $|AB| = 0$ and in this case we also have $|AB| = |A| \, |B|$.

EXAMPLE 9. Let

$$A = \begin{bmatrix} 1 & 2 \\ 3 & 4 \end{bmatrix} \quad \text{and} \quad B = \begin{bmatrix} 2 & -1 \\ 1 & 2 \end{bmatrix}.$$

Then

$$|A| = -2 \quad \text{and} \quad |B| = 5.$$

On the other hand, $AB = \begin{bmatrix} 4 & 3 \\ 10 & 5 \end{bmatrix}$ and $|AB| = -10 = |A| \, |B|$.

From Theorem 5.9 it follows that if A is nonsingular, then $|A^{-1}| = 1/|A|$. It also follows from Theorem 5.9 that if A and B are similar matrices (that is, if $B = P^{-1}AP$ for some nonsingular matrix P), then $|A| = |B|$.

The determinant of a sum of two $n \times n$ matrices A and B is, in general, not the sum of the determinants of A and B. The best result we can give along these lines is that if A, B, and C are $n \times n$ matrices all of whose entries are equal except possibly for the kth row (column), and the kth row (column) of C is the sum of the kth rows (columns) of A and B, then $|C| = |A| + |B|$. We shall not prove this result, but will consider an example.

EXAMPLE 10. Let

$$A = \begin{bmatrix} 2 & 2 & 3 \\ 0 & 3 & 4 \\ 0 & 2 & 4 \end{bmatrix}, \quad B = \begin{bmatrix} 2 & 2 & 3 \\ 0 & 3 & 4 \\ 1 & -2 & -4 \end{bmatrix},$$

and

$$C = \begin{bmatrix} 2 & 2 & 3 \\ 0 & 3 & 4 \\ 1 & 0 & 0 \end{bmatrix}.$$

Then $|A| = 8$, $|B| = -9$, and $|C| = -1$, so $|C| = |A| + |B|$.

5.2. Exercises

1. Evaluate:

   (a) $\begin{vmatrix} 4 & 2 & 0 \\ 0 & -2 & 5 \\ 0 & 0 & 3 \end{vmatrix}$.   (b) $\begin{vmatrix} 4 & 0 & 0 \\ 0 & 2 & 0 \\ 0 & 0 & 3 \end{vmatrix}$.   (c) $\begin{vmatrix} 4 & 0 & 0 & 0 \\ -1 & 2 & 0 & 0 \\ 1 & 2 & -3 & 0 \\ 1 & 5 & 3 & 5 \end{vmatrix}$.

   (d) $\begin{vmatrix} 3 & 0 \\ 2 & 1 \end{vmatrix}$.

2. Evaluate:

   (a) $\begin{vmatrix} 2 & 1 \\ 4 & 3 \end{vmatrix}$.   (b) $\begin{vmatrix} 4 & 1 & 3 \\ 2 & 3 & 0 \\ 1 & 3 & 2 \end{vmatrix}$.   (c) $\begin{vmatrix} 3 & 4 & 2 \\ 2 & 5 & 0 \\ 3 & 0 & 0 \end{vmatrix}$.

   (d) $\begin{vmatrix} 4 & 2 & 2 & 0 \\ 2 & 0 & 0 & 0 \\ 3 & 0 & 0 & 1 \\ 0 & 0 & 1 & 0 \end{vmatrix}$.   (e) $\begin{vmatrix} 4 & 2 & 3 & -4 \\ 3 & -2 & 1 & 5 \\ -2 & 0 & 1 & -3 \\ 8 & -2 & 6 & 4 \end{vmatrix}$.

3. Is $|AB| = |BA|$? Justify your answer.

4. If $|AB| = 0$, is $|A| = 0$ or $|B| = 0$? Give reasons for your answer.

5. Show that if A is skew symmetric and n is odd, then $|A| = 0$.

6. Show that if c is a scalar and A is $n \times n$, then $|cA| = c^n |A|$.

7. Compute:

   (a) $\begin{vmatrix} 2 & -2 \\ 3 & -1 \end{vmatrix}$.   (b) $\begin{vmatrix} 4 & -3 & 5 \\ 5 & 2 & 0 \\ 2 & 0 & 4 \end{vmatrix}$.   (c) $\begin{vmatrix} 2 & 0 & 1 & 4 \\ 3 & 2 & -4 & -2 \\ 2 & 3 & -1 & 0 \\ 11 & 8 & -4 & 6 \end{vmatrix}$.

8. Show that if A is a matrix such that in each row and in each column one and only one element is $\neq 0$, then $|A| \neq 0$.

9. If $\begin{vmatrix} a_1 & a_2 & a_3 \\ b_1 & b_2 & b_3 \\ c_1 & c_2 & c_3 \end{vmatrix} = 3$, find $\begin{vmatrix} a_1 + 2b_1 - 3c_1 & a_2 + 2b_2 - 3c_2 & a_3 + 2b_3 - 3c_3 \\ b_1 & b_2 & b_3 \\ c_1 & c_2 & c_3 \end{vmatrix}$.

10. Verify that $|AB| = |A| |B|$ for the following:

    (a) $A = \begin{bmatrix} 1 & -2 & 3 \\ -2 & 3 & 1 \\ 0 & 1 & 0 \end{bmatrix}$, $B = \begin{bmatrix} 1 & 0 & 2 \\ 3 & -2 & 5 \\ 2 & 1 & 3 \end{bmatrix}$.

    (b) $A = \begin{bmatrix} 2 & 3 & 6 \\ 0 & 3 & 2 \\ 0 & 0 & -4 \end{bmatrix}$, $B = \begin{bmatrix} 3 & 0 & 0 \\ 4 & 5 & 0 \\ 2 & 1 & -2 \end{bmatrix}$.

11. Show that if A is nonsingular, then $|A^{-1}| = 1/|A|$.

12. Evaluate:

(a) $\begin{vmatrix} -4 & 2 & 0 & 0 \\ 2 & 3 & 1 & 0 \\ 3 & 1 & 0 & 2 \\ 1 & 3 & 0 & 3 \end{vmatrix}.$ (b) $\begin{vmatrix} 2 & 0 & 0 & 0 \\ -5 & 3 & 0 & 0 \\ 3 & 2 & 4 & 0 \\ 4 & 2 & 1 & -5 \end{vmatrix}.$

(c) $\begin{vmatrix} t-1 & -1 & -2 \\ 0 & t-2 & 2 \\ 0 & 0 & t-3 \end{vmatrix}.$ (d) $\begin{vmatrix} t+1 & 4 \\ 2 & t-3 \end{vmatrix}.$

13. Show that if $AB = I_n$, then $|A| \neq 0$ and $|B| \neq 0$.

14. (a) Show that if $A = A^{-1}$, then $|A| = \pm 1$.
    (b) If $A' = A^{-1}$, what is $|A|$?

15. Show that if A and B are square matrices, then $\begin{vmatrix} A & O \\ O & B \end{vmatrix} = |A||B|.$

16. If A is a nonsingular matrix such that $A^2 = A$, what is $|A|$?

17. Show that if A and B are similar, then $|A| = |B|$.

18. Show that if A, B, and C are square matrices, then $\begin{vmatrix} A & O \\ C & B \end{vmatrix} = |A||B|.$

19. Show that if A and B are both $n \times n$, then
    (a) $|A'B'| = |A||B'|$.   (b) $|A'B'| = |A'||B|$.

20. Verify the result in Exercise 15 for $A = \begin{bmatrix} 1 & 2 \\ 3 & 4 \end{bmatrix}$ and $B = \begin{bmatrix} 2 & 1 \\ -3 & 2 \end{bmatrix}.$

21. Use the properties of Section 5.2 to prove that
$$\begin{vmatrix} 1 & a & a^2 \\ 1 & b & b^2 \\ 1 & c & c^2 \end{vmatrix} = (b-a)(c-a)(c-b).$$ (*Hint:* Use factorization.)
This determinant is called a **Vandermonde determinant**.

22. If $|A| = 2$, find $|A^5|$.

23. Use Theorem 5.8 to determine which of the following matrices are non-singular:

(a) $\begin{bmatrix} 1 & 2 & 3 \\ 0 & 1 & 2 \\ 2 & -3 & 1 \end{bmatrix}.$ (b) $\begin{bmatrix} 1 & 2 \\ 3 & 4 \end{bmatrix}.$ (c) $\begin{bmatrix} 1 & 3 & 2 \\ 2 & 1 & 4 \\ 1 & -7 & 2 \end{bmatrix}.$

(d) $\begin{bmatrix} 1 & 2 & 0 & 5 \\ 3 & 4 & 1 & 7 \\ -2 & 5 & 2 & 0 \\ 0 & 1 & 2 & -7 \end{bmatrix}.$

24. Use Corollary 5.1 to find out whether rank $A = 3$ for the following matrices:

(a) $A = \begin{bmatrix} 1 & 2 & 3 \\ 2 & 1 & 0 \\ -3 & 1 & 2 \end{bmatrix}.$ (b) $A = \begin{bmatrix} 1 & 3 & -4 \\ -2 & 1 & 2 \\ -9 & 15 & 0 \end{bmatrix}.$

(c) $A = \begin{bmatrix} 1 & 0 & 1 \\ 1 & 1 & 0 \\ 2 & 1 & 0 \end{bmatrix}.$

25. Use Corollary 5.2 to find out whether the following homogeneous system has a nontrivial solution (do *not* solve):

$$x_1 - 2x_2 + x_3 = 0$$
$$2x_1 + 3x_2 + x_3 = 0$$
$$3x_1 + x_2 + 2x_3 = 0.$$

26. Repeat Exercise 25 for the following homogeneous system:

$$\begin{bmatrix} 1 & 2 & 0 & 1 \\ 0 & 1 & 2 & 3 \\ 0 & 0 & 1 & 2 \\ 0 & 1 & 2 & -1 \end{bmatrix} \begin{bmatrix} x_1 \\ x_2 \\ x_3 \\ x_4 \end{bmatrix} = \begin{bmatrix} 0 \\ 0 \\ 0 \\ 0 \end{bmatrix}.$$

27. Let $A = [a_{ij}]$ be an upper triangular matrix. Prove that A is nonsingular if and only if $a_{ii} \neq 0$, for $i = 1, 2, \ldots, n$.

28. Let $A^2 = A$. Prove that either A is singular or $|A| = 1$.

29. Prove Corollary 5.2.

5.3. Cofactor Expansion

So far we have evaluated determinants by using Definition 5.2 and the properties established in Section 5.2. We now develop a method for evaluating the determinant of an $n \times n$ matrix which reduces the problem to the evaluation of determinants of matrices of order $n - 1$. We can then repeat the process for these $(n - 1) \times (n - 1)$ matrices until we get to 2×2 matrices.

Definition 5.3. Let $A = [a_{ij}]$ be an $n \times n$ matrix. Let M_{ij} be the $(n - 1) \times (n - 1)$ submatrix of A obtained by deleting the ith row and jth column of A. The determinant $|M_{ij}|$ is called the **minor** of a_{ij}.

Definition 5.4. Let $A = [a_{ij}]$ be an $n \times n$ matrix. The **cofactor** A_{ij} of a_{ij} is defined as $A_{ij} = (-1)^{i+j}|M_{ij}|$.

EXAMPLE 1. Let $A = \begin{bmatrix} 3 & -1 & 2 \\ 4 & 5 & 6 \\ 7 & 1 & 2 \end{bmatrix}$. Then

$$\left| M_{12} \right| = \begin{vmatrix} 4 & 6 \\ 7 & 2 \end{vmatrix} = 8 - 42 = -34, \qquad \left| M_{23} \right| = \begin{vmatrix} 3 & -1 \\ 7 & 1 \end{vmatrix} = 3 + 7 = 10,$$

and

$$\left| M_{31} \right| = \left| \begin{matrix} -1 & 2 \\ 5 & 6 \end{matrix} \right| = -6 - 10 = -16.$$

Also, $A_{12} = (-1)^{1+2}|M_{12}| = (-1)(-34) = 34$, $A_{23} = (-1)^{2+3}|M_{23}| = (-1)(10) = -10$, and $A_{31} = (-1)^{1+3}|M_{31}| = (1)(-16) = -16$.

If we think of the sign $(-1)^{i+j}$ as being located in position (i, j) of an $n \times n$ matrix, then the signs form a checkerboard pattern that has a $+$ in the $(1, 1)$ position. The patterns for $n = 3$ and $n = 4$ are as follows:

$$
\begin{matrix}
+ & - & + & \quad & + & - & + & - \\
- & + & - & \quad & - & + & - & + \\
+ & - & + & \quad & + & - & + & - \\
  &   &   & \quad & - & + & - & + \\
\end{matrix}
$$
$$
\begin{matrix}
n = 3 & \qquad\qquad & n = 4
\end{matrix}
$$

Theorem 5.10. *Let $A = [a_{ij}]$ be an $n \times n$ matrix. Then*

$$|A| = a_{i1}A_{i1} + a_{i2}A_{i2} + \cdots + a_{in}A_{in}$$
(expansion of $|A|$ about the ith row)

and

$$|A| = a_{1j}A_{1j} + a_{2j}A_{2j} + \cdots + a_{nj}A_{nj}$$
(expansion of $|A|$ about the jth column).

Proof: The first formula follows from the second by Theorem 5.1, that is from the fact that $|A'| = |A|$. We omit the general proof and consider the 3×3 matrix $A = [a_{ij}]$. From (1) in Section 5.1,

$$
\begin{aligned}
|A| = {} & a_{11}a_{22}a_{33} + a_{12}a_{23}a_{31} + a_{13}a_{21}a_{32} \\
& - a_{11}a_{23}a_{32} - a_{12}a_{21}a_{33} - a_{13}a_{22}a_{31}.
\end{aligned}
\tag{1}
$$

We can write this expression as

$$
\begin{aligned}
|A| = {} & a_{11}(a_{22}a_{33} - a_{23}a_{32}) + a_{12}(a_{23}a_{31} - a_{21}a_{33}) \\
& + a_{13}(a_{21}a_{32} - a_{22}a_{31}).
\end{aligned}
$$

Now,

$$A_{11} = (-1)^{1+1} \left| \begin{matrix} a_{22} & a_{23} \\ a_{32} & a_{33} \end{matrix} \right| = (a_{22}a_{33} - a_{23}a_{32}).$$

$$A_{12} = (-1)^{1+2} \left| \begin{matrix} a_{21} & a_{23} \\ a_{31} & a_{33} \end{matrix} \right| = (a_{23}a_{31} - a_{21}a_{33})$$

$$A_{13} = (-1)^{1+3} \left| \begin{matrix} a_{21} & a_{22} \\ a_{31} & a_{32} \end{matrix} \right| = (a_{21}a_{32} - a_{22}a_{31}).$$

Hence

$$|A| = a_{11}A_{11} + a_{12}A_{12} + a_{13}A_{13},$$

which is the expansion of $|A|$ about the first row.

If we now write (1) as

$$|A| = a_{13}(a_{21}a_{32} - a_{22}a_{31}) + a_{23}(a_{12}a_{31} - a_{11}a_{32})$$
$$+ a_{33}(a_{11}a_{22} - a_{12}a_{21}),$$

we can easily verify that

$$|A| = a_{13}A_{13} + a_{23}A_{23} + a_{33}A_{33},$$

which is the expansion of $|A|$ about the third column.

EXAMPLE 2. To evaluate

$$\begin{vmatrix} 1 & 2 & -3 & 4 \\ -4 & 2 & 1 & 3 \\ 3 & 0 & 0 & -3 \\ 2 & 0 & -2 & 3 \end{vmatrix},$$

it is best to expand about either the second column or the third row, because they each have two zeros. Obviously, the optimal course of action is to expand about the row or column that has the largest number of zeros, because in that case the cofactors A_{ij} of those a_{ij} which are zero do not have to be evaluated since $a_{ij}A_{ij} = (0)(A_{ij}) = 0$. Thus, expanding about the third row, we have

$$\begin{vmatrix} 1 & 2 & -3 & 4 \\ -4 & 2 & 1 & 3 \\ 3 & 0 & 0 & -3 \\ 2 & 0 & -2 & 3 \end{vmatrix}$$

$$= (-1)^{3+1}(3)\begin{vmatrix} 2 & -3 & 4 \\ 2 & 1 & 3 \\ 0 & -2 & 3 \end{vmatrix} + (-1)^{3+2}(0)\begin{vmatrix} 1 & -3 & 4 \\ -4 & 1 & 3 \\ 2 & -2 & 3 \end{vmatrix}$$

$$+ (-1)^{3+3}(0)\begin{vmatrix} 1 & 2 & 4 \\ -4 & 2 & 3 \\ 2 & 0 & 3 \end{vmatrix} + (-1)^{3+4}(-3)\begin{vmatrix} 1 & 2 & -3 \\ -4 & 2 & 1 \\ 2 & 0 & -2 \end{vmatrix}$$

$$= (+1)(3)(20) + 0 + 0 + (-1)(-3)(-4) = 48.$$

We can use the properties of Section 5.2 to introduce many zeros in a given row or column and then expand about that row or column. Consider the following example.

EXAMPLE 3. We have

$$
\begin{vmatrix}
1 & 2 & -3 & 4 \\
-4 & 2 & 1 & 3 \\
1 & 0 & 0 & -3 \\
2 & 0 & -2 & 3
\end{vmatrix}
=
\begin{vmatrix}
1 & 2 & -3 & 7 \\
-4 & 2 & 1 & -9 \\
1 & 0 & 0 & 0 \\
2 & 0 & -2 & 9
\end{vmatrix}
$$

$\mathbf{c}_4 + 3\mathbf{c}_1 \to \mathbf{c}_4$

$$
= (-1)^{3+1}(1)
\begin{vmatrix}
2 & -3 & 7 \\
2 & 1 & -9 \\
0 & -2 & 9
\end{vmatrix}
=
(-1)^4(1)
\begin{vmatrix}
0 & -4 & 16 \\
2 & 1 & -9 \\
0 & -2 & 9
\end{vmatrix}
$$

$\mathbf{r}_1 - \mathbf{r}_2 \to \mathbf{r}_1$

$$
= (-1)^4(1)(8) = 8.
$$

Here $\mathbf{c}_4 + 3\mathbf{c}_1 \to \mathbf{c}_4$ means that the fourth column $\mathbf{c}_4$ plus three times the first $\mathbf{c}_1$ replaces the fourth.

5.3. Exercises

1. Let $A = \begin{bmatrix} 1 & 0 & -2 \\ 3 & 1 & 4 \\ 5 & 2 & -3 \end{bmatrix}$. Find the following minors:

   (a) $|M_{13}|$.   (b) $|M_{22}|$.   (c) $|M_{31}|$.   (d) $|M_{32}|$.

2. Let $A = \begin{bmatrix} 1 & 0 & 3 & 0 \\ 2 & 1 & 4 & -1 \\ 3 & 2 & 4 & 0 \\ 0 & 3 & -1 & 0 \end{bmatrix}$. Find the following cofactors:

   (a) A_{12}.   (b) A_{23}.   (c) A_{33}.   (d) A_{41}.

3. Use Theorem 5.10 to evaluate the determinants in Exercise 1 of Section 5.2.

4. Show by a column (row) expansion that if $A = [a_{ij}]$ is upper (lower) triangular, then $|A| = a_{11}a_{22} \cdots a_{nn}$.

5. If $A = [a_{ij}]$ is a 3×3 matrix, develop the general expression for $|A|$ by expanding:

   (a) About the second column.

   (b) About the third row.

   Compare these answers with those obtained for Example 8 in Section 5.1.

6. Use Theorem 5.10 to evaluate the determinants in Exercise 2 of Section 5.2.

7. Find values of t for which

   (a) $\begin{vmatrix} t-2 & 2 \\ 3 & t-3 \end{vmatrix} = 0.$ (b) $\begin{vmatrix} t-1 & -4 \\ 0 & t-4 \end{vmatrix} = 0.$

8. Use Theorem 5.10 to evaluate the determinants in Exercise 7 of Section 5.2.

9. Use Theorem 5.10 to evaluate the determinants in Exercise 12 of Section 5.2.

10. Find values of t for which

$$\begin{vmatrix} t-1 & 0 & 1 \\ -2 & t+2 & -1 \\ 0 & 0 & t+1 \end{vmatrix} = 0.$$

11. Let A be an $n \times n$ matrix.

(a) Show that $f(t) = |tI_n - A|$ is a polynomial in t of degree n.

(b) What is the coefficient of t^n in $f(t)$?

(c) What is the constant term in $f(t)$?

12. Verify your answers to Exercise 11 with the following matrices:

(a) $\begin{bmatrix} 1 & 2 \\ 3 & 4 \end{bmatrix}.$
 (b) $\begin{bmatrix} 1 & 3 & 2 \\ 2 & -1 & 3 \\ 3 & 0 & 1 \end{bmatrix}.$
 (c) $\begin{bmatrix} 1 & 1 \\ 1 & 1 \end{bmatrix}.$

5.4. The Inverse of a Matrix

We saw in Section 5.3 that Theorem 5.10 provides formulas for expanding $|A|$ about either a row or a column of A. Thus $|A| = a_{i1}A_{i1} + a_{i2}A_{i2} + \cdots + a_{in}A_{in}$ is the expansion of $|A|$ about the ith row. It is interesting to ask what $a_{i1}A_{k1} + a_{i2}A_{k2} + \cdots + a_{in}A_{kn}$ is for $i \neq k$, because as soon as we answer this question we shall obtain another method for finding the inverse of a nonsingular matrix.

Theorem 5.11. *If $A = [a_{ij}]$ is an $n \times n$ matrix, then*

$$a_{i1}A_{k1} + a_{i2}A_{k2} + \cdots + a_{in}A_{kn} = 0 \qquad \text{for } i \neq k;$$
$$a_{1j}A_{1k} + a_{2j}A_{2k} + \cdots + a_{nj}A_{nk} = 0 \qquad \text{for } j \neq k.$$

Proof: We prove only the first formula. The second follows from the first by Theorem 5.1.

Consider the matrix B obtained from A by replacing the kth row of A by the ith row of A. Thus B is a matrix with two identical rows—the ith and kth, so $|B| = 0$. Now expand $|B|$ about the kth row. The elements of the kth row of B are $a_{i1}, a_{i2}, \ldots, a_{in}$. The cofactors of the kth row are $A_{k1}, A_{k2}, \ldots, A_{kn}$. Thus $0 = |B| = a_{i1}A_{k1} + a_{i2}A_{k2} + \cdots + a_{in}A_{kn}$, as we wanted to show.

EXAMPLE 1. Let $A = \begin{bmatrix} 1 & 2 & 3 \\ -2 & 3 & 1 \\ 4 & 5 & -2 \end{bmatrix}$. Then

$$A_{21} = (-1)^{2+1} \begin{vmatrix} 2 & 3 \\ 5 & -2 \end{vmatrix} = 19,$$

$$A_{22} = (-1)^{2+2} \begin{vmatrix} 1 & 3 \\ 4 & -2 \end{vmatrix} = -14, \quad \text{and} \quad A_{23} = (-1)^{2+3} \begin{vmatrix} 1 & 2 \\ 4 & 5 \end{vmatrix} = 3.$$

Now $a_{31}A_{21} + a_{32}A_{22} + a_{33}A_{23} = (4)(19) + (5)(-14) + (-2)(3) = 0$.

We may summarize our expansion results by writing

$$a_{i1}A_{k1} + a_{i2}A_{k2} + \cdots + a_{in}A_{kn} = |A| \quad \text{if } i = k$$
$$= 0 \quad \text{if } i \neq k$$

and

$$a_{1j}A_{1k} + a_{2j}A_{2k} + \cdots + a_{nj}A_{nk} = |A| \quad \text{if } j = k$$
$$= 0 \quad \text{if } j \neq k.$$

Definition 5.5. Let $A = [a_{ij}]$ be an $n \times n$ matrix. The $n \times n$ matrix adj A, called the **adjoint** of A, is the matrix whose (i, j) entry is the cofactor A_{ji} of a_{ji}. Thus

$$\text{adj } A = \begin{bmatrix} A_{11} & A_{21} & \cdots & A_{n1} \\ A_{12} & A_{22} & \cdots & A_{n2} \\ \vdots & & & \vdots \\ A_{1n} & A_{2n} & \cdots & A_{nn} \end{bmatrix}.$$

EXAMPLE 2. Let $A = \begin{bmatrix} 3 & -2 & 1 \\ 5 & 6 & 2 \\ 1 & 0 & -3 \end{bmatrix}$.

The cofactors are

$$A_{11} = (-1)^{1+1} \begin{vmatrix} 6 & 2 \\ 0 & -3 \end{vmatrix} = -18,$$

$$A_{12} = (-1)^{1+2} \begin{vmatrix} 5 & 2 \\ 1 & -3 \end{vmatrix} = 17, \quad \text{and} \quad A_{13} = (-1)^{1+3} \begin{vmatrix} 5 & 6 \\ 1 & 0 \end{vmatrix} = -6;$$

$$A_{21} = (-1)^{2+1} \begin{vmatrix} -2 & 1 \\ 0 & -3 \end{vmatrix} = -6,$$

$$A_{22} = (-1)^{2+2} \begin{vmatrix} 3 & 1 \\ 1 & -3 \end{vmatrix} = -10, \quad \text{and} \quad A_{23} = (-1)^{2+3} \begin{vmatrix} 3 & -2 \\ 1 & 0 \end{vmatrix} = -2;$$

$$A_{31} = (-1)^{3+1} \begin{vmatrix} -2 & 1 \\ 6 & 2 \end{vmatrix} = -10,$$

$$A_{32} = (-1)^{3+2} \begin{vmatrix} 3 & 1 \\ 5 & 2 \end{vmatrix} = -1, \quad \text{and} \quad A_{33} = (-1)^{3+3} \begin{vmatrix} 3 & -2 \\ 5 & 6 \end{vmatrix} = 28.$$

Hence

$$\text{adj } A = \begin{bmatrix} -18 & -6 & -10 \\ 17 & -10 & -1 \\ -6 & -2 & 28 \end{bmatrix}.$$

Theorem 5.12. *If $A = [a_{ij}]$ is an $n \times n$ matrix, then $A(\text{adj } A) = (\text{adj } A)A = |A|I_n$.*

Proof: We have

$$A(\text{adj } A) = \begin{bmatrix} a_{11} & a_{12} & \cdots & a_{1n} \\ a_{21} & a_{22} & \cdots & a_{2n} \\ \vdots & \vdots & & \vdots \\ a_{i1} & a_{i2} & \cdots & a_{in} \\ \vdots & \vdots & & \vdots \\ a_{n1} & a_{n2} & \cdots & a_{nn} \end{bmatrix} \begin{bmatrix} A_{11} & A_{21} & \cdots & A_{j1} & \cdots & A_{n1} \\ A_{12} & A_{22} & \cdots & A_{j2} & \cdots & A_{n2} \\ \vdots & \vdots & & \vdots & & \vdots \\ A_{1n} & A_{2n} & \cdots & A_{jn} & \cdots & A_{nn} \end{bmatrix}.$$

The (i, j) entry in the product matrix $A(\text{adj } A)$ is

$$a_{i1}A_{j1} + a_{i2}A_{j2} + \cdots + a_{in}A_{jn} = |A| \quad \text{if } i = j$$
$$= 0 \quad \text{if } i \neq j.$$

This means that

$$A(\text{adj } A) = \begin{bmatrix} |A| & 0 & \cdot & \cdot & \cdot & 0 \\ 0 & |A| & & & & \vdots \\ \vdots & & \cdot & & & 0 \\ 0 & \cdot & \cdot & \cdot & 0 & |A| \end{bmatrix} = |A|I_n.$$

The (i, j) entry in the product matrix $(\text{adj } A) A$ is

$$A_{1i}a_{1j} + A_{2i}a_{2j} + \cdots + A_{ni}a_{nj} = |A| \quad \text{if } i = j$$
$$= 0 \quad \text{if } i \neq j.$$

Thus $(\text{adj } A)A = |A|I_n$.

EXAMPLE 3. Consider the matrix of Example 2. Then

$$\begin{bmatrix} 3 & -2 & 1 \\ 5 & 6 & 2 \\ 1 & 0 & -3 \end{bmatrix} \begin{bmatrix} -18 & -6 & -10 \\ 17 & -10 & -1 \\ -6 & -2 & 28 \end{bmatrix}$$

$$= \begin{bmatrix} -94 & 0 & 0 \\ 0 & -94 & 0 \\ 0 & 0 & -94 \end{bmatrix} = -94 \begin{bmatrix} 1 & 0 & 0 \\ 0 & 1 & 0 \\ 0 & 0 & 1 \end{bmatrix}$$

and

$$\begin{bmatrix} -18 & -6 & -10 \\ 17 & -10 & -1 \\ -6 & -2 & 28 \end{bmatrix} \begin{bmatrix} 3 & -2 & 1 \\ 5 & 6 & 2 \\ 1 & 0 & -3 \end{bmatrix} = -94 \begin{bmatrix} 1 & 0 & 0 \\ 0 & 1 & 0 \\ 0 & 0 & 1 \end{bmatrix}.$$

We now have a new method for finding the inverse of a nonsingular matrix, and we state this result as a corollary.

Corollary 5.3. *If A is an $n \times n$ matrix and $|A| \neq 0$, then $A^{-1} = \dfrac{1}{|A|} (\text{adj } A)$.*

Proof: By Theorem 5.12, $A(\text{adj } A) = |A|I_n$, so if $|A| \neq 0$, then

$$A\left(\frac{1}{|A|} (\text{adj } A)\right) = \frac{1}{|A|} (A(\text{adj } A)) = \frac{1}{|A|} (|A|I_n) = I_n.$$

Hence $A^{-1} = \dfrac{1}{|A|} (\text{adj } A)$.

EXAMPLE 4. Again consider the matrix of Example 2. Then $|A| = -94$, and

$$A^{-1} = \frac{1}{|A|} (\text{adj } A) = \begin{bmatrix} 18/94 & 6/94 & 10/94 \\ -17/94 & 10/94 & 1/94 \\ 6/94 & 2/94 & -28/94 \end{bmatrix}.$$

We might note that the method of inverting a nonsingular matrix given in Corollary 5.3 is much less efficient than the methods given in Chapter 1. In fact, the computation of A^{-1} using determinants, as given in Corollary 5.3, becomes too expensive from a computing point of view for $n > 4$. We discuss these matters in Section 5.6, where we deal with determinants from a computational point of view. However, Corollary 5.3 is still a useful result on other grounds.

5.4. Exercises

1. Verify Theorem 5.11 for the matrix $A = \begin{bmatrix} -2 & 3 & 0 \\ 4 & 1 & -3 \\ 2 & 0 & 1 \end{bmatrix}$ by computing $a_{11}A_{12} + a_{21}A_{22} + a_{31}A_{32}$.

2. Let $A = \begin{bmatrix} 2 & 1 & 3 \\ -1 & 2 & 0 \\ 3 & -2 & 1 \end{bmatrix}$.
   (a) Find adj A.
   (b) Compute $|A|$.
   (c) Verify Theorem 5.12; that is, show that $A(\text{adj } A) = (\text{adj } A)A = |A|I_3$.

3. Let $A = \begin{bmatrix} 6 & 2 & 8 \\ -3 & 4 & 1 \\ 4 & -4 & 5 \end{bmatrix}$.
   (a) Find adj A.
   (b) Compute $|A|$.
   (c) Verify Theorem 5.12; that is, show that $(\text{adj } A)A = (\text{adj } A)A = |A|I_3$.

4. Find the inverse of the matrix in Exercise 2 by the method given in Corollary 5.3.

5. Repeat Exercise 5 of Section 1.6 by the method given in Corollary 5.3. Compare your results with those obtained earlier.

6. Prove that if A is a nonsingular symmetric matrix, then A^{-1} is symmetric.

7. Use the method given in Corollary 5.3 to find the inverse, if it exists, of
   (a) $A = \begin{bmatrix} 0 & 2 & 1 & 3 \\ 2 & -1 & 3 & 4 \\ -2 & 1 & 5 & 2 \\ 0 & 1 & 0 & 2 \end{bmatrix}$. (b) $\begin{bmatrix} 4 & 2 & 2 \\ 0 & 1 & 2 \\ 1 & 0 & 3 \end{bmatrix}$. (c) $\begin{bmatrix} 3 & 2 \\ -3 & 4 \end{bmatrix}$.

8. Prove that if A is a nonsingular upper triangular matrix, then A^{-1} is upper triangular.

9. Use the method given in Corollary 5.3 to find the inverse of
$$A = \begin{bmatrix} a & b \\ c & d \end{bmatrix} \quad \text{if } ad - bc \neq 0.$$

10. Use the method given in Corollary 5.3 to find the inverse of

$$A = \begin{bmatrix} 1 & a & a^2 \\ 1 & b & b^2 \\ 1 & c & c^2 \end{bmatrix}.$$

(*Hint:* See Exercise 21 in Section 5.2, where $|A|$ was computed.)

11. Use the method given in Corollary 5.3 to find the inverse of

$$A = \begin{bmatrix} 4 & 0 & 0 \\ 0 & -3 & 0 \\ 0 & 0 & 2 \end{bmatrix}.$$

12. Use the method given in Corollary 5.3 to find the inverse of

$$A = \begin{bmatrix} 4 & 1 & 2 \\ 0 & -3 & 3 \\ 0 & 0 & 2 \end{bmatrix}.$$

13. Prove that $|\operatorname{adj} A| = |A|^{n-1}$.

14. Prove that if A is symmetric, then adj A is symmetric.

5.5. Other Applications of Determinants

We can use the results developed in Theorem 5.12 to obtain another method for solving a linear system of n equations in n unknowns. This method is known as **Cramer's rule.**

Theorem 5.13 (Cramer's Rule). *Let*

$$a_{11}x_1 + a_{12}x_2 + \cdots + a_{1n}x_n = b_1$$
$$a_{21}x_1 + a_{22}x_2 + \cdots + a_{2n}x_n = b_2$$
$$\vdots$$
$$a_{n1}x_1 + a_{n2}x_2 + \cdots + a_{nn}x_n = b_n$$

be a linear system of n equations in n unknowns and let $A = [a_{ij}]$ be the coefficient matrix, so that we can write the given system as $AX = B$, where

$$B = \begin{bmatrix} b_1 \\ b_2 \\ \vdots \\ b_n \end{bmatrix}.$$ *If $|A| \neq 0$, then the system has the unique solution*

$$x_1 = \frac{|A_1|}{|A|}, \; x_2 = \frac{|A_2|}{|A|}, \ldots, x_n = \frac{|A_n|}{|A|},$$

where A_i is the matrix obtained from A by replacing its ith column by B.

Proof: If $|A| \neq 0$, then, by Theorem 5.8, A is nonsingular. Hence

$$
X = \begin{bmatrix} x_1 \\ x_2 \\ \vdots \\ x_n \end{bmatrix} = A^{-1}B = \begin{bmatrix} \dfrac{A_{11}}{|A|} & \dfrac{A_{21}}{|A|} & \cdots & \dfrac{A_{n1}}{|A|} \\[2mm] \dfrac{A_{12}}{|A|} & \dfrac{A_{22}}{|A|} & \cdots & \dfrac{A_{n2}}{|A|} \\[2mm] \vdots & \vdots & & \vdots \\[2mm] \dfrac{A_{1i}}{|A|} & \dfrac{A_{2i}}{|A|} & \cdots & \dfrac{A_{ni}}{|A|} \\[2mm] \vdots & \vdots & & \vdots \\[2mm] \dfrac{A_{1n}}{|A|} & \dfrac{A_{2n}}{|A|} & \cdots & \dfrac{A_{nn}}{|A|} \end{bmatrix} \begin{bmatrix} b_1 \\ b_2 \\ \vdots \\ b_n \end{bmatrix}.
$$

This means that

$$
x_i = \frac{A_{1i}}{|A|} b_1 + \frac{A_{2i}}{|A|} b_2 + \cdots + \frac{A_{ni}}{|A|} b_n \quad \text{for } i = 1, 2, \ldots, n.
$$

Now let

$$
A_i = \begin{bmatrix} a_{11} & a_{12} & \cdots & a_{1i-1} & b_1 & a_{1i+1} & \cdots & a_{1n} \\ a_{21} & a_{22} & \cdots & a_{2i-1} & b_2 & a_{2i+1} & \cdots & a_{2n} \\ \vdots & \vdots & & \vdots & \vdots & \vdots & & \vdots \\ a_{n1} & a_{n2} & \cdots & a_{ni-1} & b_n & a_{ni+1} & \cdots & a_{nn} \end{bmatrix}.
$$

If we evaluate $|A_i|$ by expanding about the cofactors of the ith column, we find that

$$
|A_i| = A_{1i}b_1 + A_{2i}b_2 + \cdots + A_{ni}b_n.
$$

Hence

$$
x_i = \frac{|A_i|}{|A|} \quad \text{for } i = 1, 2, \ldots, n.
$$

In this expression for x_i, the determinant, $|A_i|$, of A_i can be calculated by any method desired. It was only in the *derivation* of the expression for x_i that we had to evaluate $|A_i|$ by expanding about the ith column.

EXAMPLE 1. Consider the following linear system:

$$
\begin{aligned}
-2x_1 + 3x_2 - x_3 &= 1 \\
x_1 + 2x_2 - x_3 &= 4 \\
-2x_1 - x_2 + x_3 &= -3.
\end{aligned}
$$

We have $|A| = \begin{vmatrix} -1 & 3 & -1 \\ 1 & 2 & -1 \\ -2 & -1 & 1 \end{vmatrix} = -2$. Then

$$x_1 = \frac{\begin{vmatrix} 1 & 3 & -1 \\ 4 & 2 & -1 \\ -3 & -1 & 1 \end{vmatrix}}{|A|} = \frac{-4}{-2} = 2,$$

$$x_2 = \frac{\begin{vmatrix} -2 & 1 & -1 \\ 1 & 4 & -1 \\ -2 & -3 & 1 \end{vmatrix}}{|A|} = \frac{-6}{-2} = 3,$$

and

$$x_3 = \frac{\begin{vmatrix} -2 & 3 & 1 \\ 1 & 2 & 4 \\ -2 & -1 & -3 \end{vmatrix}}{|A|} = \frac{-8}{-2} = 4.$$

We note that Cramer's rule is only applicable to the case in which we have n equations in n unknowns and the coefficient matrix A is nonsingular. Cramer's rule becomes computationally inefficient if $n \geq 4$ and it is then better to use the Gaussian elimination or Gauss–Jordan reduction methods discussed in Section 1.5.

Our next application of determinants will enable us to tell whether a set of n vectors in R^n or R_n is linearly independent.

Theorem 5.14. *Let* $S = \{\alpha_1, \alpha_2, \ldots, \alpha_n\}$ *be a set of n vectors in R^n (R_n). Let A be the matrix whose columns (rows) are the elements of S. Then S is linearly independent if and only if* $|A| \neq 0$.

Proof: We shall prove the result for columns only; the proof for rows is analogous.

If S is linearly independent, then the dimension of the column space of A = rank $A = n$, and from Corollary 5.1 it follows that $|A| \neq 0$. Conversely, if $|A| \neq 0$, then, by Corollary 5.1 we know that rank A = dim column space of $A = n$. Hence S is linearly independent.

EXAMPLE 2. To find out if $S = \{[1 \quad 2 \quad 3], [0 \quad 1 \quad 2], [3 \quad 0 \quad -1]\}$ is linearly independent, we form the matrix A whose rows are the vectors in S:

$$A = \begin{bmatrix} 1 & 2 & 3 \\ 0 & 1 & 2 \\ 3 & 0 & -1 \end{bmatrix}.$$

Since $|A| = 2$, we conclude that S is linearly independent.

EXAMPLE 3. To find out if $S = \{t^2 + t, t + 1, t - 1\}$ is a basis for P_2, we note that P_2 is a three-dimensional vector space isomorphic to R^3 under the mapping $L: P_2 \to R^3$ defined by $L(at^2 + bt + c) = \begin{bmatrix} a \\ b \\ c \end{bmatrix}$. Therefore, S is a basis for P_2 if and only if $T = \{L(t^2 + t), L(t + 1), L(t - 1)\}$ is a basis for R^3. To decide whether this is so, we apply Theorem 5.14. Thus, let A be the matrix whose columns are $L(t^2 + t), L(t + 1), L(t - 1)$, respectively. Now

$$L(t^2 + t) = \begin{bmatrix} 1 \\ 1 \\ 0 \end{bmatrix}, \quad L(t + 1) = \begin{bmatrix} 0 \\ 1 \\ 1 \end{bmatrix}, \quad \text{and} \quad L(t - 1) = \begin{bmatrix} 0 \\ 1 \\ -1 \end{bmatrix},$$

so

$$A = \begin{bmatrix} 1 & 0 & 0 \\ 1 & 1 & 1 \\ 0 & 1 & -1 \end{bmatrix}.$$

Since $|A| = -2$, we conclude that T is linearly independent. Hence S is linearly independent, and since dim $P_2 = 3$, S is a basis for P_2.

We now recall that if a set S of n vectors in R^n (R_n) is linearly independent, then S spans R^n (R_n), and conversely, if S spans R^n (R_n), then S is linearly independent (Theorem 2.9 in Section 2.2). Thus the condition in Theorem 5.14 (that $|A| \neq 0$) is also necessary and sufficient for S to span R^n (R_n).

Determinants and Cross Products (Optional)

Our final application of determinants is to cross products. Recall the definition given in Section 3.1 for the cross product $\alpha \times \beta$ of the vectors $\alpha = a_1\mathbf{i} + a_2\mathbf{j} + a_3\mathbf{k}$ and $\beta = b_1\mathbf{i} + b_2\mathbf{j} + b_3\mathbf{k}$ in R^3:

$$\alpha \times \beta = (a_2b_3 - a_3b_2)\mathbf{i} + (a_3b_1 - a_1b_3)\mathbf{j} + (a_1b_2 - a_2b_1)\mathbf{k}.$$

If we formally write the matrix

$$C = \begin{bmatrix} \mathbf{i} & \mathbf{j} & \mathbf{k} \\ a_1 & a_2 & a_3 \\ b_1 & b_2 & b_3 \end{bmatrix},$$

then the determinant of C, obtained by expanding about the cofactors of the first row, is $\alpha \times \beta$, that is,

$$\alpha \times \beta = |C| = \begin{vmatrix} a_2 & b_3 \\ b_2 & b_3 \end{vmatrix}\mathbf{i} - \begin{vmatrix} a_1 & a_3 \\ b_1 & b_3 \end{vmatrix}\mathbf{j} + \begin{vmatrix} a_1 & a_2 \\ b_1 & b_2 \end{vmatrix}\mathbf{k}.$$

Of course, C is not really a matrix and $|C|$ is not really a determinant, but it is convenient to think of the computation in this way.

EXAMPLE 4. If $\alpha = 2\mathbf{i} + \mathbf{j} + 2\mathbf{k}$ and $\beta = 3\mathbf{i} - \mathbf{j} - 3\mathbf{k}$, as in Example 4 of Section 3.1, then

$$C = \begin{bmatrix} \mathbf{i} & \mathbf{j} & \mathbf{k} \\ 2 & 1 & 2 \\ 3 & -1 & -3 \end{bmatrix}$$

and $|C| = \alpha \times \beta = -\mathbf{i} + 12\mathbf{j} - 5\mathbf{k}$, when expanded about its first row.

5.5. Exercises

1. If possible, solve the following linear system by Cramer's rule:

$$\begin{aligned} 2x_1 + 4x_2 + 6x_3 &= 2 \\ x_1 \quad\quad + 2x_3 &= 0 \\ 2x_1 + 3x_2 - x_3 &= -5. \end{aligned}$$

2. Repeat Exercise 1 for the linear system

$$\begin{bmatrix} 1 & 1 & 1 & -2 \\ 0 & 2 & 1 & 3 \\ 2 & 1 & -1 & 2 \\ 1 & -1 & 0 & 1 \end{bmatrix} \begin{bmatrix} x_1 \\ x_2 \\ x_3 \\ x_4 \end{bmatrix} = \begin{bmatrix} -4 \\ 4 \\ 5 \\ 4 \end{bmatrix}.$$

3. Is $S = \left\{ \begin{bmatrix} 2 \\ 2 \\ 3 \end{bmatrix}, \begin{bmatrix} 1 \\ 0 \\ 2 \end{bmatrix}, \begin{bmatrix} 0 \\ 1 \\ 3 \end{bmatrix} \right\}$ a linearly independent set of vectors in R^3?
Use Theorem 5.14.

4. Solve the following linear system for x_3, by Cramer's rule:

$$\begin{aligned} 2x_1 + x_2 + x_3 &= 6 \\ 3x_1 + 2x_2 - 2x_3 &= -2 \\ x_1 + x_2 + 2x_3 &= 4. \end{aligned}$$

5. Repeat Exercise 5 of Section 1.5; use Cramer's rule.

6. Is $S = \{[0 \ -1 \ 4], [4 \ 1 \ 2], [2 \ 0 \ 3]\}$, a basis for R_3? Use Theorem 5.14.

7. Repeat Exercise 1 for the following linear system:

$$\begin{aligned} 2x_1 - x_2 + 3x_3 &= 0 \\ x_1 + 2x_2 - 3x_3 &= 0 \\ 4x_1 + 2x_2 + x_3 &= 0. \end{aligned}$$

8. Does the set $S = \left\{ \begin{bmatrix} 3 \\ 1 \\ 0 \end{bmatrix}, \begin{bmatrix} 0 \\ -1 \\ 1 \end{bmatrix}, \begin{bmatrix} 1 \\ 2 \\ -1 \end{bmatrix} \right\}$ span R^3? Use Theorem 5.14.

9. Repeat Exercise 6(b) of Section 1.5; use Cramer's rule.

10. Repeat Exercise 1 for the following linear system:

$$\begin{aligned} 2x_1 + 3x_2 + 7x_3 &= 0 \\ -2x_1 \qquad\quad - 4x_3 &= 0 \\ x_1 + 2x_2 + 4x_3 &= 0. \end{aligned}$$

11. Is $S = \{t^3 + t + 1, 2t^2 + 3, t - 1, 2t^3 - 2t^2\}$ a basis for P_3? Use Theorem 5.14.

12. Is $S = \left\{ \begin{bmatrix} 1 & 1 \\ 0 & 1 \end{bmatrix}, \begin{bmatrix} 0 & 2 \\ 1 & 3 \end{bmatrix}, \begin{bmatrix} 0 & 0 \\ -1 & 2 \end{bmatrix}, \begin{bmatrix} 0 & -2 \\ 0 & 3 \end{bmatrix} \right\}$ a basis for $_2R_2$? Use Theorem 5.14.

13. Let $\alpha = a_1\mathbf{i} + a_2\mathbf{j} + a_3\mathbf{k}$, $\beta = b_1\mathbf{i} + b_2\mathbf{j} + b_3\mathbf{k}$, and $\gamma = c_1\mathbf{i} + c_2\mathbf{j} + c_3\mathbf{k}$ be vectors in R^3. Show that

$$(\alpha \times \beta) \cdot \gamma = \begin{vmatrix} a_1 & a_2 & a_3 \\ b_1 & b_2 & b_3 \\ c_1 & c_2 & c_3 \end{vmatrix}.$$

14. Compute $\alpha \times \beta$ by the method of Example 4.
(a) $\alpha = 2\mathbf{i} + 3\mathbf{j} + 4\mathbf{k}$, $\beta = -\mathbf{i} + 3\mathbf{j} - \mathbf{k}$.
(b) $\alpha = \mathbf{i} + \mathbf{k}$, $\beta = 2\mathbf{i} + 3\mathbf{j} - \mathbf{k}$.
(c) $\alpha = \mathbf{i} - \mathbf{j} + 2\mathbf{k}$, $\beta = 3\mathbf{i} - 4\mathbf{j} + \mathbf{k}$.
(d) $\alpha = 2\mathbf{i} + \mathbf{j} - 2\mathbf{k}$, $\beta = \mathbf{i} + 3\mathbf{k}$.

5.6. Determinants from a Computational Point of View

So far we have developed two methods for solving a linear system: Gaussian elimination (or Gauss–Jordan reduction) and Cramer's rule. We also have two methods for inverting a nonsingular matrix: the method involving determinants and the method developed in Section 1.6, which uses elementary matrices. We must then develop some criteria for choosing one or the other method depending upon our needs.

In general, if we are seeking numerical answers, then any method involving determinants can be used for $n \le 4$. We shall compare Gaussian elimination and Cramer's rule for solving the linear system $AX = B$, when A is 25×25.

If we find X by Cramer's rule, then we must first obtain $|A|$. We can find $|A|$ by cofactor expansion, say $|A| = a_{11}A_{11} + a_{21}A_{21} + \cdots + a_{n1}A_{n1}$, where we have expanded about the first column. Note that if each cofactor is available, we require 25 multiplications. Now each cofactor A_{ij} is the determinant of a 24×24 matrix, and it can be expanded about a given row or column, requiring 24 multiplications. Thus the computation of $|A|$ requires $25 \times 24 \times 23 \times \cdots \times 2 \times 1 = 25!$ multiplications. If we were to use a futuristic computer capable of performing *100 billion* multiplications *per second*, it would take about *4 million years* to evaluate $|A|$. However, Gaussian elimination takes about 25^3 multiplications, and we would find the solution in less than *1 second*. Of course, if A is $n \times n$, we can compute $|A|$ in a much more efficient way, by using elementary row operations to reduce A to triangular form and then using Theorem 5.7 (see Example 8 in Section 5.2). When implemented this way, Cramer's rule will require approximately n^4 multiplications, compared to n^3 multiplications for Gaussian elimination.

The importance of determinants does not lie in their computational usage; determinants enable one to express the inverse of a matrix and the solution to a system of n linear equations in n unknowns by means of *expressions* or *formulas*. Gaussian elimination and the method for finding A^{-1} using elementary matrices have the property that we cannot write a *formula* for the answer; we must proceed numerically to obtain the answer. Sometimes we do not need a numerical answer but merely an expression for the answer, because we may wish to further manipulate the answer, for example, integrate it. However, the most important reason for studying determinants is that they play a key role in the further study of the properties of a linear transformation mapping a vector space V into itself. This study will be undertaken in Chapter 6.

CHAPTER **6**

Eigenvalues and Eigenvectors

6.1. Diagonalization

At this point we return to the question raised at the end of Section 4.5: If $L: V \to V$ is a linear transformation of an n-dimensional vector space V into itself (a linear operator on V), when and how can we find a basis S for V such that L is represented with respect to S by a diagonal matrix D? In this section we shall formulate this problem precisely; we also define some pertinent terminology. Later in the chapter we settle the question for symmetric matrices ($A = A'$) and briefly discuss the situation in the general case. In this chapter every matrix is square.

Definition 6.1. Let $L: V \to V$ be a linear transformation of an n-dimensional vector space V into itself. We say that L is **diagonalizable** or can be **diagonalized** if there exists a basis S for V such that L is represented with respect to S by a diagonal matrix D.

The linear transformation defined in Example 2 of Section 4.5 is a diagonalizable linear transformation.

Let $L: V \to V$ be a diagonalizable linear transformation of an n-dimensional vector space V into itself and let $S = \{\alpha_1, \alpha_2, \ldots, \alpha_n\}$ be a basis for V such that L is represented with respect to S by a diagonal matrix

$$D = \begin{bmatrix} c_1 & 0 & \cdots & & 0 \\ 0 & c_2 & \cdots & & 0 \\ \vdots & & & & 0 \\ & & & & \vdots \\ 0 & & \cdots & 0 & c_n \end{bmatrix}.$$

Now recall that if D represents L with respect to S, then the jth column of D is the coordinate vector $[L(\alpha_j)]_S$ of $L(\alpha_j)$ with respect to S. Thus we have

$$[L(\alpha_j)]_S = \begin{bmatrix} 0 \\ 0 \\ \vdots \\ c_j \\ 0 \\ \vdots \\ 0 \end{bmatrix} \leftarrow j\text{th row,}$$

which means that

$$L(\alpha_j) = 0\alpha_1 + 0\alpha_2 + \cdots + 0\alpha_{j-1} + c_j\alpha_j + 0\alpha_{j+1} + \cdots + 0\alpha_n = c_j\alpha_j.$$

Conversely, let $S = \{\alpha_1, \alpha_2, \ldots, \alpha_n\}$ be a basis for V such that

$$L(\alpha_j) = c_j\alpha_j = 0\alpha_1 + 0\alpha_2 + \cdots$$
$$+ 0\alpha_{j-1} + c_j\alpha_j + 0\alpha_{j+1} + \cdots + 0\alpha_n \qquad \text{for } j = 1, 2, \ldots, n.$$

We now determine the matrix representing L with respect to S. The jth column

of this matrix is $[L(\alpha_j)]_S = \begin{bmatrix} 0 \\ 0 \\ \vdots \\ c_j \\ 0 \\ \vdots \\ 0 \end{bmatrix}$. Hence $D = \begin{bmatrix} c_1 & 0 & \cdots & & 0 \\ 0 & c_2 & \cdots & & 0 \\ \vdots & & & & \vdots \\ & & & & 0 \\ 0 & & \cdots & 0 & c_n \end{bmatrix}$, a

diagonal matrix, represents L with respect to S. Thus L is diagonalizable.

The situation described in the preceding paragraph is extremely important, and we now formulate some terminology for it.

Definition 6.2. Let $L: V \to V$ be a linear transformation of an n-dimensional vector space V into itself (a linear operator on V). The real number c is called an **eigenvalue** of L if there exists a nonzero vector α in V such that

$$L(\alpha) = c\alpha. \tag{1}$$

Every nonzero vector α satisfying this equation is then called an **eigenvector** of L **associated with the eigenvalue** c. We might mention that the word "eigenvalue" is a hybrid one ("eigen" in German means "proper"). Eigenvalues are also called **proper, characteristic,** or **latent values,** and eigenvectors are also called **proper, characteristic,** or **latent vectors.**

Note that $\alpha = \theta_V$ always satisfies (1), but θ_V is not an eigenvector, since we insist that an eigenvector be a nonzero vector.

In some applications one encounters matrices with complex entries and vector spaces with scalars that are complex numbers. (Some remarks on these ideas are made at the beginning of the next section.) In such a setting the preceding definition of eigenvalue is modified so that an eigenvalue can be a real *or* a complex number. Such treatments are presented in more advanced books. In this book we require that an eigenvalue be a real number.

We can now state the following theorem, whose proof has been given.

Theorem 6.1. *Let $L: V \to V$ be a linear transformation of an n-dimensional vector space V into itself. Then L is diagonalizable if and only if V has a basis S of eigenvectors of L. Moreover, if D is the diagonal matrix representing L with respect to S, then the entries on the main diagonal of D are the eigenvalues of L.*

EXAMPLE 1. Let $L: V \to V$ be the linear operator defined by $L(\alpha) = 2\alpha$. We can see that the only eigenvalue of L is $c = 2$ and that every nonzero vector in V is an eigenvector of L associated with the eigenvalue $c = 2$.

Let $L: V \to V$ be a linear operator on V. If c is an eigenvalue of L and α is an eigenvector associated with c, then α has the property that L maps α into a scalar multiple of itself, the multiplier being the eigenvalue. We thus have a rather nice geometric interpretation of eigenvalues and eigenvectors.

In Figure 6.1 we show α and $L(\alpha)$ for the cases $c > 1$, $0 < c < 1$, and $c < 0$. Thus, if $c > 1$, then L is a dilation, and if $0 < c < 1$, then L is a contraction; if $c < 0$, then L reverses the direction of α.

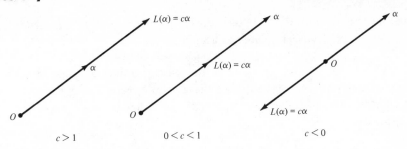

FIGURE 6.1

EXAMPLE 2. Let $L: R^2 \to R^2$ be the linear transformation defined by $L\left(\begin{bmatrix} a_1 \\ a_2 \end{bmatrix}\right) = \begin{bmatrix} a_2 \\ a_1 \end{bmatrix}$. Then we can see that

$$L\left(\begin{bmatrix} a \\ a \end{bmatrix}\right) = 1\begin{bmatrix} a \\ a \end{bmatrix} \quad \text{and} \quad L\left(\begin{bmatrix} a \\ -a \end{bmatrix}\right) = -1\begin{bmatrix} a \\ -a \end{bmatrix}.$$

Thus any vector of the form $\begin{bmatrix} a \\ a \end{bmatrix}$, for example, $\alpha_1 = \begin{bmatrix} 1 \\ 1 \end{bmatrix}$, where a is any nonzero real number, is an eigenvector of L associated with the eigenvalue $c = 1$; any vector of the form $\begin{bmatrix} a \\ -a \end{bmatrix}$, such as $\alpha_2 = \begin{bmatrix} 1 \\ -1 \end{bmatrix}$, where a is any nonzero real number, is an eigenvector of L associated with the eigenvalue $c = -1$ (see Figure 6.2).

Example 2 shows that an eigenvalue c can have associated with it many different eigenvectors. In fact, if α is an eigenvector of L associated with the eigenvalue c [that is, $L(\alpha) = c\alpha$], then $L(r\alpha) = rL(\alpha) = r(c\alpha) = c(r\alpha)$, for any

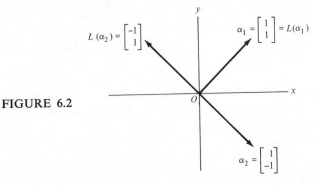

FIGURE 6.2

real number r. Thus, if $r \neq 0$, then $r\alpha$ is also an eigenvector of L associated with c so that eigenvectors are never unique.

EXAMPLE 3. Let $L: R^2 \to R^2$ be the linear transformation performing a counterclockwise $90°$ rotation. Note that the only vector α mapped by L into a multiple of itself is the zero vector. Hence L has no eigenvalues and no eigenvectors.

EXAMPLE 4. Let $L: R_2 \to R_2$ be defined by $L([a_1 \quad a_2]) = [0 \quad a_2]$. We can then see that $L([a \quad 0]) = 0[a \quad 0]$ so that a vector of the form $[a \quad 0]$, where a is any nonzero real number (such as $[2 \quad 0]$), is an eigenvector of L associated with the eigenvalue $c = 0$. Also, $L([0 \quad a]) = 1[0 \quad a]$ so that a vector of the form $[0 \quad a]$ (such as $[0 \quad 1]$), where a is any nonzero real number, is an eigenvector of L associated with the eigenvalue $c = 1$.

Example 4 points out the fact that although the zero *vector* cannot be an eigenvector, zero can be an eigenvalue.

If A is an $n \times n$ matrix, we can consider, as in Section 4.2, the linear transformation $L: R^n \to R^n$ defined by $L(\alpha) = A\alpha$ for α in R^n. If c is a scalar and $\alpha \neq \theta$ a vector in R^n such that

$$A\alpha = c\alpha,$$

then we shall say that c is an **eigenvalue** of A and α is an **eigenvector** of A **associated with** c. That is, c is an eigenvalue of L and α is an eigenvector of L associated with c. We also say that A is **diagonalizable**, or can be **diagonalized**, if A is similar to a diagonal matrix D. Recall from Theorem 4.11 that if A and D are similar, then they both represent the same linear transformation $L: R^n \to R^n$, with respect to corresponding ordered bases S and T for R^n. Since L is represented by D with respect to T, we see that L is diagonalizable. Thus we can restate Theorem 6.1 as follows.

Theorem 6.2. *An $n \times n$ matrix A is similar to a diagonal matrix D if and only if R^n has a basis of eigenvectors of A. Moreover, the elements on the main diagonal of D are the eigenvalues of A.*

To use Theorem 6.2, we need only show that there is a set of n eigenvectors of A that are linearly independent, since n linearly independent vectors in R^n form a basis for R^n.

EXAMPLE 5. Let $A = \begin{bmatrix} 1 & 1 \\ -2 & 4 \end{bmatrix}$. We wish to find the eigenvalues of A

and their associated eigenvectors. Thus we wish to find all scalars c and all nonzero vectors $\alpha = \begin{bmatrix} a_1 \\ a_2 \end{bmatrix}$ that satisfy (1), obtaining

$$\begin{bmatrix} 1 & 1 \\ -2 & 4 \end{bmatrix}\begin{bmatrix} a_1 \\ a_2 \end{bmatrix} = c\begin{bmatrix} a_1 \\ a_2 \end{bmatrix},$$

which yields

$$\begin{matrix} a_1 + a_2 = ca_1 \\ -2a_1 + 4a_2 = ca_2 \end{matrix} \quad \text{or} \quad \begin{matrix} (c-1)a_1 - a_2 = 0 \\ 2a_1 + (c-4)a_2 = 0. \end{matrix} \qquad (2)$$

This homogeneous system of two equations in two unknowns has a nontrivial solution if and only if the determinant of the coefficient matrix is zero. Thus

$$\begin{vmatrix} c-1 & -1 \\ 2 & c-4 \end{vmatrix} = 0.$$

This means that $c^2 - 5c + 6 = 0 = (c-3)(c-2)$, and so $c = 2$ and $c = 3$ are the eigenvalues of A. To find an eigenvector of A associated with $c = 2$, we set up $A\alpha = 2\alpha$, or

$$\begin{bmatrix} 1 & 1 \\ -2 & 4 \end{bmatrix}\begin{bmatrix} a_1 \\ a_2 \end{bmatrix} = 2\begin{bmatrix} a_1 \\ a_2 \end{bmatrix}.$$

This gives

$$\begin{matrix} a_1 + a_2 = 2a_1 \\ -2a_1 + 4a_2 = 2a_2 \end{matrix} \quad \text{or} \quad \begin{matrix} (2-1)a_1 - a_2 = 0 \\ 2a_1 + (2-4)a_2 = 0 \end{matrix}$$

$$\text{or} \quad \begin{matrix} a_1 - a_2 = 0. \\ 2a_1 - 2a_2 = 0. \end{matrix}$$

Note that we could have obtained this last homogeneous system by merely substituting $c = 2$ in (2). Then the vector $\begin{bmatrix} 1 \\ 1 \end{bmatrix}$ is a nontrivial solution to this last homogeneous system and is, therefore, an eigenvector associated with $c = 2$. Similarly, for $c = 3$ we get, from (2),

$$\begin{matrix} (3-1)a_1 - a_2 = 0 \\ 2a_1 + (3-4)a_2 = 0 \end{matrix} \quad \text{or} \quad \begin{matrix} 2a_1 - a_2 = 0 \\ 2a_1 - a_2 = 0. \end{matrix}$$

Thus $\begin{bmatrix} 1 \\ 2 \end{bmatrix}$ is an eigenvector associated with $c = 3$.

Since $S = \left\{ \begin{bmatrix} 1 \\ 1 \end{bmatrix}, \begin{bmatrix} 1 \\ 2 \end{bmatrix} \right\}$ is linearly independent, R^2 has a basis of two eigen-vectors of A and hence A can be diagonalized. From Theorem 6.2 we conclude that A is similar to $D = \begin{bmatrix} 2 & 0 \\ 0 & 3 \end{bmatrix}$.

EXAMPLE 6. Let $A = \begin{bmatrix} 1 & 1 \\ 0 & 1 \end{bmatrix}$. To find out if A can be diagonalized, we set up (1), obtaining

$$\begin{bmatrix} 1 & 1 \\ 0 & 1 \end{bmatrix} \begin{bmatrix} a_1 \\ a_2 \end{bmatrix} = c \begin{bmatrix} a_1 \\ a_2 \end{bmatrix}.$$

This means that

$$a_1 + a_2 = ca_1$$
$$a_2 = ca_2.$$

This can only hold for $c = 1$. If $c = 1$, then $a_2 = 0$. Thus $\begin{bmatrix} 1 \\ 0 \end{bmatrix}$ is an eigenvector of A associated with the eigenvalue $c = 1$. In fact, every eigenvector associated with $c = 1$ is of the form $\begin{bmatrix} a \\ 0 \end{bmatrix}$, with $a \neq 0$. Since A does not have two linearly independent eigenvectors, we conclude that A cannot be diagonalized.

If an $n \times n$ matrix A is similar to a diagonal matrix D, then $P^{-1}AP = D$ for some nonsingular matrix P. This means that $AP = PD$. Let

$$D = \begin{bmatrix} c_1 & 0 & \cdots & & 0 \\ 0 & c_2 & \cdots & & 0 \\ \vdots & & \cdots & & \vdots \\ & & & & 0 \\ 0 & & \cdots & 0 & c_n \end{bmatrix},$$

and let $\alpha_j, j = 1, 2, \ldots, n$, be the jth column of P. Note that the jth column of AP is $A\alpha_j$, and the jth column of PD is $c_j\alpha_j$ (see Exercise 15 in Section 1.2). Thus we have $A\alpha_j = c_j\alpha_j$, which means that c_j is an eigenvalue of A and α_j is a corresponding eigenvector.

Conversely, if $c_1, c_2, \ldots, c_n$ are n eigenvalues of an $n \times n$ matrix A and $\alpha_1, \alpha_2, \ldots, \alpha_n$ are corresponding eigenvectors forming a linearly independent set in R^n, we let P be the matrix whose jth column is α_j. From Theorem 5.14

P is nonsingular. Since $A\alpha_j = c_j\alpha_j$, $j = 1, 2, \ldots, n$, we have $AP = PD$, or $P^{-1}AP = D$, which means that A is diagonalizable. Thus, if the eigenvectors $\alpha_1, \alpha_2, \ldots, \alpha_n$ of the $n \times n$ matrix A form a linearly independent set in R^n, we can diagonalize A by letting P be the matrix whose jth column is α_j, and we find that $P^{-1}AP = D$, a diagonal matrix whose entries on the main diagonal are the corresponding eigenvalues of A.

EXAMPLE 7. Let A be as in Example 5. The eigenvalues of A are $c_1 = 2$ and $c_2 = 3$, and corresponding eigenvectors are $\alpha_1 = \begin{bmatrix} 1 \\ 1 \end{bmatrix}$ and $\alpha_2 = \begin{bmatrix} 1 \\ 2 \end{bmatrix}$. Thus $P = \begin{bmatrix} 1 & 1 \\ 1 & 2 \end{bmatrix}$ and $P^{-1} = \begin{bmatrix} 2 & -1 \\ -1 & 1 \end{bmatrix}$ (verify). Hence

$$P^{-1}AP = \begin{bmatrix} 2 & -1 \\ -1 & 1 \end{bmatrix} \begin{bmatrix} 1 & 1 \\ -2 & 4 \end{bmatrix} \begin{bmatrix} 1 & 1 \\ 1 & 2 \end{bmatrix} = \begin{bmatrix} 2 & 0 \\ 0 & 3 \end{bmatrix}.$$

On the other hand, if we let $c_1 = 3$ and $c_2 = 2$, then $\alpha_1 = \begin{bmatrix} 1 \\ 2 \end{bmatrix}$ and $\alpha_2 = \begin{bmatrix} 1 \\ 1 \end{bmatrix}$, $P = \begin{bmatrix} 1 & 1 \\ 2 & 1 \end{bmatrix}$ and $P^{-1} = \begin{bmatrix} -1 & 1 \\ 2 & -1 \end{bmatrix}$, and

$$P^{-1}AP = \begin{bmatrix} -1 & 1 \\ 2 & -1 \end{bmatrix} \begin{bmatrix} 1 & 1 \\ -2 & 4 \end{bmatrix} \begin{bmatrix} 1 & 1 \\ 2 & 1 \end{bmatrix} = \begin{bmatrix} 3 & 0 \\ 0 & 2 \end{bmatrix}.$$

In Examples 1 through 4 we found eigenvalues and eigenvectors by inspection. In Example 5 we proceeded in a more systematic fashion. We can use the procedure of Example 5 as our standard method, as follows.

Definition 6.3. Let $A = \begin{bmatrix} a_{11} & a_{12} & \cdots & a_{1n} \\ a_{21} & a_{22} & \cdots & a_{2n} \\ \vdots & & & \vdots \\ a_{n1} & a_{n2} & \cdots & a_{nn} \end{bmatrix}$ be an $n \times n$ matrix. Then

the determinant of the matrix

$$tI_n - A = \begin{bmatrix} t - a_{11} & -a_{12} & \cdots & -a_{1n} \\ -a_{21} & t - a_{22} & \cdots & -a_{2n} \\ \vdots & & & \vdots \\ -a_{n1} & -a_{n2} & \cdots & t - a_{nn} \end{bmatrix}$$

is called the **characteristic polynomial** of A. The equation $f(t) = |tI_n - A|$ $= 0$ is called the **characteristic equation** of A.

Recall from Chapter 5 that each term in the expansion of the determinant of an $n \times n$ matrix is a product of n entries of the matrix, containing exactly one entry from each row and column. Thus, if we expand $|tI_n - A|$, we obtain a polynomial of degree n. The expression involving t^n in the characteristic polynomial of A comes from the product $(t - a_{11})(t - a_{22}) \cdots (t - a_{nn})$, and so the coefficient of t^n is 1. We can then write

$$|tI_n - A| = f(t) = t^n + a_1 t^{n-1} + a_2 t^{n-2} + \cdots + a_{n-1}t + a_n.$$

Note that if we let $t = 0$, then we get $|-A| = a_n$, and thus the constant term of the characteristic polynomial of A is $a_n = (-1)^n |A|$.

EXAMPLE 8. Let $A = \begin{bmatrix} 1 & 2 & -1 \\ 1 & 0 & 1 \\ 4 & -4 & 5 \end{bmatrix}$. The characteristic polynomial of A is

$$f(t) = |tI_3 - A| = \begin{vmatrix} t-1 & -2 & 1 \\ -1 & t & -1 \\ -4 & 4 & t-5 \end{vmatrix} = t^3 - 6t^2 + 11t - 6$$

(verify).

We now connect the characteristic polynomial of a matrix with its eigenvalues in the following theorem.

Theorem 6.3. *Let A be an $n \times n$ matrix. The eigenvalues of A are the real roots of the characteristic polynomial of A.*

Proof: Let α in R^n be an eigenvector of A associated with the eigenvalue c. Then

$$A\alpha = c\alpha, \quad \text{or} \quad A\alpha = (cI_n)\alpha, \quad \text{or} \quad (cI_n - A)\alpha = O.$$

This is a homogeneous system of n equations in n unknowns; a nontrivial solution exists if and only if $|cI_n - A| = 0$. Hence c is a root of the characteristic polynomial of A.

Conversely, if c is a real root of the characteristic polynomial of A, then $|cI_n - A| = 0$, so the homogeneous system $(cI_n - A)\alpha = O$ has a nontrivial solution. Hence c is an eigenvalue of A.

Thus, to find the eigenvalues of a given matrix A, we must find the real roots of the characteristic polynomial $f(t)$. There are many methods for finding approximations to the roots of a polynomial, some of them more effective than others. Two results that are sometimes useful in this connection are: (1) the product of all the roots of the polynomial

$$f(t) = t^n + a_1 t^{n-1} + \cdots + a_{n-1} t + a_n$$

is $(-1)^n a_n$, and (2) if $a_1, a_2, \ldots, a_n$ are integers, then $f(t)$ cannot have a rational root that is not already an integer. Thus as possible real roots of $f(t)$ one can try the factors of a_n. Of course, $f(t)$ might well have irrational roots.

To minimize the computational effort and as a convenience to the reader, all the characteristic polynomials to be solved in the rest of this chapter have only integer roots, and each of these roots is a factor of the constant term of the characteristic polynomial of A. The corresponding eigenvectors are obtained by substituting for c in

$$(cI_n - A)\alpha = O \tag{3}$$

and solving the resulting homogeneous system.

EXAMPLE 9. The characteristic polynomial of the matrix A defined in Example 8 is

$$f(t) = t^3 - 6t^2 + 11t - 6 = (t - 1)(t - 2)(t - 3).$$

The eigenvalues of A are $c_1 = 1$, $c_2 = 2$, and $c_3 = 3$. To find an eigenvector α_1 associated with $c_1 = 1$, we substitute $c = 1$ in (3), obtaining

$$\begin{bmatrix} 1-1 & -2 & 1 \\ -1 & 1 & -1 \\ -4 & 4 & 1-5 \end{bmatrix} \begin{bmatrix} a_1 \\ a_2 \\ a_3 \end{bmatrix} = \begin{bmatrix} 0 \\ 0 \\ 0 \end{bmatrix}$$

or

$$\begin{bmatrix} 0 & -2 & 1 \\ -1 & 1 & -1 \\ -4 & 4 & -4 \end{bmatrix} \begin{bmatrix} a_1 \\ a_2 \\ a_3 \end{bmatrix} = \begin{bmatrix} 0 \\ 0 \\ 0 \end{bmatrix}.$$

The vector $\begin{bmatrix} -\dfrac{a}{2} \\ \dfrac{a}{2} \\ a \end{bmatrix}$ is a solution for any real number a. Thus $\alpha_1 = \begin{bmatrix} -1 \\ 1 \\ 2 \end{bmatrix}$ is an

eigenvector of A associated with $c_1 = 1$.

To find an eigenvector α_2 associated with $c_2 = 2$, we substitute $c = 2$ in (3), obtaining

$$\begin{bmatrix} 2-1 & -2 & 1 \\ -1 & 2 & -1 \\ -4 & 4 & 2-5 \end{bmatrix} \begin{bmatrix} a_1 \\ a_2 \\ a_3 \end{bmatrix} = \begin{bmatrix} 0 \\ 0 \\ 0 \end{bmatrix}$$

or

$$\begin{bmatrix} 1 & -2 & 1 \\ -1 & 2 & -1 \\ -4 & 4 & -3 \end{bmatrix} \begin{bmatrix} a_1 \\ a_2 \\ a_3 \end{bmatrix} = \begin{bmatrix} 0 \\ 0 \\ 0 \end{bmatrix}.$$

The vector $\begin{bmatrix} -\dfrac{a}{2} \\ \dfrac{a}{4} \\ a \end{bmatrix}$ is a solution for any real number a. Thus $\alpha_2 = \begin{bmatrix} -2 \\ 1 \\ 4 \end{bmatrix}$ is an

eigenvector.

To find an eigenvector α_3 associated with $c_3 = 3$, we substitute $c = 3$ in (3), obtaining

$$\begin{bmatrix} 3-1 & -2 & 1 \\ -1 & 3 & -1 \\ -4 & 4 & 3-5 \end{bmatrix} \begin{bmatrix} a_1 \\ a_2 \\ a_3 \end{bmatrix} = \begin{bmatrix} 0 \\ 0 \\ 0 \end{bmatrix}$$

or

$$\begin{bmatrix} 2 & -2 & 1 \\ -1 & 3 & -1 \\ -4 & 4 & -2 \end{bmatrix} \begin{bmatrix} a_1 \\ a_2 \\ a_3 \end{bmatrix} = \begin{bmatrix} 0 \\ 0 \\ 0 \end{bmatrix}.$$

The vector $\begin{bmatrix} -\dfrac{a}{4} \\ \dfrac{a}{4} \\ a \end{bmatrix}$ is a solution for any real number a. Thus $\alpha_3 = \begin{bmatrix} -1 \\ 1 \\ 4 \end{bmatrix}$ is an

eigenvector.

Of course, the characteristic polynomial of a matrix may have complex roots and it may even have no real roots. However, in the important case of symmetric matrices, all the roots of the characteristic polynomial are real. We shall prove this in Section 6.2 (Theorem 6.6).

EXAMPLE 10. The characteristic polynomial of $A = \begin{bmatrix} 0 & 1 \\ -1 & 0 \end{bmatrix}$ is $f(t) = t^2 + 1$, which has no real roots. Thus A has no eigenvalues and cannot be diagonalized.

Now that we know how to find the eigenvalues of A in a systematic fashion, the following is a useful theorem, because it identifies a large class of matrices that can be diagonalized.

Theorem 6.4. *An $n \times n$ matrix A is diagonalizable if all the roots of its characteristic polynomial are real and distinct.*

Proof: Let $\{c_1, c_2, \ldots, c_n\}$ be the set of distinct eigenvalues of A, and let $S = \{\alpha_1, \alpha_2, \ldots, \alpha_n\}$ be a set of associated eigenvectors. We wish to prove that S is a basis for R^n, and it suffices to show that S is linearly independent.

Suppose that S is linearly dependent. Then Theorem 2.4 implies that some vector α_j is a linear combination of the preceding vectors in S. We can assume that $S_1 = \{\alpha_1, \alpha_2, \ldots, \alpha_{j-1}\}$ is linearly independent, for otherwise one of the vectors in S_1 is a linear combination of the preceding ones, and we can choose a new set S_2, and so on. We thus have that S_1 is linearly independent and that

$$\alpha_j = a_1\alpha_1 + a_2\alpha_2 + \cdots + a_{j-1}\alpha_{j-1}, \tag{4}$$

where $a_1, a_2, \ldots, a_{j-1}$ are real numbers. This means that

$$\begin{aligned} A\alpha_j &= A(a_1\alpha_1 + a_2\alpha_2 + \cdots + a_{j-1}\alpha_{j-1}) \\ &= a_1 A\alpha_1 + a_2 A\alpha_2 + \cdots + a_{j-1}A\alpha_{j-1}. \end{aligned} \tag{5}$$

Since $c_1, c_2, \ldots, c_j$ are eigenvalues and $\alpha_1, \alpha_2, \ldots, \alpha_j$ are associated eigenvectors, we know that $A\alpha_i = c_i\alpha_i$, for $i = 1, 2, \ldots, n$. Substituting in (5), we have

$$c_j\alpha_j = a_1 c_1\alpha_1 + a_2 c_2\alpha_2 + \cdots + a_{j-1}c_{j-1}\alpha_{j-1}. \tag{6}$$

Multiplying (4) by c_j, we get

$$c_j\alpha_j = c_j a_1\alpha_1 + c_j a_2\alpha_2 + \cdots + c_j a_{j-1}\alpha_{j-1}. \tag{7}$$

Subtracting (7) from (6), we have

$$\theta = c_j\alpha_j - c_j\alpha_j = a_1(c_1 - c_j)\alpha_1 + a_2(c_2 - c_j)\alpha_2 + \cdots + a_{j-1}(c_{j-1} - c_j)\alpha_{j-1}.$$

Since S_1 is linearly independent, we must have

$$a_1(c_1 - c_j) = 0, \qquad a_2(c_2 - c_j) = 0, \ldots, a_{j-1}(c_{j-1} - c_j) = 0.$$

Now $(c_1 - c_j) \neq 0, (c_2 - c_j) \neq 0, \ldots, (c_{j-1} - c_j) \neq 0$, which implies that

$$a_1 = a_2 = \cdots = a_{j-1} = 0.$$

This means that $\alpha_j = \theta$, which is impossible if α_j is an eigenvector. Hence S is linearly independent, and so A is diagonalizable.

In the proof of Theorem 6.4 we have actually proved that eigenvectors associated with distinct eigenvalues form a linearly independent set (Exercise 28).

If all the roots of the characteristic polynomial of A are real and not all distinct, then A may or may not be diagonalizable. The characteristic polynomial of A can be written as the product of n factors, each of the form $t - c_0$, where c_0 is a root of the characteristic polynomial. Now the eigenvalues of A are the real roots of the characteristic polynomial of A. Thus the characteristic polynomial can be written as

$$(t - c_1)^{k_1}(t - c_2)^{k_2} \cdots (t - c_r)^{k_r},$$

where $c_1, c_2, \ldots, c_r$ are the distinct eigenvalues of A, and $k_1, k_2, \ldots, k_r$ are integers whose sum is n. The integer k_i is called the **multiplicity** of c_i. Thus, in Example 6, $c = 1$ is an eigenvalue of $A = \begin{bmatrix} 1 & 1 \\ 0 & 1 \end{bmatrix}$ of multiplicity 2. It can be shown that if the roots of the characteristic polynomial of A are all real, then A can be diagonalized if, for each eigenvalue of multiplicity k, we can find k linearly independent eigenvectors. This means that the solution space of the homogeneous system $(cI_n - A)\alpha = O$ has dimension k. It can also be shown that if c is an eigenvalue of A of multiplicity k, then we can never find more than k linearly independent eigenvectors associated with c.

EXAMPLE 11. Let $A = \begin{bmatrix} 0 & 0 & 1 \\ 0 & 1 & 2 \\ 0 & 0 & 1 \end{bmatrix}$. The characteristic polynomial of A is $f(t) = t(t - 1)^2$ (verify), so the eigenvalues of A are $c_1 = 0, c_2 = 1$, and

$c_3 = 1$; $c_2 = 1$ is an eigenvalue of multiplicity 2. We now consider the eigenvectors associated with the eigenvalues $c_2 = c_3 = 1$. They are obtained by solving the homogeneous system (3):

$$\begin{bmatrix} 1 & 0 & -1 \\ 0 & 0 & -2 \\ 0 & 0 & 0 \end{bmatrix} \begin{bmatrix} a_1 \\ a_2 \\ a_3 \end{bmatrix} = \begin{bmatrix} 0 \\ 0 \\ 0 \end{bmatrix}.$$

The solutions are the vectors of the form $\begin{bmatrix} 0 \\ r \\ 0 \end{bmatrix}$, where r is any real number, so the dimension of the solution space of $(1I_3 - A)\alpha = O$ is 1 (why?), and we cannot find two linearly independent eigenvectors. Thus A cannot be diagonalized.

EXAMPLE 12. Let $A = \begin{bmatrix} 0 & 0 & 0 \\ 0 & 1 & 0 \\ 1 & 0 & 1 \end{bmatrix}$. The characteristic polynomial of A is $f(t) = t(t-1)^2$ (verify), so the eigenvalues of A are $c_1 = 0$, $c_2 = 1$, and $c_3 = 1$; and $c_2 = 1$ is again an eigenvalue of multiplicity 2. Now we consider the solution space of (3):

$$\begin{bmatrix} 1 & 0 & 0 \\ 0 & 0 & 0 \\ -1 & 0 & 0 \end{bmatrix} \begin{bmatrix} a_1 \\ a_2 \\ a_3 \end{bmatrix} = \begin{bmatrix} 0 \\ 0 \\ 0 \end{bmatrix}.$$

The solutions are the vectors of the form $\begin{bmatrix} 0 \\ r \\ s \end{bmatrix}$ for any real numbers r and s.

Thus $\alpha_2 = \begin{bmatrix} 0 \\ 1 \\ 0 \end{bmatrix}$ and $\alpha_3 = \begin{bmatrix} 0 \\ 0 \\ 1 \end{bmatrix}$ are eigenvectors.

Next we look for an eigenvector associated with $c_1 = 0$. We have to solve (3):

$$\begin{bmatrix} 0 & 0 & 0 \\ 0 & -1 & 0 \\ -1 & 0 & -1 \end{bmatrix} \begin{bmatrix} a_1 \\ a_2 \\ a_3 \end{bmatrix} = \begin{bmatrix} 0 \\ 0 \\ 0 \end{bmatrix}.$$

The solutions are the vectors of the form $\begin{bmatrix} t \\ 0 \\ -t \end{bmatrix}$ for any real number t. Thus

$\alpha_1 = \begin{bmatrix} 1 \\ 0 \\ -1 \end{bmatrix}$ is an eigenvector associated with $c_1 = 0$. Now $S = \{\alpha_1, \alpha_2, \alpha_3\}$

is linearly independent and so A can be diagonalized.

Thus a matrix may fail to be diagonalizable either because all the roots of its characteristic polynomial are not real numbers, or because its eigenvectors do not form a basis for R^n.

We next prove an important result as Theorem 6.5.

Theorem 6.5. *Similar matrices have the same characteristic polynomial.*

Proof: Let A and B be similar. Then $B = P^{-1}AP$, for some nonsingular matrix P. Let $f_A(t)$ and $f_B(t)$ be the characteristic polynomials of A and B, respectively. Now

$$\begin{aligned} f_B(t) &= |tI_n - B| = |tI_n - P^{-1}AP| = |P^{-1}tI_nP - P^{-1}AP| \\ &= |P^{-1}(tI_n - A)P| = |P^{-1}| \, |tI_n - A| \, |P| \\ &= |P^{-1}| \, |P| \, |tI_n - A| = |tI_n - A| = f_A(t). \end{aligned}$$

Note that we have used the fact that $|P^{-1}| \, |P| = 1$, and that determinants are real numbers, so their order as factors in multiplication does not matter.

We now define the **characteristic polynomial** of a linear operator $L: V \to V$, as the characteristic polynomial of any matrix representing L; by Theorem 6.5 all representations of L will give the same characteristic polynomial. Of course, a scalar c is an **eigenvalue** of L if and only if c is a real root of the characteristic polynomial of L.

Eigenvalues and eigenvectors satisfy many important and interesting properties. For example, if A is an upper (lower) triangular matrix, then the eigenvalues of A are the elements on the main diagonal of A, since the determinant of such a matrix is the product of the elements on its main diagonal. Other properties are developed in the exercises for this section.

It must be pointed out that the method for finding the eigenvalues of a linear transformation or matrix by obtaining the roots of the characteristic polynomial is not practical for $n > 4$, owing to the need for evaluating a determinant. Efficient numerical methods for finding eigenvalues are studied in numerical analysis courses.

6.1. Exercises

1. Let $L: R^2 \to R^2$ be counterclockwise rotation through an angle π. Find the eigenvalues and associated eigenvectors of L.

2. Find the characteristic polynomial of the following matrices.

(a) $\begin{bmatrix} 1 & 2 & 1 \\ 0 & 1 & 2 \\ -1 & 3 & 2 \end{bmatrix}$.
(b) $\begin{bmatrix} 2 & 1 \\ -1 & 3 \end{bmatrix}$.
(c) $\begin{bmatrix} 4 & -1 & 3 \\ 0 & 2 & 1 \\ 0 & 0 & 3 \end{bmatrix}$.

3. Find the characteristic polynomial, the eigenvalues, and their associated eigenvectors for each of the following matrices.

(a) $\begin{bmatrix} 0 & 1 & 2 \\ 0 & 0 & 3 \\ 0 & 0 & 0 \end{bmatrix}$.
(b) $\begin{bmatrix} 1 & 0 & 0 \\ -1 & 3 & 0 \\ 3 & 2 & -2 \end{bmatrix}$.
(c) $\begin{bmatrix} 1 & 1 \\ 1 & 1 \end{bmatrix}$.

4. Prove that if A is an upper (lower) triangular matrix, then the eigenvalues of A are the elements on the main diagonal of A.

5. Find the characteristic polynomial, the eigenvalues and their associated eigenvectors for each of the following matrices.

(a) $\begin{bmatrix} 1 & -1 \\ 2 & 4 \end{bmatrix}$.
(b) $\begin{bmatrix} 2 & -2 & 3 \\ 0 & 3 & -2 \\ 0 & -1 & 2 \end{bmatrix}$.
(c) $\begin{bmatrix} 2 & 2 & 3 \\ 1 & 2 & 1 \\ 2 & -2 & 1 \end{bmatrix}$.

6. Prove that A and A' have the same eigenvalues. What, if anything, can we say about the associated eigenvectors of A and A'?

7. Let $L: P_2 \to P_2$ be the linear transformation defined by $L(p(u)) = p'(u)$ for $p(u)$ in P_2. Is L diagonalizable? If it is, find a basis S for P_2 with respect to which L is represented by a diagonal matrix.

8. Which of the following matrices are diagonalizable?

(a) $\begin{bmatrix} 1 & 4 \\ 1 & -2 \end{bmatrix}$.
(b) $\begin{bmatrix} 1 & 0 \\ -2 & 1 \end{bmatrix}$.
(c) $\begin{bmatrix} 1 & 1 & -2 \\ 4 & 0 & 4 \\ 1 & -1 & 4 \end{bmatrix}$.

(d) $\begin{bmatrix} 1 & 2 & 3 \\ 0 & -1 & 2 \\ 0 & 0 & 2 \end{bmatrix}$.
(e) $\begin{bmatrix} 3 & 1 & 0 \\ 0 & 3 & 1 \\ 0 & 0 & 3 \end{bmatrix}$.

9. Let $L: V \to V$ be a linear transformation, where V is an n-dimensional vector space. Let c be an eigenvalue of L. Prove that the subset of V consisting of θ_V and of all eigenvectors of L associated with c is a subspace of V. This subspace is called the **eigenspace** associated with c.

10. Prove that if c is an eigenvalue of a matrix A with associated eigenvector α, and k is a positive integer, then c^k is an eigenvalue of the matrix $A^k = A \cdot A \cdots A$ (k factors) with associated eigenvector α.

11. Let $A = \begin{bmatrix} 1 & 4 \\ 1 & -2 \end{bmatrix}$ be the matrix of Exercise 8(a). Find the eigenvalues and eigenvectors of A^2 and verify Exercise 10.

12. An $n \times n$ matrix A is called **nilpotent** if $A^k = 0$ for some positive integer k. Prove that if A is nilpotent, then 0 is the only eigenvalue of A. (*Hint:* Use Exercise 10.)

13. Let V the vector space of continuous functions with basis $\{\sin u, \cos u\}$, and let $L: V \to V$ be defined as $L(\alpha(u)) = \alpha'(u)$; that is, L is differentiation. Is L diagonalizable?

14. For each of the following matrices find, if possible, a nonsingular matrix P such that $P^{-1}AP$ is diagonal.

(a) $\begin{bmatrix} 4 & 2 & 3 \\ 2 & 1 & 2 \\ -1 & -2 & 0 \end{bmatrix}$. (b) $\begin{bmatrix} 1 & 1 & 2 \\ 0 & 1 & 0 \\ 0 & 1 & 3 \end{bmatrix}$. (c) $\begin{bmatrix} 1 & 2 & 3 \\ 0 & 1 & 0 \\ 2 & 1 & 2 \end{bmatrix}$.

(d) $\begin{bmatrix} 0 & -1 \\ 2 & 3 \end{bmatrix}$. (e) $\begin{bmatrix} 3 & -2 & 1 \\ 0 & 2 & 0 \\ 0 & 0 & 0 \end{bmatrix}$.

15. Let c be an eigenvalue of the $n \times n$ matrix A. Prove that the subset of R^n consisting of the zero vector and of all eigenvectors of A associated with c is a subspace of R^n. This subspace is called the **eigenspace** associated with c. (This result is a corollary to the result in Exercise 9.)

16. Let $L: R_3 \to R_3$ be defined by

$$L[a_1 \quad a_2 \quad a_3] = [2a_1 + 3a_2 \quad -a_2 + 4a_3 \quad 3a_3].$$

Find the characteristic polynomial and the eigenvalues and eigenvectors of L.

17. Let A be an $n \times n$ matrix.
    (a) Show that $|A|$ is the product of all the roots of the characteristic polynomial of A.
    (b) Show that A is singular if and only if 0 is an eigenvalue of A.
    (c) Also prove the analogous statement for a linear transformation: If $L: V \to V$ is a linear transformation, show that L is not one-one if and only if 0 is an eigenvalue of L.
    (d) Show that if A is nilpotent, then A is singular.

18. Let $L: V \to V$ be an invertible linear operator and let c be an eigenvalue of L with associated eigenvector α.
    (a) Show that $1/c$ is an eigenvalue of L^{-1} with associated eigenvector α.
    (b) State and prove the analogous statement for matrices.

19. Let $L: P_2 \to P_2$ be the linear transformation defined by $L(au^2 + bu + c) = (2a + b + c)u^2 + (2c - 3b)u + 4c$. Find the eigenvalues and eigenvectors of L. Is L diagonalizable?

20. Prove that if A and B are similar matrices, then A and B have the same eigenvalues.

21. Let $A = \begin{bmatrix} a & b \\ c & d \end{bmatrix}$. Find necessary and sufficient conditions for A to be diagonalizable.

22. Let $A = \begin{bmatrix} 1 & 2 & 3 & 4 \\ 0 & -1 & 3 & 2 \\ 0 & 0 & 3 & 3 \\ 0 & 0 & 0 & 2 \end{bmatrix}$ represent the linear transformation

$L: {}_2R_2 \to {}_2R_2$ with respect to the basis

$$S = \left\{ \begin{bmatrix} 1 & 0 \\ 0 & 0 \end{bmatrix}, \begin{bmatrix} 0 & 1 \\ 0 & 0 \end{bmatrix}, \begin{bmatrix} 0 & 0 \\ 1 & 0 \end{bmatrix}, \begin{bmatrix} 0 & 0 \\ 0 & 1 \end{bmatrix} \right\}.$$

Find the eigenvalues and eigenvectors of L.

23. Let A and B be nonsingular $n \times n$ matrices. Prove that AB and BA have the same eigenvalues.

24. Let $A = \begin{bmatrix} 2 & 2 & 3 & 4 \\ 0 & 2 & 3 & 2 \\ 0 & 0 & 1 & 1 \\ 0 & 0 & 0 & 1 \end{bmatrix}$.

(a) Find a basis for the eigenspace (see Exercise 15) associated with the eigenvalue $c_1 = 1$.

(b) Find a basis for the eigenspace associated with the eigenvalue $c_2 = 2$.

25. Which of the following matrices are similar to a diagonal matrix?

(a) $\begin{bmatrix} 2 & 3 & 0 \\ 0 & 1 & 0 \\ 0 & 0 & 2 \end{bmatrix}$.   (b) $\begin{bmatrix} 2 & 3 & 1 \\ 0 & 1 & 0 \\ 0 & 0 & 2 \end{bmatrix}$.   (c) $\begin{bmatrix} -3 & 0 \\ 1 & 2 \end{bmatrix}$.

26. Let V be the vector space of continuous functions with basis $\{e^u, e^{-u}\}$. Let $L: V \to V$ be defined by $L(f(u)) = f'(u)$ for $f(u)$ in V. Show that L is diagonalizable.

27. Prove that if A is diagonalizable, then (a) A' is diagonalizable, and (b) A^k is diagonalizable, where k is a positive integer.

28. Let $c_1, c_2, \ldots, c_k$ be distinct eigenvalues of a matrix A with associated eigenvectors $\alpha_1, \alpha_2, \ldots, \alpha_k$. Prove that $\{\alpha_1, \alpha_2, \ldots, \alpha_k\}$ is linearly independent. (*Hint:* See the proof of Theorem 6.4.)

29. Let A and B be $n \times n$ matrices such that $A\alpha = c\alpha$ and $B\alpha = d\alpha$. Show that
(a) $(A + B)\alpha = (c + d)\alpha$.
(b) $(AB)\alpha = (cd)\alpha$.

30. The **Cayley–Hamilton theorem** states that a matrix satisfies its characteristic equation; that is, if A is an $n \times n$ matrix with a characteristic polynomial

$$f(t) = t^n + a_1 t^{n-1} + \cdots + a_{n-1} t + a_n,$$

then

$$A^n + a_1 A^{n-1} + \cdots + a_{n-1} A + a_n I_n = O.$$

The proof and applications of this result, unfortunately, lie beyond the scope of this book. Verify the Cayley–Hamilton theorem for the following matrices:

(a) $\begin{bmatrix} 1 & 2 & 3 \\ 2 & -1 & 5 \\ 3 & 2 & 1 \end{bmatrix}$. (b) $\begin{bmatrix} 1 & 2 & 3 \\ 0 & 2 & 2 \\ 0 & 0 & -3 \end{bmatrix}$. (c) $\begin{bmatrix} 3 & 3 \\ 2 & 4 \end{bmatrix}$.

31. Let A be an $n \times n$ matrix whose characteristic polynomial is $f(t) = t^n + a_1 t^{n-1} + \cdots + a_{n-1} t + a_n$. If A is nonsingular, show that

$$A^{-1} = -\frac{1}{a_n}(A^{n-1} + a_1 A^{n-2} + \cdots + a_{n-2}A + a_{n-1}I_n).$$

[*Hint:* Use the Cayley–Hamilton theorem (Exercise 30).]

6.2. Diagonalization of Symmetric Matrices

In this section we consider the diagonalization of symmetric matrices ($A = A'$). We restrict our attention to symmetric matrices because they are easier to handle than general matrices and because they arise in many applied problems.

Theorem 6.4 assures us that an $n \times n$ matrix A is diagonalizable if it has n distinct eigenvalues; if this is not so, then A may fail to be diagonalizable. However, every symmetric matrix can be diagonalized; that is, if A is symmetric, there exists a nonsingular matrix P such that $P^{-1}AP = D$ where D is a diagonal matrix. Moreover, P has some noteworthy properties that we remark upon. We thus turn to the study of symmetric matrices in this section.

We first prove that all the roots of the characteristic polynomial of a symmetric matrix are real. This provides us with an opportunity for a brief digression on complex numbers and their place in linear algebra; most readers will have some familiarity with complex numbers.

A **complex number** c is a number of the form $c = a + bi$, where a and b are real numbers and where $i^2 = -1$; a is called the **real part** of c and b is called the **imaginary part** of c. We say that two complex numbers $c_1 = a_1 + b_1 i$ and $c_2 = a_2 + b_2 i$ are **equal** if their real and imaginary parts are equal, that is, if $a_1 = a_2$ and $b_1 = b_2$. Of course, every real number is a complex number— its imaginary part is 0. If $c_1 = a_1 + b_1 i$ and $c_2 = a_2 + b_2 i$ are complex numbers, then their **sum** is

$$c_1 + c_2 = (a_1 + a_2) + (b_1 + b_2)i$$

and their **product** is

$$c_1 c_2 = (a_1 a_2 - b_1 b_2) + (a_1 b_2 + a_2 b_1)i.$$

If $c = a + bi$ is a complex number, then the **conjugate** of c is the complex number $\bar{c} = a - bi$. It is easy to show that if c and d are complex numbers, then the following basic properties of complex arithmetic hold:

1. $\bar{\bar{c}} = c$.
2. $\overline{c + d} = \bar{c} + \bar{d}$.
3. $\overline{cd} = \bar{c}\bar{d}$.
4. c is a real number if and only if $c = \bar{c}$.
5. $c\bar{c}$ is a nonnegative (real) number and $c\bar{c} = 0$ if and only if $c = 0$.

We prove property 4 here. Let $c = a + bi$ so that $\bar{c} = a - bi$. If $c = \bar{c}$, then $a + bi = a - bi$, so $b = 0$ and c is real. On the other hand, if c is real, then $c = a$ and $\bar{c} = a$, so $c = \bar{c}$.

A polynomial of degree n with real coefficients has n complex roots, some, all, or none of which may be real numbers. Thus the polynomial $f_1(t) = t^4 - 1$ has the roots i, $-i$, 1, and -1; the polynomial $f_2(t) = t^2 - 1$ has the roots 1 and -1; the polynomial $f_3(t) = t^2 + 1$ has the roots i and $-i$.

Almost everything in the first five chapters of this book remains true if we cross out the word *real* and replace it by the word *complex*. Thus we have matrices with complex entries; we can solve linear systems using complex arithmetic; and we define complex vector spaces, complex inner product spaces, and linear transformations of one complex vector space into another. We also evaluate the determinant of a complex matrix using complex arithmetic. It might further be noted that there are computer codes to solve linear algebra problems, for example, to solve linear systems using complex numbers. Such codes carry out the complex arithmetic following the properties given above. As a matter of fact, we can do most of the first five chapters not only over the real numbers and complex numbers, but also over any *field* (the real numbers and the complex numbers are familiar examples of a field). However, we have decided to limit ourselves to the real numbers (rather than also deal with the complex numbers) for the following reasons:

1. The results would not look markedly different.
2. For a good many applications the real number situation is adequate; the transition to the complex numbers is, as we have just remarked, not too difficult.
3. The computational effort when using complex numbers is, in most cases, more than double that resulting with only real numbers.

The following example will illustrate how to solve linear systems with complex coefficients and how to evaluate the determinant of a matrix with complex entries.

EXAMPLE 1. Consider the linear system

$$\begin{bmatrix} 2+i & 1+i \\ 3-i & 2-2i \end{bmatrix} \begin{bmatrix} a_1 + b_1 i \\ a_2 + b_2 i \end{bmatrix} = \begin{bmatrix} 3+6i \\ 7-i \end{bmatrix}.$$

We wish to find a solution $x_1 = a_1 + b_1 i$ and $x_2 = a_2 + b_2 i$. We first rewrite the given linear system as

$$\left(\begin{bmatrix} 2 & 1 \\ 3 & 2 \end{bmatrix} + i \begin{bmatrix} 1 & 1 \\ -1 & -2 \end{bmatrix} \right) \left(\begin{bmatrix} a_1 \\ a_2 \end{bmatrix} + i \begin{bmatrix} b_1 \\ b_2 \end{bmatrix} \right) = \begin{bmatrix} 3 \\ 7 \end{bmatrix} + i \begin{bmatrix} 6 \\ -1 \end{bmatrix}.$$

Multiplying, we have

$$\left(\begin{bmatrix} 2 & 1 \\ 3 & 2 \end{bmatrix} \begin{bmatrix} a_1 \\ a_2 \end{bmatrix} - \begin{bmatrix} 1 & 1 \\ -1 & -2 \end{bmatrix} \begin{bmatrix} b_1 \\ b_2 \end{bmatrix} \right)$$
$$+ i \left(\begin{bmatrix} 2 & 1 \\ 3 & 2 \end{bmatrix} \begin{bmatrix} b_1 \\ b_2 \end{bmatrix} + \begin{bmatrix} 1 & 1 \\ -1 & -2 \end{bmatrix} \begin{bmatrix} a_1 \\ a_2 \end{bmatrix} \right) = \begin{bmatrix} 3 \\ 7 \end{bmatrix} + i \begin{bmatrix} 6 \\ -1 \end{bmatrix}.$$

The real and imaginary parts on both sides of the equation must agree, respectively, and so we have

$$\begin{bmatrix} 2 & 1 \\ 3 & 2 \end{bmatrix} \begin{bmatrix} a_1 \\ a_2 \end{bmatrix} - \begin{bmatrix} 1 & 1 \\ -1 & -2 \end{bmatrix} \begin{bmatrix} b_1 \\ b_2 \end{bmatrix} = \begin{bmatrix} 3 \\ 7 \end{bmatrix}$$

and

$$\begin{bmatrix} 2 & 1 \\ 3 & 2 \end{bmatrix} \begin{bmatrix} b_1 \\ b_2 \end{bmatrix} + \begin{bmatrix} 1 & 1 \\ -1 & -2 \end{bmatrix} \begin{bmatrix} a_1 \\ a_2 \end{bmatrix} = \begin{bmatrix} 6 \\ -1 \end{bmatrix}.$$

This leads to the linear system

$$\begin{aligned} 2a_1 + a_2 - b_1 - b_2 &= 3 \\ 3a_1 + 2a_2 + b_1 + 2b_2 &= 7 \\ a_1 + a_2 + 2b_1 + b_2 &= 6 \\ -a_1 - 2a_2 + 3b_1 + 2b_2 &= -1, \end{aligned}$$

which can be written as

$$\begin{bmatrix} 2 & 1 & -1 & -1 \\ 3 & 2 & 1 & 2 \\ 1 & 1 & 2 & 1 \\ -1 & -2 & 3 & 2 \end{bmatrix} \begin{bmatrix} a_1 \\ a_2 \\ b_1 \\ b_2 \end{bmatrix} = \begin{bmatrix} 3 \\ 7 \\ 6 \\ -1 \end{bmatrix}.$$

This linear system of four equations in four unknowns is now solved as in Chapter 1. The solution is $a_1 = 1$, $a_2 = 2$, $b_1 = 2$, and $b_2 = -1$. Thus $x_1 = 1 + 2i$ and $x_2 = 2 - i$ is the solution to the given linear system.

We might also compute the determinant of the coefficient matrix. We have

$$\begin{vmatrix} 2 + i & 1 + i \\ 3 - i & 2 - 2i \end{vmatrix} = (2 + i)(2 - 2i) - (3 - i)(1 + i)$$
$$= (6 - 2i) - (4 + 2i) = 2 - 4i.$$

We have made all these remarks about complex numbers to prove the following important theorem.

Theorem 6.6. *All the roots of the characteristic polynomial of a real symmetric matrix are real numbers.*

Proof: Let $c = a + bi$ be any root of the characteristic polynomial of A. We shall prove that $b = 0$, so that c is a real number. Now $|cI_n - A| = 0 = |(a + bi)I_n - A|$. This means that the homogeneous system

$$((a + bi)I_n - A)(\alpha + \beta i) = O = O + iO \tag{1}$$

has a nontrivial solution $\alpha + \beta i$, where α and β are vectors in R^n that are not both the zero vector. Carrying out the multiplication in (1), we obtain

$$(aI_n\alpha - A\alpha - bI_n\beta) + i(aI_n\beta + bI_n\alpha - A\beta) = O + Oi. \tag{2}$$

Setting the real and imaginary parts equal to O, we have

$$\begin{aligned} aI_n\alpha - A\alpha - bI_n\beta &= O \\ aI_n\beta - A\beta + bI_n\alpha &= O. \end{aligned} \tag{3}$$

Forming the inner products of both sides of the first equation in (3) with β and of both sides of the second equation of (3) with α, we have

$$\begin{aligned} (\beta, aI_n\alpha - A\alpha - bI_n\beta) &= (\beta, O) = 0 \\ (aI_n\beta - A\beta + bI_n\alpha, \alpha) &= (O, \alpha) = 0 \end{aligned}$$

or

$$\begin{aligned} a(\beta, I_n\alpha) - (\beta, A\alpha) - b(\beta, I_n\beta) &= 0 \\ a(I_n\beta, \alpha) - (A\beta, \alpha) + b(I_n\alpha, \alpha) &= 0. \end{aligned} \tag{4}$$

Now, by Equation (1) in Section 3.2, we see that $(I_n\beta, \alpha) = (\beta, I_n'\alpha) = (\beta, I_n\alpha)$ and that $(A\beta, \alpha) = (\beta, A'\alpha) = (\beta, A\alpha)$. Note that we have used the fact that $I_n' = I_n$ and that since A is symmetric, we have $A' = A$. Subtracting the two equations in (4), we now get

$$-b(\beta, I_n\beta) - b(I_n\alpha, \alpha) = 0 \tag{5}$$

or

$$-b[(\beta, \beta) + (\alpha, \alpha)] = 0. \tag{6}$$

Since α and β are not both the zero vector, $(\alpha, \alpha) > 0$ or $(\beta, \beta) > 0$. From (6) we conclude that $b = 0$. Hence every root of the characteristic polynomial of A is a real number.

Once we have established this result, we know that complex numbers do not enter into the study of the diagonalization problem for real symmetric matrices. Thus throughout the remainder of this book we again deal only with the real numbers.

Corollary 6.1. *If A is a symmetric $n \times n$ matrix and all the eigenvalues of A are distinct, then A is diagonalizable.*

Proof: Since A is symmetric, all the roots of its characteristic polynomial are real. From Theorem 6.4 it now follows that A can be diagonalized. Moreover, if D is the diagonal matrix that is similar to A, then the elements on the main diagonal of D are the eigenvalues of A.

Theorem 6.7. *If A is a symmetric $n \times n$ matrix, then eigenvectors that belong to distinct eigenvalues of A are orthogonal.*

Proof: Let α_1 and α_2 be eigenvectors of A which are associated with the distinct eigenvalues c_1 and c_2 of A. We then have $A\alpha_1 = c_1\alpha_1$ and $A\alpha_2 = c_2\alpha_2$. Now $c_1(\alpha_1, \alpha_2) = (c_1\alpha_1, \alpha_2) = (A\alpha_1, \alpha_2) = (\alpha_1, A'\alpha_2) = (\alpha_1, A\alpha_2) = (\alpha_1, c_2\alpha_2) = c_2(\alpha_1, \alpha_2)$. Hence

$$0 = c_1(\alpha_1, \alpha_2) - c_2(\alpha_1, \alpha_2) = (c_1 - c_2)(\alpha_1, \alpha_2).$$

Since $c_1 \neq c_2$, we conclude that $(\alpha_1, \alpha_2) = 0$.

EXAMPLE 2. Let $A = \begin{bmatrix} 0 & 0 & -2 \\ 0 & -2 & 0 \\ -2 & 0 & 3 \end{bmatrix}$. The characteristic polynomial of A is

$$f(t) = (t + 2)(t - 4)(t + 1)$$

(verify), so the eigenvalues of A are $c_1 = -2$, $c_2 = 4$, $c_3 = -1$. Associated eigenvectors are the nontrivial solutions of the homogeneous system [equation (3) in Section 6.1]

$$\begin{bmatrix} c & 0 & 2 \\ 0 & c+2 & 0 \\ 2 & 0 & c-3 \end{bmatrix} \begin{bmatrix} a_1 \\ a_2 \\ a_3 \end{bmatrix} = \begin{bmatrix} 0 \\ 0 \\ 0 \end{bmatrix}.$$

For $c_1 = -2$, we find (verify) that α_1 is any vector of the form $\begin{bmatrix} 0 \\ r \\ 0 \end{bmatrix}$, where

r is any nonzero real number. Thus we may take $\alpha_1 = \begin{bmatrix} 0 \\ 1 \\ 0 \end{bmatrix}$. For $c_2 = 4$ we

find (verify) that α_2 is any vector of the form $\begin{bmatrix} -s/2 \\ 0 \\ s \end{bmatrix}$, where s is any nonzero

real number. Thus we may take $\alpha_2 = \begin{bmatrix} -1 \\ 0 \\ 2 \end{bmatrix}$. For $c_3 = -1$, we find (verify)

that α_3 is any vector of the form $\begin{bmatrix} 2t \\ 0 \\ t \end{bmatrix}$, where t is any nonzero real number.

Thus we may take $\alpha_3 = \begin{bmatrix} 2 \\ 0 \\ 1 \end{bmatrix}$. It is clear that $\{\alpha_1, \alpha_2, \alpha_3\}$ is orthogonal and

linearly independent. Thus A is similar to $D = \begin{bmatrix} -2 & 0 & 0 \\ 0 & 4 & 0 \\ 0 & 0 & -1 \end{bmatrix}$.

If A can be diagonalized, then there exists a nonsingular matrix P such that $P^{-1}AP$ is diagonal. Moreover, the columns of P are eigenvectors of A. Now, if the eigenvectors of A form an orthogonal set S, as happens when A is symmetric and the eigenvalues of A are distinct, then since any scalar multiple of an eigenvector of A is also an eigenvector of A, we can normalize S to obtain an orthonormal set $T = \{\alpha_1, \alpha_2, \ldots, \alpha_n\}$ of eigenvectors of A. The jth column of P is the eigenvector α_j and we now see what type of matrix P must be. We can write P as a partitioned matrix in the form $P = [\alpha_1 \, \alpha_2 \, \cdots \, \alpha_n]$. Then

$$P' = \begin{bmatrix} \alpha_1' \\ \alpha_2' \\ \vdots \\ \alpha_n' \end{bmatrix}, \text{ where } \alpha_i' \text{ is the transpose of the } n \times 1 \text{ matrix (or vector) } \alpha_i.$$

We find that the (i, j) entry in $P'P$ is (α_i, α_j); also $(\alpha_i, \alpha_j) = 1$ if $i = j$ and $(\alpha_i, \alpha_j) = 0$ if $i \neq j$. Hence $P'P = I_n$, which means that $P' = P^{-1}$. Such matrices are important enough to have a special name.

Definition 6.4. A square matrix A is called **orthogonal** if $A^{-1} = A'$.

Of course, we can also say that A is orthogonal if $A'A = I_n$.

EXAMPLE 3. Let $A = \begin{bmatrix} \frac{2}{3} & -\frac{2}{3} & \frac{1}{3} \\ \frac{2}{3} & \frac{1}{3} & -\frac{2}{3} \\ \frac{1}{3} & \frac{2}{3} & \frac{2}{3} \end{bmatrix}$. It is easy to check that $A'A = I_n$.

EXAMPLE 4. Let A be the matrix defined in Example 2. We already know that the set of eigenvectors $\left\{ \begin{bmatrix} 0 \\ 1 \\ 0 \end{bmatrix}, \begin{bmatrix} -1 \\ 0 \\ 2 \end{bmatrix}, \begin{bmatrix} 2 \\ 0 \\ 1 \end{bmatrix} \right\}$ is orthogonal. If we normalize these vectors, we find that $T = \left\{ \begin{bmatrix} 0 \\ 1 \\ 0 \end{bmatrix}, \begin{bmatrix} -1/\sqrt{5} \\ 0 \\ 2/\sqrt{5} \end{bmatrix}, \begin{bmatrix} 2/\sqrt{5} \\ 0 \\ 1/\sqrt{5} \end{bmatrix} \right\}$ is an orthonormal basis for R^3. The matrix P such that $P^{-1}AP$ is diagonal is the matrix whose columns are the vectors in T. Thus $P = \begin{bmatrix} 0 & -1/\sqrt{5} & 2/\sqrt{5} \\ 1 & 0 & 0 \\ 0 & 2/\sqrt{5} & 1/\sqrt{5} \end{bmatrix}$.

We leave it to the reader to verify that P is an orthogonal matrix and that

$$P^{-1}AP = P'AP = \begin{bmatrix} -2 & 0 & 0 \\ 0 & 4 & 0 \\ 0 & 0 & -1 \end{bmatrix}.$$

The following theorem is not difficult to prove.

Theorem 6.8. *The $n \times n$ matrix A is orthogonal if and only if the columns (and rows) of A form an orthonormal set.*

Proof: Exercise.

If A is an orthogonal matrix, then it is easy to show that $|A| = \pm 1$. We now look at the geometrical properties of orthogonal matrices. If A is an orthogonal $n \times n$ matrix, then associated with A is the linear transformation $L: R^n \to R^n$ (recall Chapter 4) defined by $L(\alpha) = A\alpha$ for α in R^n. We now compute $(L(\alpha), L(\beta))$ for any α and β in R^n, using the standard inner product on R^n. We have

$$(L(\alpha), L(\beta)) = (A\alpha, A\beta) = (\alpha, A'A\beta)(\alpha, A^{-1}A\beta) = (\alpha, I_n\beta) = (\alpha, \beta).$$

This means that L preserves length. It is, of course, clear that if ϕ is the angle between vectors α and β in R^n, then the angle between $L(\alpha)$ and $L(\beta)$ is also ϕ. Conversely, let $L: R^n \to R^n$ be a linear transformation which preserves length; that is, $(L(\alpha), (L(\beta)) = (\alpha, \beta)$ for any α and β in R^n. Let A be the matrix representing L with respect to the natural basis for R^n. Then $L(\alpha) = A\alpha$. We now have

$$(\alpha, \beta) = (L(\alpha), L(\beta)) = (A\alpha, A\beta) = (\alpha, A'A\beta).$$

Since this holds for all α in R^n, then, by Exercise 7(f) in Section 3.2, we conclude that $A'A\beta = \beta$ for any β in R^n. It follows that $A'A = I_n$, so A is an orthogonal matrix. Other properties of orthogonal matrices are examined in the exercises.

We now turn to the general situation for a symmetric matrix; even if A has eigenvalues whose multiplicities are greater than one, it turns out that we can still diagonalize A. We omit the proof of the following theorem.

Theorem 6.9. *If A is a symmetric $n \times n$ matrix, then there exists an orthogonal matrix P such that $P^{-1}AP = P'AP = D$, a diagonal matrix. The eigenvalues of A lie on the main diagonal of D.*

It can be shown that if A has an eigenvalue c of multiplicity k, then the solution space of the homogeneous system $(cI_n - A)\alpha = O$ [equation (3) in Section 6.1] has dimension k. This means that there exist k linearly independent eigenvectors of A associated with the eigenvalue c. We can, of course, choose an orthonormal basis for this solution space. Thus we obtain a set of k orthonormal eigenvectors associated with the eigenvalue c. Since eigenvectors associated with distinct eigenvalues are orthogonal, if we form the set of all eigenvectors we get an orthonormal set. Hence the matrix P whose columns are the eigenvectors is orthogonal.

EXAMPLE 5. Let

$$A = \begin{bmatrix} 0 & 2 & 2 \\ 2 & 0 & 2 \\ 2 & 2 & 0 \end{bmatrix}.$$

The characteristic polynomial of A is $f(t) = (t + 2)^2(t - 4)$ (verify), so its eigenvalues are

$$c_1 = -2, \quad c_2 = -2, \quad \text{and} \quad c_3 = 4.$$

That is, -2 is an eigenvalue of multiplicity 2. To find the eigenvectors associated with -2, we solve the homogeneous system $(-2I_3 - A)\alpha = O$:

$$\begin{bmatrix} -2 & -2 & -2 \\ -2 & -2 & -2 \\ -2 & -2 & -2 \end{bmatrix} \begin{bmatrix} a_1 \\ a_2 \\ a_3 \end{bmatrix} = \begin{bmatrix} 0 \\ 0 \\ 0 \end{bmatrix}. \tag{7}$$

A basis for the solution space of (7) consists of the eigenvectors $\alpha_1 = \begin{bmatrix} -1 \\ 1 \\ 0 \end{bmatrix}$

and $\alpha_2 = \begin{bmatrix} -1 \\ 0 \\ 1 \end{bmatrix}$. Now α_1 and α_2 are not orthogonal, since $(\alpha_1, \alpha_2) \neq 0$.

We can use the Gram–Schmidt process to obtain an orthonormal basis for the solution space of (7) (the eigenspace associated with -2) as follows. Let $\beta_1 = \alpha_1$ and

$$\beta_2 = \alpha_2 - \frac{(\alpha_2, \beta_1)}{(\beta_1, \beta_1)} \alpha_1 = \begin{bmatrix} -\frac{1}{2} \\ -\frac{1}{2} \\ 1 \end{bmatrix}.$$

To eliminate fractions, we let $\beta_2^* = 2\beta_2 = \begin{bmatrix} -1 \\ -1 \\ 2 \end{bmatrix}$. The set $\{\beta_1, \beta_2^*\}$ is an

orthogonal set of vectors. Normalizing, we obtain

$$\gamma_1 = \frac{\beta_1}{|\beta_1|} = \frac{1}{\sqrt{2}} \begin{bmatrix} -1 \\ 1 \\ 0 \end{bmatrix} \quad \text{and} \quad \gamma_2 = \frac{\beta_2^*}{|\beta_2^*|} = \frac{1}{\sqrt{6}} \begin{bmatrix} -1 \\ -1 \\ 2 \end{bmatrix}.$$

The set $\{\gamma_1, \gamma_2\}$ is an orthonormal basis for the eigenspace associated with $c = -2$. Now we find a basis for the eigenspace associated with $c = 4$ by solving the homogeneous system $(4I_3 - A)\alpha = O$:

$$\begin{bmatrix} 4 & -2 & -2 \\ -2 & 4 & -2 \\ -2 & -2 & 4 \end{bmatrix} \begin{bmatrix} a_1 \\ a_2 \\ a_3 \end{bmatrix} = \begin{bmatrix} 0 \\ 0 \\ 0 \end{bmatrix}.$$

A basis for this eigenspace consists (verify) of the vector $\alpha_3 = \begin{bmatrix} 1 \\ 1 \\ 1 \end{bmatrix}$. Normal-

izing this vector, we have the eigenvector $\gamma_3 = \dfrac{1}{\sqrt{3}} \begin{bmatrix} 1 \\ 1 \\ 1 \end{bmatrix}$, as a basis for the

eigenspace associated with $c = 4$. Since eigenvectors associated with distinct eigenvalues are orthogonal, γ_3 is orthogonal to both γ_1 and γ_2. Thus the set $\{\gamma_1, \gamma_2, \gamma_3\}$ is an orthonormal basis of R^3 consisting of eigenvectors of A. The matrix P is the matrix whose jth column is γ_j:

$$P = \begin{bmatrix} -1/\sqrt{2} & -1/\sqrt{6} & 1/\sqrt{3} \\ 1/\sqrt{2} & -1/\sqrt{6} & 1/\sqrt{3} \\ 0 & 2/\sqrt{6} & 1/\sqrt{3} \end{bmatrix}.$$

We leave it to the reader to verify that

$$P^{-1}AP = P'AP = \begin{bmatrix} -2 & 0 & 0 \\ 0 & -2 & 0 \\ 0 & 0 & 4 \end{bmatrix}.$$

EXAMPLE 6. Let $A = \begin{bmatrix} 1 & 2 & 0 & 0 \\ 2 & 1 & 0 & 0 \\ 0 & 0 & 1 & 2 \\ 0 & 0 & 2 & 1 \end{bmatrix}$. Either by straightforward com-

putation, or by Exercise 15 in Section 5.2, we find that the characteristic polynomial of A is $f(t) = (t + 1)^2(t - 3)^2$, so its eigenvalues are $c_1 = -1$, $c_2 = -1$, $c_3 = 3$, and $c_4 = 3$. We now compute the associated eigenvectors and the orthogonal matrix P. The eigenspace associated with the eigenvalue

-1, of multiplicity 2, is the solution space of the homogeneous system $(-1I_4 - A)\alpha = O$:

$$\begin{bmatrix} -2 & -2 & 0 & 0 \\ -2 & -2 & 0 & 0 \\ 0 & 0 & -2 & -2 \\ 0 & 0 & -2 & -2 \end{bmatrix} \begin{bmatrix} a_1 \\ a_2 \\ a_3 \\ a_4 \end{bmatrix} = \begin{bmatrix} 0 \\ 0 \\ 0 \\ 0 \end{bmatrix},$$

which is the set of all vectors of the form

$$\begin{bmatrix} r \\ -r \\ s \\ -s \end{bmatrix} = r \begin{bmatrix} 1 \\ -1 \\ 0 \\ 0 \end{bmatrix} + s \begin{bmatrix} 0 \\ 0 \\ 1 \\ -1 \end{bmatrix},$$

where r and s are any real numbers. Thus the eigenvectors

$$\begin{bmatrix} 1 \\ -1 \\ 0 \\ 0 \end{bmatrix} \quad \text{and} \quad \begin{bmatrix} 0 \\ 0 \\ 1 \\ -1 \end{bmatrix}$$

form a basis for the eigenspace associated with -1, and the dimension of this eigenspace is 2. Note that the eigenvectors

$$\begin{bmatrix} 1 \\ -1 \\ 0 \\ 0 \end{bmatrix} \quad \text{and} \quad \begin{bmatrix} 0 \\ 0 \\ 1 \\ -1 \end{bmatrix}$$

happen to be orthogonal. Since we are looking for an orthonormal basis for this eigenspace, we take

$$\alpha_1 = \begin{bmatrix} 1/\sqrt{2} \\ -1/\sqrt{2} \\ 0 \\ 0 \end{bmatrix} \quad \text{and} \quad \alpha_2 = \begin{bmatrix} 0 \\ 0 \\ 1/\sqrt{2} \\ -1/\sqrt{2} \end{bmatrix}$$

as eigenvectors associated with c_1 and c_2, respectively. Then $\{\alpha_1, \alpha_2\}$ is an orthonormal basis for the eigenspace associated with -1. The eigenspace associated with the eigenvalue 3, of multiplicity 2, is the solution space of the homogeneous system $(3I_4 - A)\alpha = O$:

$$\begin{bmatrix} 2 & -2 & 0 & 0 \\ -2 & 2 & 0 & 0 \\ 0 & 0 & 2 & -2 \\ 0 & 0 & -2 & 2 \end{bmatrix} \begin{bmatrix} a_1 \\ a_2 \\ a_3 \\ a_4 \end{bmatrix} = \begin{bmatrix} 0 \\ 0 \\ 0 \\ 0 \end{bmatrix},$$

which is the set of all vectors of the form

$$\begin{bmatrix} r \\ r \\ s \\ s \end{bmatrix} = r \begin{bmatrix} 1 \\ 1 \\ 0 \\ 0 \end{bmatrix} + s \begin{bmatrix} 0 \\ 0 \\ 1 \\ 1 \end{bmatrix},$$

where r and s are any real numbers. Thus the eigenvectors $\begin{bmatrix} 1 \\ 1 \\ 0 \\ 0 \end{bmatrix}$ and $\begin{bmatrix} 0 \\ 0 \\ 1 \\ 1 \end{bmatrix}$

form a basis for the eigenspace associated with 3, and the dimension of this eigenspace is 2. Since these eigenvectors are orthogonal, we normalize them and let

$$\alpha_3 = \begin{bmatrix} 1/\sqrt{2} \\ 1/\sqrt{2} \\ 0 \\ 0 \end{bmatrix} \quad \text{and} \quad \alpha_4 = \begin{bmatrix} 0 \\ 0 \\ 1/\sqrt{2} \\ 1/\sqrt{2} \end{bmatrix}$$

be eigenvectors associated with c_3 and c_4, respectively. Then $\{\alpha_3, \alpha_4\}$ is an orthonormal basis for the eigenspace associated with 3. Now eigenvectors associated with distinct eigenvalues are orthogonal, so $\{\alpha_1, \alpha_2, \alpha_3, \alpha_4\}$ is an orthonormal basis for R^4. The matrix P is the matrix whose jth column is α_j, $j = 1, 2, 3, 4$. Thus

$$P = \begin{bmatrix} 1/\sqrt{2} & 0 & 1/\sqrt{2} & 0 \\ -1/\sqrt{2} & 0 & 1/\sqrt{2} & 0 \\ 0 & 1/\sqrt{2} & 0 & 1/\sqrt{2} \\ 0 & -1/\sqrt{2} & 0 & 1/\sqrt{2} \end{bmatrix}.$$

We leave it to the reader to verify that P is an orthogonal matrix and that

$$P^{-1}AP = P'AP = \begin{bmatrix} -1 & 0 & 0 & 0 \\ 0 & -1 & 0 & 0 \\ 0 & 0 & 3 & 0 \\ 0 & 0 & 0 & 3 \end{bmatrix}.$$

If A is an $n \times n$ symmetric matrix, we know that we can find an orthogonal matrix P such that $P^{-1}AP$ is diagonal. Conversely, suppose that A is a matrix for which we can find an orthogonal matrix P such that $P^{-1}AP = D$ is a diagonal matrix. What type of matrix is A? Since $P^{-1}AP = D$, $A = PDP^{-1}$. Also, $P^{-1} = P'$, since P is orthogonal. Then

$$A' = (PDP')' = (P')'D'P' = PDP' = A,$$

which means that A is symmetric.

Some remarks about nonsymmetric matrices are in order at this point. Theorem 6.4 assures us that A is diagonalizable if all the roots of its characteristic polynomial are real and distinct. We also studied examples, in Section 6.1, of nonsymmetric matrices that had repeated eigenvalues which were diagonalizable and others which were not diagonalizable. There are some striking differences between the symmetric and nonsymmetric cases, which we now summarize. Thus, if A is nonsymmetric, then the roots of its characteristic polynomial need not all be real numbers; if an eigenvalue c has multiplicity k, then the solution space of $(cI_n - A)\alpha = O$ may have dimension less than k; if the roots of the characteristic polynomial of A are all real, it is still possible for the eigenvectors not to form a basis for R^n; eigenvectors associated with distinct eigenvalues need not be orthogonal. Thus, in Example 12 of Section 6.1, the eigenvectors α_1 and α_3 associated with the eigenvalues $c_1 = 0$ and $c_3 = 1$ are not orthogonal. If a matrix A cannot be diagonalized, then we can often find a matrix B similar to A which is "nearly diagonal." The matrix B is said to be in **Jordan canonical form**; the study of such matrices lies beyond the scope of this book but is studied in advanced books on linear algebra [for example, K. Hoffman and R. Kunze, *Linear Algebra*, 2nd ed., Prentice-Hall, Inc. (Englewood Cliffs, N.J.: 1971)]; it plays a key role in many applications of linear algebra.

It should also be noted that, in many applications, we need only find a diagonal matrix D which is similar to the given matrix A; that is, we do not require the orthogonal matrix P such that $P^{-1}AP = D$.

Eigenvalue problems arise in all applications involving vibrations; they arise in aerodynamics, elasticity, nuclear physics, mechanics, chemical engineering, biology, differential equations, and so on. Many of the matrices to be diagonalized in applied problems are either symmetric or all the roots of their characteristic polynomial are real. Of course, the methods for finding eigenvalues that have been presented in this chapter are not recommended for matrices of large order because of the need to evaluate determinants.

6.2. Exercises

1. Prove properties 1, 2, 3, and 5 of the complex numbers discussed in the text.

2. (a) Solve the linear system

$$2x_1 + (3 - i)x_2 = 11 + 7i$$
$$(3 - i)x_1 - 3x_2 = -3 - 10i.$$

(b) Compute $\begin{vmatrix} 1 + i & 3 \\ 2i & 2 - i \end{vmatrix}$.

3. Verify that $P = \begin{bmatrix} \frac{2}{3} & -\frac{2}{3} & \frac{1}{3} \\ \frac{2}{3} & \frac{1}{3} & -\frac{2}{3} \\ \frac{1}{3} & \frac{2}{3} & \frac{2}{3} \end{bmatrix}$ is an orthogonal matrix.

4. Find the inverse of each of the following orthogonal matrices.

(a) $A = \begin{bmatrix} 1 & 0 & 0 \\ 0 & \cos \phi & \sin \phi \\ 0 & -\sin \phi & \cos \phi \end{bmatrix}$. (b) $B = \begin{bmatrix} 1 & 0 & 0 \\ 0 & 1/\sqrt{2} & -1/\sqrt{2} \\ 0 & -1/\sqrt{2} & -1/\sqrt{2} \end{bmatrix}$.

5. Show that if A and B are orthogonal matrices, then AB is an orthogonal matrix.

6. Show that if A is an orthogonal matrix, then A^{-1} is orthogonal.

7. Prove Theorem 6.8.

8. Verify Theorem 6.8 for the matrices in Exercise 4.

9. Verify that the matrix P in Example 4 is an orthogonal matrix and that

$$P^{-1}AP = P'AP = \begin{bmatrix} -2 & 0 & 0 \\ 0 & 4 & 0 \\ 0 & 0 & -1 \end{bmatrix}.$$

10. Show that if A is an orthogonal matrix, then $|A| = \pm 1$.

11. (a) Verify that the matrix $\begin{bmatrix} \cos \phi & \sin \phi \\ -\sin \phi & \cos \phi \end{bmatrix}$ is orthogonal.

(b) Prove that if A is an orthogonal 2×2 matrix, then there exists a real number ϕ such that $A = \pm \begin{bmatrix} \cos \phi & \sin \phi \\ -\sin \phi & \cos \phi \end{bmatrix}$.

12. For the orthogonal matrix $A = \begin{bmatrix} 1/\sqrt{2} & -1/\sqrt{2} \\ -1/\sqrt{2} & -1/\sqrt{2} \end{bmatrix}$, verify that $(A\alpha, A\beta) = (\alpha, \beta)$ for any α and β in R^2.

13. Let A be an $n \times n$ orthogonal matrix, and let $L: R^n \to R^n$ be the linear transformation associated with A; that is, $L(\alpha) = A\alpha$ for α in R^n. Let ϕ be the angle between vectors α and β in R^n. Prove that if A is orthogonal, then the angle between $L(\alpha)$ and $L(\beta)$ is also ϕ.

14. A linear transformation $L: V \to V$, where V is an n-dimensional Euclidean space is called **orthogonal** if $(L(\alpha), L(\beta)) = (\alpha, \beta)$. Let S be an orthonormal basis for V and let the matrix A represent L with respect to S. Prove that A is orthogonal.

15. Let $L: R^2 \to R^2$ be the linear transformation performing a counterclockwise rotation through $45°$, and let A be the matrix representing L with respect to the natural basis for R^2. Prove that A is orthogonal.

16. Let A be an $n \times n$ matrix and let $B = P^{-1}AP$ be similar to A. Prove that if α is an eigenvector of A associated with the eigenvalue c of A, then $P^{-1}\alpha$ is an eigenvector of B associated with the eigenvalue c of B.

In Exercises 17 through 22, diagonalize each given matrix and find an orthogonal matrix P such that $P^{-1}AP$ is diagonal.

17. $A = \begin{bmatrix} 2 & 2 \\ 2 & 2 \end{bmatrix}$.

18. $A = \begin{bmatrix} 0 & 0 & 1 \\ 0 & 0 & 0 \\ 1 & 0 & 0 \end{bmatrix}$.

19. $A = \begin{bmatrix} 0 & 0 & 0 \\ 0 & 2 & 2 \\ 0 & 2 & 2 \end{bmatrix}$.

20. $A = \begin{bmatrix} 0 & 0 & 0 & 0 \\ 0 & 0 & 0 & 0 \\ 0 & 0 & 0 & 1 \\ 0 & 0 & 1 & 0 \end{bmatrix}$.

21. $A = \begin{bmatrix} 0 & -1 & -1 \\ -1 & 0 & -1 \\ -1 & -1 & 0 \end{bmatrix}$.

22. $A = \begin{bmatrix} -1 & 2 & 2 \\ 2 & -1 & 2 \\ 2 & 2 & -1 \end{bmatrix}$.

In Exercises 23 through 30, diagonalize each given matrix.

23. $A = \begin{bmatrix} 2 & 1 \\ 1 & 2 \end{bmatrix}$.

24. $A = \begin{bmatrix} 2 & 2 & 0 & 0 \\ 2 & 2 & 0 & 0 \\ 0 & 0 & 2 & 2 \\ 0 & 0 & 2 & 2 \end{bmatrix}.$

25. $A = \begin{bmatrix} 1 & 1 & 0 \\ 1 & 1 & 0 \\ 0 & 0 & 1 \end{bmatrix}.$

26. $A = \begin{bmatrix} 1 & 0 & 0 \\ 0 & 3 & -2 \\ 0 & -2 & 3 \end{bmatrix}.$

27. $A = \begin{bmatrix} 1 & 0 & 0 \\ 0 & 1 & 1 \\ 0 & 1 & 1 \end{bmatrix}.$

28. $A = \begin{bmatrix} 0 & 0 & 0 & 1 \\ 0 & 0 & 0 & 0 \\ 0 & 0 & 0 & 0 \\ 1 & 0 & 0 & 0 \end{bmatrix}.$

29. $A = \begin{bmatrix} 1 & -1 & 2 \\ -1 & 1 & 2 \\ 2 & 2 & 2 \end{bmatrix}.$

30. $A = \begin{bmatrix} -3 & 0 & -1 \\ 0 & -2 & 0 \\ -1 & 0 & -3 \end{bmatrix}.$

31. Prove Theorem 6.9 for the 2 × 2 case by studying the two possible cases for the roots of the characteristic polynomial of A.

32. Let $L: V \to V$ be an orthogonal linear transformation (see Exercise 14), where V is an n-dimensional Euclidean space. Show that if c is an eigenvalue of L, then $|c| = 1$.

6.3. Real Quadratic Forms

Let

$$ax^2 + 2bxy + cy^2 = d, \tag{1}$$

where a, b, c, and d are real numbers, be the equation of a conic centered at the origin of a rectangular Cartesian coordinate system in two-dimensional space. In many applications we are interested in identifying the type of conic represented by the equation; that is, we have to decide whether the equation represents an ellipse, a hyperbola, a parabola, or any degenerate forms of the first two of these, such as circles and straight lines. The approach frequently used is to rotate the x- and y-axes (or change coordinates) so as to obtain a new set of axes, x' and y', in which the $x'y'$ term is no longer present; we can then identify the type of conic rather easily. The analogous problem can be studied for quadric surfaces in three-dimensional space, that is, surfaces whose equation is

$$ax^2 + 2dxy + 2exz + by^2 + 2fyz + cz^2 = g, \tag{2}$$

where $a, b, c, d, e, f,$ and g are real numbers. The solution is to again change the x-, y-, and z-axes to x'-, y'-, and z'-axes so as to eliminate the $x'y'$, $x'z'$, and $y'z'$ terms. The expressions on the left sides of (1) and (2) are quadratic forms. Quadratic forms arise in statistics, mechanics, and in other problems in physics; in quadratic programming; in the study of maxima and minima of functions of several variables; and in other applied problems. In this section we apply our results on eigenvalues and eigenvectors of matrices to give a brief treatment of real quadratic forms in n variables, generalizing the techniques used for quadratic forms in two and three variables.

Definition 6.5. If A is a symmetric matrix, then the function $g: R^n \to R^1$ (a real-valued function on R^n) defined by

$$g(X) = X'AX,$$

where

$$X = \begin{bmatrix} x_1 \\ x_2 \\ \vdots \\ x_n \end{bmatrix},$$

is called a **real quadratic form in the n variables** $x_1, x_2, \ldots, x_n$. The matrix A is called the **matrix of the quadratic form** g. We shall also denote the quadratic form by $g(X)$.

EXAMPLE 1. The left side of (1) is the quadratic form in the variables x and y:

$$g(X) = X'AX,$$

where

$$X = \begin{bmatrix} x \\ y \end{bmatrix} \quad \text{and} \quad A = \begin{bmatrix} a & b \\ b & c \end{bmatrix}.$$

EXAMPLE 2. The left side of (2) is the quadratic form

$$g(X) = X'AX,$$

where

$$X = \begin{bmatrix} x \\ y \\ z \end{bmatrix} \quad \text{and} \quad A = \begin{bmatrix} a & d & e \\ d & b & f \\ e & f & c \end{bmatrix}.$$

EXAMPLE 3. The following expressions are quadratic forms:

(a) $3x^2 - 5xy - 7y^2 = \begin{bmatrix} x & y \end{bmatrix} \begin{bmatrix} 3 & -\frac{5}{2} \\ -\frac{5}{2} & -7 \end{bmatrix} \begin{bmatrix} x \\ y \end{bmatrix}.$

(b) $3x^2 - 7xy + 5xz + 4y^2 - 4yz - 3z^2$

$$= \begin{bmatrix} x & y & z \end{bmatrix} \begin{bmatrix} 3 & -\frac{7}{2} & \frac{5}{2} \\ -\frac{7}{2} & 4 & -2 \\ \frac{5}{2} & -2 & -3 \end{bmatrix} \begin{bmatrix} x \\ y \\ z \end{bmatrix}.$$

Suppose now that $g(X) = X'AX$ is a quadratic form. To simplify the quadratic form, we change from the variables $x_1, x_2, \ldots, x_n$ to the variables $y_1, y_2, \ldots, y_n$. Suppose that the old variables are related to the new variables by $X = PY$ for some orthogonal matrix P. Then

$$g(X) = X'AX = (PY)'A(PY) = Y'(P'AP)Y = Y'BY,$$

where $B = P'AP$. We shall let the reader verify the fact that if A is a symmetric matrix, then $P'AP$ is also symmetric (Exercise 3). Thus

$$h(Y) = Y'BY$$

is another quadratic form and $g(X) = h(Y)$.

This situation is important enough to formulate the following definitions.

Definition 6.6. Two $n \times n$ matrices A and B are said to be **congruent*** if $B = P'AP$ for an orthogonal matrix P.

Definition 6.7. Two quadratic forms g and h with matrices A and B, respectively, are said to be **equivalent*** if A and B are congruent.

EXAMPLE 4. Consider the quadratic form in the variables x and y defined by

$$g(X) = 2x^2 + 2xy + 2y^2 = [x \quad y]\begin{bmatrix} 2 & 1 \\ 1 & 2 \end{bmatrix}\begin{bmatrix} x \\ y \end{bmatrix}. \tag{3}$$

We now change from the variables x and y to the variables x' and y'. Suppose that the old variables are related to the new variables by the equations

$$x = \frac{1}{\sqrt{2}}x' + \frac{1}{\sqrt{2}}y' \quad \text{and} \quad y = \frac{1}{\sqrt{2}}x' - \frac{1}{\sqrt{2}}y', \tag{4}$$

which can be written in matrix form as

$$X = \begin{bmatrix} x \\ y \end{bmatrix} = \begin{bmatrix} 1/\sqrt{2} & 1/\sqrt{2} \\ 1/\sqrt{2} & -1/\sqrt{2} \end{bmatrix}\begin{bmatrix} x' \\ y' \end{bmatrix} = PY,$$

where the orthogonal matrix

$$P = \begin{bmatrix} 1/\sqrt{2} & 1/\sqrt{2} \\ 1/\sqrt{2} & -1/\sqrt{2} \end{bmatrix} \quad \text{and} \quad Y = \begin{bmatrix} x' \\ y' \end{bmatrix}.$$

We shall soon see why and how this particular matrix P was selected. Substituting in (3), we obtain

$$g(X) = X'AX = (PY)'A(PY) = Y'P'APY$$

$$= [x' \quad y']\begin{bmatrix} 1/\sqrt{2} & 1/\sqrt{2} \\ 1/\sqrt{2} & -1/\sqrt{2} \end{bmatrix}'\begin{bmatrix} 2 & 1 \\ 1 & 2 \end{bmatrix}\begin{bmatrix} 1/\sqrt{2} & 1/\sqrt{2} \\ 1/\sqrt{2} & -1/\sqrt{2} \end{bmatrix}\begin{bmatrix} x' \\ y' \end{bmatrix}$$

$$= [x' \quad y']\begin{bmatrix} 3 & 0 \\ 0 & 1 \end{bmatrix}\begin{bmatrix} x' \\ y' \end{bmatrix} = h(Y).$$

---

*Some other authors use the terms **orthogonally congruent** and **orthogonally equivalent** for our **congruent** and **equivalent**.

Thus the matrices

$$\begin{bmatrix} 2 & 1 \\ 1 & 2 \end{bmatrix} \quad \text{and} \quad \begin{bmatrix} 3 & 0 \\ 0 & 1 \end{bmatrix}$$

are congruent and the quadratic forms g and h are equivalent.

The equation

$$g(X) = 2x^2 + 2xy + 2y^2 = 9 \tag{5}$$

represents a conic section. Since g is the quadratic form defined in Example 4, it is equivalent to the quadratic form

$$h(Y) = 3x'^2 + y'^2.$$

Now the equation

$$h(Y) = 3x'^2 + y'^2 = 9 \tag{6}$$

is the equation of an ellipse. We can then conclude that the conic section whose equation is (5) is the same ellipse as that represented by (6).

We now turn to the question of how to select the matrix P.

Theorem 6.10 (Principal Axis Theorem). *Any quadratic form in n variables* $g(X) = X'AX$ *is equivalent to a quadratic form,* $h(Y) = c_1 y_1^2 + c_2 y_2^2 + \cdots + c_n y_n^2,$ *where*

$$Y = \begin{bmatrix} y_1 \\ y_2 \\ \vdots \\ y_n \end{bmatrix}$$

and $c_1, c_2, \ldots, c_n$ *are the eigenvalues of the matrix A of g.*

Proof: If A is the matrix of g, then, since A is symmetric, we know, by Theorem 6.9, that A can be diagonalized by an orthogonal matrix. This means that there exists an orthogonal matrix P such that $D = P^{-1}AP$ is a diagonal matrix. Since P is orthogonal, $P^{-1} = P'$, so $D = P'AP$. Moreover, the elements on the main diagonal of D are the eigenvalues $c_1, c_2, \ldots, c_n$ of A. The quadratic form h with matrix D is given by

$$h(Y) = c_1 y_1^2 + c_2 y_2^2 + \cdots + c_n y_n^2;$$

g and h are equivalent.

EXAMPLE 5. Consider the quadratic form g in the variables x, y, and z, defined by

$$g(X) = 2x^2 + 4y^2 + 6yz - 4z^2.$$

The matrix of g is

$$A = \begin{bmatrix} 2 & 0 & 0 \\ 0 & 4 & 3 \\ 0 & 3 & -4 \end{bmatrix}$$

and the eigenvalues of A are (verify)

$$c_1 = 2, \qquad c_2 = 5, \qquad \text{and} \qquad c_3 = -5.$$

Let h be the quadratic form in the variables x', y', and z' defined by

$$h(Y) = 2x'^2 + 5y'^2 - 5z'^2.$$

Then g and h are equivalent. Now

$$D = \begin{bmatrix} 2 & 0 & 0 \\ 0 & 5 & 0 \\ 0 & 0 & -5 \end{bmatrix}$$

is the matrix of h; A and D are congruent matrices.

Note that to apply Theorem 6.10 to diagonalize a given quadratic form, as shown in Example 5, we do not need to know the eigenvectors of A (nor the matrix P); we only require the eigenvalues of A.

We now return to quadratic forms in two and three variables. The equation $X'AX = 1$, where X is a vector in R^2, is the locus of a central conic in the plane. From Theorem 6.10 it follows that there is a Cartesian coordinate system in the plane with respect to which the equation of the central conic is $ax^2 + by^2 = 1$, where a and b are real numbers. Similarly, the equation $X'AX = 1$, where X is a vector in R^3, is the locus of a central quadric surface in R^3. From Theorem 6.10 it follows that there is a Cartesian coordinate system in 3-space with respect to which the equation of the quadric surface is $ax^2 + by^2 + cz^2 = 1$, where a, b, and c are real numbers. The principal axes of the conic or surface lie along the coordinate axes, and this is the reason for calling Theorem 6.10 the **principal axis theorem**.

EXAMPLE 6. Consider the conic section whose equation is (5),

$$g(X) = 2x^2 + 2xy + 2y^2 = 9.$$

This conic section can also be described by the equation

$$h(Y) = 3x'^2 + y'^2 = 9,$$

which can be rewritten as

$$\frac{x'^2}{3} + \frac{y'^2}{9} = 1.$$

This is an ellipse (Figure 6.3) whose major axis is along the y'-axis. The major axis is of length 3; the minor axis is of length $\sqrt{3}$. We now note that there is a very close connection between the eigenvectors of the matrix of (5) and the location of the x'- and y'-axes.

Since $X = PY$, we have $Y = P^{-1}X = P'X = PX$ (P is orthogonal and, in this example, also symmetric). Thus

$$x' = \frac{1}{\sqrt{2}}x + \frac{1}{\sqrt{2}}y \quad \text{and} \quad y' = \frac{1}{\sqrt{2}}x - \frac{1}{\sqrt{2}}y.$$

This means that, in terms of the x- and y-axes, the x'-axis lies along the vector

$$\alpha_1 = \begin{bmatrix} 1/\sqrt{2} \\ 1/\sqrt{2} \end{bmatrix}$$

and the y'-axis lies along the vector

$$\alpha_2 = \begin{bmatrix} 1/\sqrt{2} \\ -1/\sqrt{2} \end{bmatrix}.$$

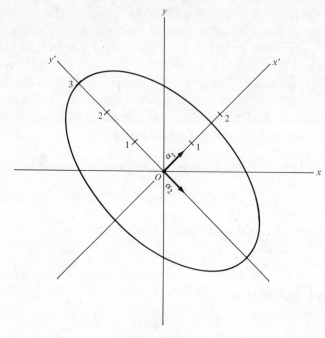

FIGURE 6.3

Now α_1 and α_2 are the columns of the matrix

$$P = \begin{bmatrix} 1/\sqrt{2} & 1/\sqrt{2} \\ 1/\sqrt{2} & -1/\sqrt{2} \end{bmatrix},$$

which in turn are eigenvectors of the matrix of (5). Thus the x'- and y'-axes lie along the eigenvectors of the matrix of (5) (see Figure 6.3).

The situation described in Example 6 is true in general. Thus the principal axes of a conic or surface lie along the eigenvectors of the matrix of the quadratic form.

Let $g(X) = X'AX$ be a quadratic form in n variables. Then we know that g is equivalent to the quadratic form $h(Y) = c_1 y_1^2 + c_2 y_2^2 + \cdots + c_n y_n^2$, where $c_1, c_2, \ldots, c_n$ are the eigenvalues of the matrix A of g. We can label the eigenvalues so that all the positive eigenvalues of A, if any, are listed first, followed by all the negative eigenvalues, if any, followed by the zero eigenvalues, if any. Thus let $c_1, c_2, \ldots, c_p$ be positive, $c_{p+1}, c_{p+2}, \ldots, c_r$ be negative, and

$c_{r+1}, c_{r+2}, \ldots, c_n$ be zero. We now define the diagonal matrix H whose entries on the main diagonal are

$$\frac{1}{\sqrt{c_1}}, \frac{1}{\sqrt{c_2}}, \ldots, \frac{1}{\sqrt{c_p}}, \frac{1}{\sqrt{-c_{p+1}}}, \frac{1}{\sqrt{-c_{p+2}}}, \ldots, \frac{1}{\sqrt{-c_r}}, 1, 1, \ldots, 1,$$

with $n - r$ ones. Let D be the diagonal matrix whose entries on the main diagonal are $c_1, c_2, \ldots, c_p, c_{p+1}, \ldots, c_r, c_{r+1}, \ldots, c_n$; A and D are congruent. Let $D_1 = H'DH$ be the matrix whose diagonal elements are $1, 1, \ldots, 1, -1, \ldots, -1, 0, 0, \ldots, 0$ (p ones, $n - r$ zeros); D and D_1 are then congruent. From Exercise 4 it follows that A and D_1 are congruent. In terms of quadratic forms, we have established Theorem 6.11.

Theorem 6.11. *A quadratic form $g(X) = X'AX$ in n variables is equivalent to a quadratic form $h(Y) = y_1^2 + y_2^2 + \cdots + y_p^2 - y_{p+1}^2 - y_{p+2}^2 - \cdots - y_r^2$.*

It is clear that the rank of the matrix D_1 is r, the number of nonzero entries on its main diagonal. Now it can be shown that congruent matrices have equal ranks. Since the rank of D_1 is r, the rank of A is also r. We also refer to r as the **rank** of the quadratic form g whose matrix is A. It can be shown that the number p of positive terms in the quadratic form h of Theorem 6.11 is unique; that is, no matter how we simplify the given quadratic form g to obtain an equivalent quadratic form, the latter will always have p positive terms. Hence the quadratic form h in Theorem 6.11 is unique; it is often called the **canonical form** of a quadratic form in n variables. The difference between the number of positive eigenvalues and the number of negative eigenvalues is $s = p - (r - p) = 2p - r$ and is called the **signature** of the quadratic form. Thus, if g and h are equivalent quadratic forms, then they have equal ranks and signatures. However, it can also be shown that if g and h have equal ranks and signatures, then they are equivalent.

EXAMPLE 7. Consider the quadratic form

$$g(X) = 3x_2^2 + 8x_2x_3 - 3x_3^2 = X'AX$$

$$= [x_1 \quad x_2 \quad x_3] \begin{bmatrix} 0 & 0 & 0 \\ 0 & 3 & 4 \\ 0 & 4 & -3 \end{bmatrix} \begin{bmatrix} x_1 \\ x_2 \\ x_3 \end{bmatrix}.$$

The eigenvalues of A are (verify)

$$c_1 = 5, \qquad c_2 = -5, \qquad \text{and} \qquad c_3 = 0.$$

In this case A is congruent to

$$D = \begin{bmatrix} 5 & 0 & 0 \\ 0 & -5 & 0 \\ 0 & 0 & 0 \end{bmatrix}.$$

If we let

$$H = \begin{bmatrix} 1/\sqrt{5} & 0 & 0 \\ 0 & 1/\sqrt{5} & 0 \\ 0 & 0 & 1 \end{bmatrix},$$

then

$$D_1 = H'DH = \begin{bmatrix} 1 & 0 & 0 \\ 0 & -1 & 0 \\ 0 & 0 & 0 \end{bmatrix}$$

and A are congruent, and the given quadratic form is equivalent to the canonical form

$$h(Y) = y_1^2 - y_2^2.$$

The rank of g is 2, and since $p = 1$, the signature $s = 2p - r = 0$.

As a final application of quadratic forms we consider positive definite, symmetric matrices. We recall that, in Section 3.2, a symmetric $n \times n$ matrix A is called positive definite if $X'AX > 0$ for every nonzero vector X in R^n.

If A is a symmetric matrix, then $X'AX$ is a quadratic form $g(X) = X'AX$ and, by Theorem 6.10, g is equivalent to h, where $h(Y) = c_1 y_1^2 + c_2 y_2^2 + \cdots + c_p y_p^2 + c_{p+1} y_{p+1}^2 + c_{p+2} y_{p+2}^2 + \cdots + c_r y_r^2$. Now A is positive definite if and only if $h(Y) > 0$ for each $Y \neq O$. However, this can happen if and only if all summands in $h(Y)$ are positive. These remarks have established the following theorem.

Theorem 6.12. *A symmetric matrix A is positive definite if and only if all the eigenvalues of A are positive.*

A quadratic form is then called **positive definite** if its matrix is positive definite.

6.3. Exercises

1. Write each of the following quadratic forms as $X'AX$.
   (a) $-3x^2 + 5xy - 2y^2$.
   (b) $2x_1^2 + 3x_1x_2 - 5x_1x_3 + 7x_2x_3$.
   (c) $3x_1^2 + x_1x_2 - 2x_1x_3 + x_2^2 - 4x_2x_3 - 2x_3^2$.

2. Write each of the following quadratic forms as $X'AX$.
   (a) $x^2 - 4xy - 3y^2 + 6yz + 4z^2$.
   (b) $4x^2 - 6xy + 2y^2$.
   (c) $-2x_1x_2 + 4x_1x_3 + 6x_2x_3$.

3. Prove that if A is a symmetric matrix, then $P'AP$ is also symmetric.

4. If A, B, and C are $n \times n$ symmetric matrices, prove the following.
   (a) A and A are congruent.
   (b) If A and B are congruent, then B and A are congruent.
   (c) If A and B are congruent and if B and C are congruent, then A and C are congruent.

5. For each of the following symmetric matrices A find a diagonal matrix D that is congruent to A.

   (a) $A = \begin{bmatrix} -1 & 0 & 0 \\ 0 & 1 & 1 \\ 0 & 1 & 1 \end{bmatrix}$.   (b) $A = \begin{bmatrix} 1 & 1 & 1 \\ 1 & 1 & 1 \\ 1 & 1 & 1 \end{bmatrix}$.

6. Prove that if A is symmetric, then A is congruent to a diagonal matrix D.

   In Exercises 7 through 11 find a quadratic form of the type in Theorem 6.10 that is equivalent to the given quadratic form.

7. $2x^2 - 4xy - y^2$.
8. $x_1^2 + x_2^2 + 2x_2x_3 + x_3^2$.
9. $2xz$.
10. $2x_2^2 + 4x_2x_3 + 2x_3^2$.
11. $-2x_1^2 - 4x_2^2 - 6x_2x_3 + 4x_3^2$.

   In Exercises 12 through 16 find a quadratic form of the type in Theorem 6.11 that is equivalent to the given quadratic form.

12. $2x^2 + 4xy + 2y^2$.
13. $x_1^2 + 2x_1x_2 + x_2^2 + x_3^2$.
14. $2x_1^2 + 4x_2^2 + 10x_2x_3 + 4x_3^2$.
15. $2x_1^2 + 3x_2^2 + 4x_2x_3 + 3x_3^2$.
16. $-3x_1^2 + 2x_2^2 + 4x_2x_3 + 2x_3^2$.
17. Find all quadratic forms $g(X) = X'AX$ in two variables of the type described in Theorem 6.11. What conics do the equations $X'AX = 1$ represent?
18. Find all quadratic forms $g(X) = X'AX$ in two variables of rank 1 of the type described in Theorem 6.11. What conics do the equations $X'AX = 1$ represent?

19. Let $g(X) = 4x^2 - 10xy + 4y^2 = 1$ be the equation of a central conic. Identify the conic by finding a quadratic form of the type in Theorem 6.10 that is equivalent to g.

20. Let $g(X) = x^2 + 2y^2 + 2yz + 2z^2 = 1$ be the equation of a central quadric surface. Identify the quadric surface by finding a quadratic form of the type in Theorem 6.10 that is equivalent to g.

21. Let $g(X) = 4x_2^2 - 10x_2x_3 + 4x_3^2$ be a quadratic form in three variables. Find a quadratic form of the type in Theorem 6.11 that is equivalent to g. What is the rank of g? What is the signature of g?

22. Which of the following quadratic forms in three variables are equivalent?

$$g_1(X) = x_1^2 + 2x_1x_2 + x_2^2 + x_3^2$$
$$g_2(X) = 2x_2^2 + 2x_2x_3 + 2x_3^2$$
$$g_3(X) = 3x_2^2 + 8x_2x_3 - 3x_3^2$$
$$g_4(X) = 3x_2^2 - 4x_2x_3 + 3x_3^2.$$

23. Which of the following matrices are positive definite?

(a) $\begin{bmatrix} 2 & -1 \\ -1 & 2 \end{bmatrix}$. (b) $\begin{bmatrix} 2 & 1 \\ 1 & 2 \end{bmatrix}$. (c) $\begin{bmatrix} 0 & -1 \\ -1 & 0 \end{bmatrix}$.

(d) $\begin{bmatrix} 1 & 1 \\ 1 & 1 \end{bmatrix}$. (e) $\begin{bmatrix} 2 & 2 \\ 2 & 2 \end{bmatrix}$.

24. Let $A = \begin{bmatrix} a & b \\ b & d \end{bmatrix}$ be a 2 × 2 symmetric matrix. Prove that A is positive definite if and only if $|A| > 0$ and $a > 0$.

25. Prove that a symmetric matrix A is positive definite if and only if $A = P'P$ for a nonsingular matrix P.

CHAPTER 7

Differential Equations

Differential equations occur often in all branches of science and engineering; linear algebra is helpful in the formulation and solution of differential equations. In this section we provide only a brief survey of the approach; books on differential equations deal with the subject in much greater detail and several suggestions for further reading are given at the end of this chapter.

We consider the homogeneous linear system of differential equations

$$
\begin{aligned}
\dot{x}_1(t) &= a_{11}x_1(t) + a_{12}x_2(t) + \cdots + a_{1n}x_n(t) \\
\dot{x}_2(t) &= a_{21}x_1(t) + a_{22}x_2(t) + \cdots + a_{2n}x_n(t) \\
&\;\;\vdots \\
\dot{x}_n(t) &= a_{n1}x_1(t) + a_{n2}x_2(t) + \cdots + a_{nn}x_n(t),
\end{aligned}
\tag{1}
$$

where the a_{ij} are constants. We seek functions $x_1(t)$, $x_2(t)$, ..., $x_n(t)$ satisfying (1). The expression

$$
b_1 x_1(t) + b_2 x_2(t) + \cdots + b_n x_n(t),
$$

where $b_1, b_2, \ldots, b_n$ are arbitrary constants, is called the **general solution** to (1). If we assign values to these constants, we obtain a **specific solution** to (1).

Often a specific solution is determined by specifying the **initial conditions** $x_1(0) = f_1, x_2(0) = f_2, \ldots, x_n(0) = f_n$, where $f_1, f_2, \ldots, f_n$ are given constants.

We can write (1) in matrix form by letting

$$X(t) = \begin{bmatrix} x_1(t) \\ x_2(t) \\ \vdots \\ x_n(t) \end{bmatrix}, \qquad A = \begin{bmatrix} a_{11} & a_{12} & \cdots & a_{1n} \\ a_{21} & a_{22} & \cdots & a_{2n} \\ \vdots & \vdots & & \vdots \\ a_{n1} & a_{n2} & \cdots & a_{nn} \end{bmatrix},$$

and defining

$$\dot{X}(t) = \begin{bmatrix} \dot{x}_1(t) \\ \dot{x}_2(t) \\ \vdots \\ \dot{x}_n(t) \end{bmatrix}.$$

Then (1) can be written as

$$\dot{X}(t) = AX(t). \tag{2}$$

We shall often write (2) as

$$\dot{X} = AX.$$

EXAMPLE 1. The simplest homogeneous linear system of differential equations is the system consisting of a single equation $\dfrac{dx}{dt} = rx$, where r is a constant. We find, using calculus, that $x(t) = be^{rt}$, where b is an arbitrary constant, is the general solution. If $x(0) = b_0$ is an initial value, then we can determine the constant in the general solution by writing $b_0 = x(0) = be^{r(0)} = b$. Thus the specific solution satisfying the given initial condition is $x(t) = b_0 e^{rt}$.

EXAMPLE 2. Consider the following linear system of three differential equations:

$$\begin{aligned} \dot{x}_1 &= 3x_1 \\ \dot{x}_2 &= -2x_2 \\ \dot{x}_3 &= 4x_3. \end{aligned}$$

This is a very easy system to solve since the equations can be integrated separately. Thus the general solution is given as

$$x_1(t) = b_1 e^{3t}$$
$$x_2(t) = b_2 e^{-2t}$$
$$x_3(t) = b_3 e^{4t},$$

where b_1, b_2, and b_3 are arbitrary constants. If we are given the initial conditions $x_1(0) = 3$, $x_2(0) = 5$, and $x_3(0) = 7$, then the specific solution determined by these initial conditions is

$$x_1(t) = 3e^{3t}$$
$$x_2(t) = 5e^{-2t}$$
$$x_3(t) = 7e^{4t}.$$

We can write this system of equations in matrix form as

$$\dot{X} = \begin{bmatrix} \dot{x}_1 \\ \dot{x}_2 \\ \dot{x}_3 \end{bmatrix} = \begin{bmatrix} 3 & 0 & 0 \\ 0 & -2 & 0 \\ 0 & 0 & 4 \end{bmatrix} \begin{bmatrix} x_1 \\ x_2 \\ x_3 \end{bmatrix}.$$

The general solution is

$$X(t) = \begin{bmatrix} b_1 e^{3t} \\ b_2 e^{-2t} \\ b_3 e^{4t} \end{bmatrix} = b_1 \begin{bmatrix} 1 \\ 0 \\ 0 \end{bmatrix} e^{3t} + b_2 \begin{bmatrix} 0 \\ 1 \\ 0 \end{bmatrix} e^{-2t} + b_3 \begin{bmatrix} 0 \\ 0 \\ 1 \end{bmatrix} e^{4t}.$$

We ask the reader to show that the set of all solutions to a homogeneous linear system of differential equations $\dot{X} = AX$ is a vector space V (this situation is analogous to that for a homogeneous system of linear equations $AX = O$), which is a subspace of the vector space of all real-valued functions. Thus, if x_1 and x_2 are in V, we define $x_1 + x_2$ by $(x_1 + x_2)(t) = x_1(t) + x_2(t)$, and if x is in V and k is a scalar, we define kx by $(kx)(t) = kx(t)$. The space V is called the **solution space** of the linear system. It can be shown (a more difficult problem) that if A is $n \times n$, then the dimension of this vector space is n.

EXAMPLE 3. Consider the linear system of differential equations

$$\begin{matrix} \dot{x}_1 = x_1 - x_2 \\ \dot{x}_2 = 2x_1 + 4x_2 \end{matrix} \quad \text{or} \quad \dot{X} = \begin{bmatrix} 1 & -1 \\ 2 & 4 \end{bmatrix} X. \tag{3}$$

The system is not diagonal, so we cannot integrate each equation individually. However, let us make the following change of variables:

$$x_1 = u_1 + u_2 \quad \text{and} \quad x_2 = -u_1 - 2u_2.$$

Of course, we shall soon show how we chose this particular change of variables. Then

$$\dot{x}_1 = \dot{u}_1 + \dot{u}_2 \quad \text{and} \quad \dot{x}_2 = -\dot{u}_1 - 2\dot{u}_2.$$

Substituting in (3), we have

$$\begin{aligned} \dot{u}_1 + \dot{u}_2 &= 2u_1 + 3u_2 \\ -\dot{u}_1 - 2\dot{u}_2 &= -2u_1 - 6u_2. \end{aligned} \tag{4}$$

First add the equations in (4) and then add twice the first equation in (4) to the second one to obtain the equivalent linear system

$$\begin{aligned} \dot{u}_1 &= 2u_1 \\ \dot{u}_2 &= 3u_2. \end{aligned} \tag{5}$$

Now (5) is a system in which each equation can be integrated individually. Thus the general solution to (5) is

$$\begin{aligned} u_1(t) &= b_1 e^{2t} \\ u_2(t) &= b_2 e^{3t}. \end{aligned}$$

Then the general solution to (3) is

$$\begin{aligned} x_1(t) &= b_1 e^{2t} + b_2 e^{3t} \\ x_2(t) &= -b_1 e^{2t} - 2b_2 e^{3t} \end{aligned} \quad \text{or} \quad X(t) = b_1 \begin{bmatrix} 1 \\ -1 \end{bmatrix} e^{2t} + b_2 \begin{bmatrix} 1 \\ -2 \end{bmatrix} e^{3t}.$$

If we look at the solution just given to Example 3 in matrix form, we shall see how to proceed in general. Thus the given linear system is

$$\dot{X} = \begin{bmatrix} \dot{x}_1 \\ \dot{x}_2 \end{bmatrix} = \begin{bmatrix} 1 & -1 \\ 2 & 4 \end{bmatrix} \begin{bmatrix} x_1 \\ x_2 \end{bmatrix}. \tag{6}$$

We made the change of variables

$$X = \begin{bmatrix} x_1 \\ x_2 \end{bmatrix} = \begin{bmatrix} 1 & 1 \\ -1 & -2 \end{bmatrix} \begin{bmatrix} u_1 \\ u_2 \end{bmatrix}, \tag{7}$$

and then let

$$\dot{X} = \begin{bmatrix} \dot{x}_1 \\ \dot{x}_2 \end{bmatrix} = \begin{bmatrix} 1 & 1 \\ -1 & -2 \end{bmatrix} \begin{bmatrix} \dot{u}_1 \\ \dot{u}_2 \end{bmatrix}. \tag{8}$$

Substituting (7) and (8) in (6), we have

$$\begin{bmatrix} 1 & 1 \\ -1 & -2 \end{bmatrix} \begin{bmatrix} \dot{u}_1 \\ \dot{u}_2 \end{bmatrix} = \begin{bmatrix} 1 & -1 \\ 2 & 4 \end{bmatrix} \begin{bmatrix} 1 & 1 \\ -1 & -2 \end{bmatrix} \begin{bmatrix} u_1 \\ u_2 \end{bmatrix}$$

or

$$\begin{bmatrix} \dot{u}_1 \\ \dot{u}_2 \end{bmatrix} = \begin{bmatrix} 1 & 1 \\ -1 & -2 \end{bmatrix}^{-1} \begin{bmatrix} 1 & -1 \\ 2 & 4 \end{bmatrix} \begin{bmatrix} 1 & 1 \\ -1 & -2 \end{bmatrix} \begin{bmatrix} u_1 \\ u_2 \end{bmatrix} = \begin{bmatrix} 2 & 0 \\ 0 & 3 \end{bmatrix} \begin{bmatrix} u_1 \\ u_2 \end{bmatrix}. \tag{9}$$

Solving (9) as in Example 2 and using (7), we find that the general solution to the given linear system is then

$$X(t) = b_1 \begin{bmatrix} 1 \\ -1 \end{bmatrix} e^{2t} + b_2 \begin{bmatrix} 1 \\ -2 \end{bmatrix} e^{3t}.$$

We now can see how to carry out this procedure for the case when A can be diagonalized. Consider the linear system $\dot{X} = AX$, where A is $n \times n$. Suppose that A is similar to a diagonal matrix D, so that there exists a nonsingular matrix P such that $D = P^{-1}AP$. Now let $X = PU$, where

$$U = \begin{bmatrix} u_1 \\ u_2 \\ \vdots \\ u_n \end{bmatrix},$$

and $\dot{X} = P\dot{U}$, where

$$\dot{U} = \begin{bmatrix} \dot{u}_1 \\ \dot{u}_2 \\ \vdots \\ \dot{u}_n \end{bmatrix}.$$

Then the given linear system $\dot{X} = AX$ becomes $P\dot{U} = A(PU)$, or $\dot{U} = (P^{-1}AP)U$, or $\dot{U} = DU$. We can write the latter equation as

$$
\begin{bmatrix} \dot{u}_1 \\ \dot{u}_2 \\ \vdots \\ \dot{u}_n \end{bmatrix} = \begin{bmatrix} c_1 & 0 & \cdots & & 0 \\ 0 & c_2 & \cdots & & 0 \\ \vdots & & \cdots & & \vdots \\ & & & & 0 \\ 0 & & \cdots & 0 & c_n \end{bmatrix} \begin{bmatrix} u_1 \\ u_2 \\ \vdots \\ u_n \end{bmatrix},
$$

where $c_1, c_2, \ldots, c_n$ are the eigenvalues of A.

This system of equations can be solved easily. Its general solution is

$$
U(t) = \begin{bmatrix} b_1 e^{c_1 t} \\ b_2 e^{c_2 t} \\ \vdots \\ b_n e^{c_n t} \end{bmatrix}.
$$

The general solution X to the original system $\dot{X} = AX$ is

$$
X(t) = \begin{bmatrix} x_1(t) \\ x_2(t) \\ \vdots \\ x_n(t) \end{bmatrix} = PU = \begin{bmatrix} p_{11} & p_{22} & \cdots & p_{1n} \\ p_{21} & p_{22} & \cdots & p_{2n} \\ \vdots & \vdots & & \vdots \\ p_{n1} & p_{n2} & \cdots & p_{nn} \end{bmatrix} \begin{bmatrix} b_1 e^{c_1 t} \\ b_2 e^{c_2 t} \\ \vdots \\ b_n e^{c_n t} \end{bmatrix}
$$

$$
= b_1 P_1 e^{c_1 t} + b_2 P_2 e^{c_2 t} + \cdots + b_n P_n e^{c_n t},
$$

where $P_1, P_2, \ldots, P_n$ are the columns of the matrix P, and $b_1, b_2, \ldots, b_n$ are arbitrary constants. We recall from Chapter 6 that the columns $P_1, P_2, \ldots, P_n$ of the matrix P are eigenvectors of A associated with the eigenvalues $c_1, c_2, \ldots, c_n$ of A. Thus, if the matrix A can be diagonalized, as happens when $c_1, c_2, \ldots, c_n$ are real and distinct, then we can always solve the linear system $\dot{X} = AX$ by integrating each equation of the corresponding diagonal system individually.

EXAMPLE 4. Consider the following linear system of differential equations:

$$
\dot{X} = \begin{bmatrix} \dot{x}_1 \\ \dot{x}_2 \\ \dot{x}_3 \end{bmatrix} = \begin{bmatrix} 0 & 1 & 0 \\ 0 & 0 & 1 \\ 8 & -14 & 7 \end{bmatrix} \begin{bmatrix} x_1 \\ x_2 \\ x_3 \end{bmatrix}.
$$

The characteristic polynomial of A is

$$f(t) = t^3 - 7t^2 + 14t - 8 = (t - 1)(t - 2)(t - 4),$$

so the eigenvalues of A are $c_1 = 1, c_2 = 2,$ and $c_3 = 4$. Associated eigenvectors are (verify)

$$\begin{bmatrix} 1 \\ 1 \\ 1 \end{bmatrix}, \begin{bmatrix} 1 \\ 2 \\ 4 \end{bmatrix}, \begin{bmatrix} 1 \\ 4 \\ 16 \end{bmatrix},$$

respectively. The general solution is

$$X(t) = b_1 \begin{bmatrix} 1 \\ 1 \\ 1 \end{bmatrix} e^t + b_2 \begin{bmatrix} 1 \\ 2 \\ 4 \end{bmatrix} e^{2t} + b_3 \begin{bmatrix} 1 \\ 4 \\ 16 \end{bmatrix} e^{4t},$$

where b_1, b_2, and b_3 are arbitrary constants.

EXAMPLE 5. We consider the linear system of Example 4 and now wish to find the specific solution determined by the initial conditions $x_1(0) = 4$, $x_2(0) = 6$, and $x_3(0) = 8$.

We write our general solution as

$$X(t) = \begin{bmatrix} 1 & 1 & 1 \\ 1 & 2 & 4 \\ 1 & 4 & 16 \end{bmatrix} \begin{bmatrix} b_1 e^t \\ b_2 e^{2t} \\ b_3 e^{4t} \end{bmatrix}.$$

Now

$$X(0) = \begin{bmatrix} 4 \\ 6 \\ 8 \end{bmatrix} = \begin{bmatrix} 1 & 1 & 1 \\ 1 & 2 & 4 \\ 1 & 4 & 16 \end{bmatrix} \begin{bmatrix} b_1 e^0 \\ b_2 e^0 \\ b_3 e^0 \end{bmatrix}.$$

Thus

$$\begin{bmatrix} b_1 \\ b_2 \\ b_3 \end{bmatrix} = \begin{bmatrix} 1 & 1 & 1 \\ 1 & 2 & 4 \\ 1 & 4 & 16 \end{bmatrix}^{-1} \begin{bmatrix} 4 \\ 6 \\ 8 \end{bmatrix} = \begin{bmatrix} \frac{8}{3} & -2 & \frac{1}{3} \\ -2 & \frac{5}{2} & -\frac{1}{2} \\ \frac{1}{3} & -\frac{1}{2} & \frac{1}{6} \end{bmatrix} \begin{bmatrix} 4 \\ 6 \\ 8 \end{bmatrix} = \begin{bmatrix} \frac{4}{3} \\ 3 \\ -\frac{1}{3} \end{bmatrix}.$$

The specific solution is

$$X(t) = \tfrac{4}{3}\begin{bmatrix} 1 \\ 1 \\ 1 \end{bmatrix} e^t + 3\begin{bmatrix} 1 \\ 2 \\ 4 \end{bmatrix} e^{2t} - \tfrac{1}{3}\begin{bmatrix} 1 \\ 4 \\ 16 \end{bmatrix} e^{4t}.$$

If A does not have distinct eigenvalues, then we may or may not be able to diagonalize A. Let c be an eigenvalue of A of multiplicity k. Then A can be diagonalized if and only if the dimension of the eigenspace associated with c is k, that is, if and only if the rank of the matrix $(cI_n - A)$ is $n - k$ (verify). If the rank of $(cI_n - A)$ is $n - k$, then we can find k linearly independent eigenvectors of A associated with c.

EXAMPLE 6. Consider the linear system

$$\dot{X} = AX = \begin{bmatrix} 1 & 0 & 0 \\ 0 & 3 & -2 \\ 0 & -2 & 3 \end{bmatrix} X.$$

The eigenvalues of A are $c_1 = c_2 = 1$ and $c_3 = 5$. The rank of the matrix $(1I_3 - A) = \begin{bmatrix} 0 & 0 & 0 \\ 0 & -2 & 2 \\ 0 & 2 & -2 \end{bmatrix}$ is 1, so the linearly independent eigenvectors $\begin{bmatrix} 1 \\ 0 \\ 0 \end{bmatrix}$ and $\begin{bmatrix} 0 \\ 1 \\ 1 \end{bmatrix}$ are associated with the eigenvalue 1. The eigenvector $\begin{bmatrix} 0 \\ 1 \\ -1 \end{bmatrix}$ is associated with the eigenvalue 5. The general solution to the given system is then

$$X(t) = b_1 \begin{bmatrix} 1 \\ 0 \\ 0 \end{bmatrix} e^t + b_2 \begin{bmatrix} 0 \\ 1 \\ 1 \end{bmatrix} e^t + b_3 \begin{bmatrix} 0 \\ 1 \\ -1 \end{bmatrix} e^{5t},$$

where b_1, b_2, and b_3 are arbitrary constants.

On the other hand, if we cannot diagonalize A, we are in a considerably more difficult situation. Suppose that the rank of $(cI_n - A)$ is greater than

$n - k$. The following example indicates a procedure followed in a specific case. Other cases of this type are not considered here.

EXAMPLE 7. Consider the following linear system of differential equations:

$$\dot{X} = AX = \begin{bmatrix} 0 & 1 & 0 \\ 0 & 0 & 1 \\ 12 & -16 & 7 \end{bmatrix} \begin{bmatrix} x_1 \\ x_2 \\ x_3 \end{bmatrix}.$$

The characteristic polynomial of A is

$$f(t) = t^3 - 7t^2 + 16t - 12 = (t - 2)^2(t - 3).$$

The distinct eigenvalues of A are $c_1 = 2$ and $c_2 = 3$, with respective multiplicities $k_1 = 2$ and $k_2 = 1$ (verify).

We now observe that the rank of the matrix

$$2I_3 - A = \begin{bmatrix} 2 & -1 & 0 \\ 0 & 2 & -1 \\ -12 & 16 & -5 \end{bmatrix}$$

is 2. For $c_1 = 2$ we find that $P_1 = \begin{bmatrix} 1 \\ 2 \\ 4 \end{bmatrix}$ is an eigenvector of A associated with c_1. Now let P_2 satisfy the equation $(A - 2I_3)P_2 = P_1$. We are thus solving the system of linear equations

$$\begin{bmatrix} -2 & 1 & 0 \\ 0 & -2 & 1 \\ 12 & -16 & 5 \end{bmatrix} \begin{bmatrix} a_1 \\ a_2 \\ a_3 \end{bmatrix} = \begin{bmatrix} 1 \\ 2 \\ 4 \end{bmatrix}; \quad \text{a solution is } P_2 = \begin{bmatrix} -1 \\ -1 \\ 0 \end{bmatrix}.$$

Then $P_3 = \begin{bmatrix} 1 \\ 3 \\ 9 \end{bmatrix}$ is an eigenvector of A associated with $c_2 = 3$ (verify). We can also easily verify that the general solution is

$$X = b_1 P_1 e^{2t} + b_2(P_1 t e^{2t} + P_2 e^{2t}) + b_3 P_3 e^{3t}$$

or

$$
X = b_1 \begin{bmatrix} 1 \\ 2 \\ 4 \end{bmatrix} e^{2t} + b_2 \left(\begin{bmatrix} 1 \\ 2 \\ 4 \end{bmatrix} te^{2t} + \begin{bmatrix} -1 \\ -1 \\ 0 \end{bmatrix} e^{2t} \right) + b_3 \begin{bmatrix} 1 \\ 3 \\ 9 \end{bmatrix} e^{3t}
$$

$$
= \left(b_1 \begin{bmatrix} 1 \\ 2 \\ 4 \end{bmatrix} + b_2 \begin{bmatrix} -1 \\ -1 \\ 0 \end{bmatrix} \right) e^{2t} + b_2 \begin{bmatrix} 1 \\ 2 \\ 4 \end{bmatrix} te^{2t} + b_3 \begin{bmatrix} 1 \\ 3 \\ 9 \end{bmatrix} e^{3t}
$$

$$
= (b_1 P_1 + b_2 P_2)e^{2t} + b_2 P_1 te^{2t} + b_3 P_3 e^{3t}.
$$

We now consider an **nth-order homogeneous linear differential equation with constant coefficients**:

$$
\frac{d^n x(t)}{dt^n} + a_1 \frac{d^{n-1}x(t)}{dt^{n-1}} + \cdots + a_{n-1}\frac{dx(t)}{dt} + a_n x(t) = 0, \tag{10}
$$

where $a_1, a_2, \ldots, a_n$ are real numbers. What we seek is a function x of t satisfying (10). This is called the **general solution** and again, since we are integrating, there are constants of integration in the general solution; it can be shown that there are n such constants of integration. If we assign values to these constants, we obtain a **specific solution** to (10). Often, a specific solution is determined by specifying the **initial** conditions $x(0) = f_1$, $\dot{x}(0) = f_2$, $\ldots, x^{(n-1)}(0) = f_n$, where $f_1, f_2, \ldots, f_n$ are given constants.

We first show that every nth-order homogeneous linear differential equation with constant coefficients is equivalent to a system of n first-order homogeneous linear differential equations with constant coefficients that can then be expressed in terms of matrices. If $x(t)$ is the unknown function, we let

$$
x_1(t) = x(t), \dot{x}(t) = \frac{dx(t)}{dt} = x_2(t), \ddot{x}(t) = \frac{d^2 x(t)}{dt^2} = x_3(t), \ldots, \frac{d^{n-1}x(t)}{dt^{n-1}} = x_n(t).
$$

We now obtain the following system of n first-order homogeneous linear differential equations with constant coefficients:

$$
\dot{x}_1 = \frac{dx_1(t)}{dt} = x_2(t)
$$

$$
\dot{x}_2 = \frac{dx_2(t)}{dt} = x_3(t)
$$

$$
\vdots \tag{11}
$$

$$
\dot{x}_{n-1} = \frac{d\dot{x}_{n-1}(t)}{dt} = x_n(t)
$$

$$
\dot{x}_n = \frac{dx_n(t)}{dt} = -a_n x_1(t) - a_{n-1}x_2(t) - \cdots - a_1 x_n(t).
$$

It is clear that if $x(t)$ is a solution to (10), then $x_1(t), x_2(t), \ldots, x_n(t)$ is a solution to (11); conversely, any solution to (11) also gives a solution $x_1(t) = x(t)$ to (10). Thus (10) and (11) are equivalent. We can write (11) in matrix form as $\dot{X}(t) = AX(t)$, where

$$X(t) = \begin{bmatrix} x_1(t) \\ x_2(t) \\ \vdots \\ x_n(t) \end{bmatrix}, \quad \dot{X}(t) = \begin{bmatrix} \dot{x}_1(t) \\ \dot{x}_2(t) \\ \vdots \\ \dot{x}_n(t) \end{bmatrix},$$

and

$$A = \begin{bmatrix} 0 & 1 & 0 & . & . & . & 0 \\ 0 & 0 & 1 & 0 & \cdots & & 0 \\ \vdots & \vdots & \vdots & \vdots & & & \vdots \\ 0 & 0 & . & . & . & 0 & 1 \\ -a_n & -a_{n-1} & . & . & . & & -a_1 \end{bmatrix}.$$

EXAMPLE 8. Consider the differential equation

$$\dddot{x} - 7\ddot{x} + 14\dot{x} - 8x = 0.$$

We let $x_1 = x$, $x_2 = \dot{x}_1$, $x_3 = \ddot{x}$. Then (10) becomes

$$\dot{x}_1 = \frac{dx_1}{dt} = x_2$$

$$\dot{x}_2 = \frac{dx_2}{dt} = x_3$$

$$\dot{x}_3 = \frac{dx_3}{dt} = 8x_1 - 14x_2 + 7x_3.$$

Writing this in matrix form, we get (11) as

$$\dot{X} = \begin{bmatrix} \dot{x}_1 \\ \dot{x}_2 \\ \dot{x}_3 \end{bmatrix} = \begin{bmatrix} 0 & 1 & 0 \\ 0 & 0 & 1 \\ 8 & -14 & 7 \end{bmatrix} \begin{bmatrix} x_1 \\ x_2 \\ x_3 \end{bmatrix}.$$

This is the linear system considered in Example 4. Thus the general solution is

$$x(t) = x_1(t) = b_1 e^t + b_2 e^{2t} + b_3 e^{4t},$$

where b_1, b_2, and b_3 are arbitrary constants.

EXAMPLE 9. Consider the differential equation $\dddot{x} - 7\ddot{x} + 16\dot{x} - 12x = 0$. We can write this in matrix form as

$$\dot{X} = \begin{bmatrix} \dot{x}_1 \\ \dot{x}_2 \\ \dot{x}_3 \end{bmatrix} = \begin{bmatrix} 0 & 1 & 0 \\ 0 & 0 & 1 \\ 12 & -16 & 7 \end{bmatrix} \begin{bmatrix} x_1 \\ x_2 \\ x_3 \end{bmatrix} = AX.$$

This is precisely the linear system considered in Example 7. Thus the general solution is

$$x(t) = x_1(t) = (b_1 - b_2)e^{2t} + b_2 t e^{2t} + b_3 e^{3t} = d_1 e^{2t} + d_2 t e^{2t} + d_3 e^{3t},$$

where d_1, d_2, and d_3 are arbitrary constants.

It can be shown that the nth-order homogeneous linear differential equation with constant coefficients when written in matrix form as $\dot{X} = AX$ has the property that if c is an eigenvalue of A of multiplicity $k > 1$, then the rank of the matrix $(cI_n - A)$ is always $n - k$. It can also be shown that the general solution will include the terms

$$b_1 e^{ct}, b_2 t e^{ct}, b_3 t^2 e^{ct}, \ldots, b_k t^{k-1} e^{ct},$$

where $b_1, b_2, \ldots$, and b_k are arbitrary constants. Note that to obtain the general solution to (10), we merely need $x(t) = x_1(t)$, which is the first row of P times the solution $X(t)$ to the resulting system. We do not need to know the first row of P, because the general solution involves arbitrary constants. Thus we merely need the eigenvalues of the resulting matrix A. We may also note that it is quite easy to write down the characteristic polynomial $f(t)$ of the matrix A involved in writing (10) as $\dot{X} = AX$; we can show that

$$f(t) = t^n + a_1 t^{n-1} + \cdots + a_{n-1}t + a_n.$$

Thus, in Example 9, the characteristic polynomial associated with $\dddot{x} - 7\ddot{x} + 16\dot{x} - 12x = 0$ is $f(t) = t^3 - 7t^2 + 16t - 12$. Also, the characteristic polynomial of $\dddot{x} - 7\ddot{x} + 14\dot{x} - 8x = 0$ is $f(t) = t^3 - 7t^2 + 14t - 8$.

EXAMPLE 10. Consider the differential equation $x^{(v)} - 15\ddddot{x} + 10\dddot{x} + 60\dot{x} - 72x = 0$, which we can write in matrix form as

$$\dot{X} = \begin{bmatrix} \dot{x}_1 \\ \dot{x}_2 \\ \dot{x}_3 \\ \dot{x}_4 \\ \dot{x}_5 \end{bmatrix} = \begin{bmatrix} 0 & 1 & 0 & 0 & 0 \\ 0 & 0 & 1 & 0 & 0 \\ 0 & 0 & 0 & 1 & 0 \\ 0 & 0 & 0 & 0 & 1 \\ 72 & -60 & -10 & 15 & 0 \end{bmatrix} \begin{bmatrix} x_1 \\ x_2 \\ x_3 \\ x_4 \\ x_5 \end{bmatrix} = AX.$$

The characteristic polynomial of A is $f(t) = t^5 - 15t^3 + 10t^2 + 60t - 72 = (t - 2)^3(t + 3)^2$, so the distinct eigenvalues are $c_1 = 2$ and $c_2 = -3$ with multiplicities $k_1 = 3$ and $k_2 = 2$, respectively. The general solution is then

$$x(t) = b_1 e^{2t} + b_2 t e^{2t} + b_3 t^2 e^{2t} + b_4 e^{3t} + b_5 t e^{3t},$$

where b_1, b_2, b_3, b_4, and b_5 are arbitrary constants.

It should also be pointed out that many differential equations cannot be solved in the sense that we can write a formula for the solution. Numerical methods, some of which are studied in numerical analysis, exist for obtaining numerical solutions to differential equations; computer codes for some of these methods are widely available.

Further Reading

Bentley, Donald L., and Kenneth L. Cooke. *Linear Algebra with Differential Equations.* New York: Holt, Rinehart and Winston, Inc., 1973.

Dettman, John H. *Introduction to Linear Algebra and Differential Equations.* New York: McGraw-Hill Book Company, 1974.

Rabenstein, Albert L. *Elementary Differential Equations with Linear Algebra,* 2nd ed. New York: Academic Press, Inc., 1975.

Wolfenstein, Samuel. *Introduction to Linear Algebra and Differential Equations.* San Francisco: Holden-Day, Inc., 1969.

7. Exercises

1. Consider the linear system of differential equations

$$\begin{bmatrix} \dot{x}_1 \\ \dot{x}_2 \\ \dot{x}_3 \end{bmatrix} = \begin{bmatrix} -3 & 0 & 0 \\ 0 & 4 & 0 \\ 0 & 0 & 2 \end{bmatrix} \begin{bmatrix} x_1 \\ x_2 \\ x_3 \end{bmatrix}.$$

(a) Find the general solution.

(b) Find the specific solution satisfying the initial conditions $x_1(0) = 3$, $x_2(0) = 4$, $x_3(0) = 5$.

2. Consider the linear system of differential equations

$$\begin{bmatrix} \dot{x}_1 \\ \dot{x}_2 \\ \dot{x}_3 \end{bmatrix} = \begin{bmatrix} 1 & 0 & 0 \\ 0 & -2 & 1 \\ 0 & 0 & 3 \end{bmatrix} \begin{bmatrix} x_1 \\ x_2 \\ x_3 \end{bmatrix}.$$

(a) Find the general solution.

(b) Find the specific solution satisfying $x_1(0) = 2$, $x_2(0) = 7$, $x_3(0) = 20$.

3. Find the general solution to the linear system of differential equations

$$\begin{bmatrix} \dot{x}_1 \\ \dot{x}_2 \\ \dot{x}_3 \end{bmatrix} = \begin{bmatrix} 4 & 0 & 0 \\ 3 & -5 & 0 \\ 2 & 1 & 2 \end{bmatrix} \begin{bmatrix} x_1 \\ x_2 \\ x_3 \end{bmatrix}.$$

4. Prove that the set of all solutions to a linear system of differential equations $\dot{X} = AX$ is a vector space.

5. Find the general solution to the linear system of differential equations

$$\begin{bmatrix} \dot{x}_1 \\ \dot{x}_2 \\ \dot{x}_3 \end{bmatrix} = \begin{bmatrix} 5 & 0 & 0 \\ 0 & -4 & 3 \\ 0 & 3 & 4 \end{bmatrix} \begin{bmatrix} x_1 \\ x_2 \\ x_3 \end{bmatrix}.$$

6. Find the general solution to the linear system of differential equations

$$\begin{bmatrix} \dot{x}_1 \\ \dot{x}_2 \end{bmatrix} = \begin{bmatrix} 3 & -2 \\ -2 & 3 \end{bmatrix} \begin{bmatrix} x_1 \\ x_2 \end{bmatrix}.$$

7. Find the general solution to the linear system of differential equations

$$\begin{bmatrix} \dot{x}_1 \\ \dot{x}_2 \\ \dot{x}_3 \end{bmatrix} = \begin{bmatrix} -3 & -2 & 3 \\ 0 & -5 & -2 \\ 0 & 1 & -2 \end{bmatrix} \begin{bmatrix} x_1 \\ x_2 \\ x_3 \end{bmatrix}.$$

8. Find the general solution to the linear system of differential equations

$$\begin{bmatrix} \dot{x}_1 \\ \dot{x}_2 \\ \dot{x}_3 \end{bmatrix} = \begin{bmatrix} 1 & 1 & 2 \\ 0 & 1 & 0 \\ 0 & 1 & 3 \end{bmatrix} \begin{bmatrix} x_1 \\ x_2 \\ x_3 \end{bmatrix}.$$

9. For each of the following nth-order homogeneous linear differential equations with constant coefficients: (1) find an equivalent system of n first-order

homogeneous linear differential equations with constant coefficients, and (2) write the system obtained in (1) in matrix form:

(a) $\ddot{x} - 2\dot{x} + x = 0.$ (b) $\dddot{x} - 2\ddot{x} + \dot{x} - x = 0.$ (c) $\dddot{x} - x = 0.$

10. Repeat Exercise 9 for each of the following:

   (a) $\dddot{x} + 2\ddot{x} + 3\dot{x} - 5x = 0.$ (b) $\ddot{x} - x = 0.$ (c) $\dddot{x} + \ddot{x} - x = 0.$

11. Consider the differential equation $\ddot{x} - 2\dot{x} - 3x = 0.$

   (a) Find the general solution.

   (b) Find the specific solution satisfying the initial conditions $x(0) = \dot{x}(0) = 2.$

12. Find the general solution to the differential equation $\dddot{x} - 5\ddot{x} + 7\dot{x} - 3x = 0.$

13. Find the general solution to the differential equation $\ddot{x} - 2\dot{x} + x = 0.$

14. Find the general solution to the differential equation $x^{(iv)} - 5\ddot{x} + 4x = 0.$

15. Find the general solution to the differential equation $x^{(iv)} - 10\dddot{x} + 37\ddot{x} - 60\dot{x} + 36x = 0$ (the set of eigenvalues of the associated matrix is $\{2, 2, 3, 3\}$).

16. Find the general solution to the differential equation $\dddot{x} - 3\ddot{x} - 4\dot{x} + 12x = 0.$

17. Find the general solution to the differential equation $\dddot{x} - 3\ddot{x} - 10\dot{x} + 24x = 0.$

18. Find the general solution to the differential equation $x^{(vi)} - 9x^{(v)} + 33x^{(iv)} - 63\dddot{x} + 66\ddot{x} - 36\dot{x} + 8x = 0.$ (The set of eigenvalues is $\{1, 1, 1, 2, 2, 2\}$.)

Answers to Selected Exercises

Chapter 1

Section 1.2, p. 16

1. (a) $\begin{bmatrix} 5 & -5 & 8 \\ 4 & 2 & 9 \\ 5 & 3 & 4 \end{bmatrix}$.

(b) $AB = \begin{bmatrix} 14 & 8 \\ 16 & 9 \end{bmatrix}$, $BA = \begin{bmatrix} 1 & 2 & 3 \\ 4 & 5 & 10 \\ 7 & 8 & 17 \end{bmatrix}$.

(c) $\begin{bmatrix} 0 & 10 & -9 \\ 8 & -1 & -2 \\ -5 & -4 & 3 \end{bmatrix}$.

(d) Impossible.

(e) $\begin{bmatrix} 19 & -6 \\ 30 & 21 \end{bmatrix}$.

(f) $3(2A) = 6A = \begin{bmatrix} 6 & 12 & 18 \\ 12 & 6 & 24 \end{bmatrix}$.

3. (a) $\begin{bmatrix} 1 & 2 \\ 2 & 1 \\ 3 & 4 \end{bmatrix}$.

(b) $(A')' = A$.

(c) $\begin{bmatrix} 14 & 16 \\ 8 & 9 \end{bmatrix}$.

(d) Same as (c).

(e) $(C + E)' = C' + E' = \begin{bmatrix} 5 & 4 & 5 \\ -5 & 2 & 3 \\ 8 & 9 & 4 \end{bmatrix}$.

(f) $A(2B) = 2(AB) = \begin{bmatrix} 28 & 16 \\ 32 & 18 \end{bmatrix}$.

6. (a) $\begin{bmatrix} 2 & 3 & -3 & 1 & 1 \\ 3 & 0 & 2 & 0 & 3 \\ 2 & 3 & 0 & -4 & 0 \\ 0 & 0 & 1 & 1 & 1 \end{bmatrix}$.

(b) $\begin{bmatrix} 2 & 3 & -3 & 1 & 1 \\ 3 & 0 & 2 & 0 & 3 \\ 2 & 3 & 0 & -4 & 0 \\ 0 & 0 & 1 & 1 & 1 \end{bmatrix} \begin{bmatrix} x_1 \\ x_2 \\ x_3 \\ x_4 \\ x_5 \end{bmatrix} = \begin{bmatrix} 7 \\ -2 \\ 3 \\ 5 \end{bmatrix}$.

(c) $\begin{bmatrix} 2 & 3 & -3 & 1 & 1 & \vdots & 7 \\ 3 & 0 & 2 & 0 & 3 & \vdots & -2 \\ 2 & 3 & 0 & -4 & 0 & \vdots & 3 \\ 0 & 0 & 1 & 1 & 1 & \vdots & 5 \end{bmatrix}$.

7. $\begin{aligned} -2x_1 - x_2 \qquad\quad + 4x_4 &= 5 \\ -3x_1 + 2x_2 + 7x_3 + 8x_4 &= 3 \\ x_1 \qquad\qquad\quad + 2x_4 &= 4 \\ 3x_1 + \qquad + x_3 + 3x_4 &= 6. \end{aligned}$

8. $a = 3, b = 1, c = 8, d = -2$.

11. They are equivalent.

13. (a) 4.

(b) 2.

(c) 3.

Section 1.3, p. 24

3. $A(BC) = (AB)C = \begin{bmatrix} -2 & 34 \\ 24 & -9 \end{bmatrix}$.

5. $A(rB) = r(AB) = \begin{bmatrix} -6 & 18 & -42 \\ 9 & -27 & 0 \end{bmatrix}$.

7. $C(A + B) = \begin{bmatrix} -10 & -8 & 16 \\ 10 & 14 & -28 \end{bmatrix}.$

11. $(a + b)A = \begin{bmatrix} 4 & -6 \\ 8 & 4 \end{bmatrix}.$

14. $a(A + B) = \begin{bmatrix} -12 & -12 \\ -15 & 0 \\ -3 & -9 \end{bmatrix}.$

19. $(A + B)' = \begin{bmatrix} 5 & 0 \\ 5 & 2 \\ 1 & 2 \end{bmatrix}, (cA)' = \begin{bmatrix} -4 & -8 \\ -12 & -4 \\ -8 & 12 \end{bmatrix}.$

20. $(AB)' = \begin{bmatrix} 11 & 5 \\ 15 & -4 \end{bmatrix}.$

Section 1.4, p. 32

7. $S = \begin{bmatrix} 3 & \frac{3}{2} & 0 \\ \frac{3}{2} & 2 & \frac{9}{2} \\ 0 & \frac{9}{2} & 2 \end{bmatrix}$ and $K = \begin{bmatrix} 0 & -\frac{7}{2} & 1 \\ \frac{7}{2} & 0 & -\frac{3}{2} \\ -1 & \frac{3}{2} & 0 \end{bmatrix}.$

9. $A^{-1} = \begin{bmatrix} -\frac{2}{13} & \frac{3}{13} \\ \frac{5}{13} & -\frac{1}{13} \end{bmatrix}.$

10. $A + B = \begin{bmatrix} 4 & 5 & \vdots & 0 \\ 0 & 4 & \vdots & 1 \\ \cdots & \cdots & \vdots & \cdots \\ 6 & -2 & \vdots & 6 \end{bmatrix}$ is one possible answer.

13. (a) $\begin{bmatrix} \frac{6}{13} \\ \frac{11}{13} \end{bmatrix}.$ (b) $\begin{bmatrix} \frac{8}{13} \\ \frac{19}{13} \end{bmatrix}.$

21. $A + B = \begin{bmatrix} 9 & -1 & 1 \\ 0 & -2 & 7 \\ 0 & 0 & 8 \end{bmatrix}$ and $AB = \begin{bmatrix} 18 & -5 & 11 \\ 0 & -8 & -7 \\ 0 & 0 & 15 \end{bmatrix}.$

24. $A = \begin{bmatrix} -\frac{1}{2} & \frac{1}{2} \\ 2 & -1 \end{bmatrix}.$

Section 1.5, p. 47

2. (a) $B = \begin{bmatrix} 0 & 1 & \frac{3}{2} & 2 & \frac{5}{2} \\ 0 & 0 & 1 & -2 & -3 \\ 0 & 0 & 0 & 1 & 2 \\ 0 & 0 & 0 & 0 & 1 \end{bmatrix}$ is a possible answer.

(b) $C = \begin{bmatrix} 0 & 1 & 0 & 0 & 0 \\ 0 & 0 & 1 & 0 & 0 \\ 0 & 0 & 0 & 1 & 0 \\ 0 & 0 & 0 & 0 & 1 \end{bmatrix}.$

5. $x_1 = 1, x_2 = 2, x_3 = -2.$

7. (a) $x_1 = -1, x_2 = 4, x_3 = -3.$

   (b) $x_1 = x_2 = x_3 = 0.$

   (c) $x_1 = r, x_2 = -2r, x_3 = r$, where $r =$ any real number.

   (d) $x_1 = -2r, x_2 = r, x_3 = 0$, where $r =$ any real number.

9. (a) $x_1 = 1, x_2 = 2, x_3 = 2.$

   (b) $x_1 = x_2 = x_3 = 0.$

12. $a = -2$, no solution; $a = 2$, infinitely many solutions; $a \neq \pm 2$, unique solution.

16. $a = -3$, no solution; $a = 3$, infinitely many solutions; $a \neq \pm 3$ unique solution.

17. (a) A possible answer is $\begin{bmatrix} 1 & 0 & 0 & 0 & 0 \\ -3 & 1 & 0 & 0 & 0 \\ -3 & \frac{1}{2} & 1 & 0 & 0 \\ -2 & \frac{3}{2} & -\frac{7}{2} & 1 & 0 \end{bmatrix}.$

   (b) $\begin{bmatrix} 1 & 0 & 0 & 0 & 0 \\ 0 & 1 & 0 & 0 & 0 \\ 0 & 0 & 1 & 0 & 0 \\ 0 & 0 & 0 & 1 & 0 \end{bmatrix}.$

18. $a = \pm\sqrt{6}$, no solution; $a \neq \pm\sqrt{6}$, unique solution.

19. (a) A possible answer is $\begin{bmatrix} 1 & 0 & 0 & 0 & 0 \\ 2 & 1 & 0 & 0 & 0 \\ 3 & \frac{5}{3} & 1 & 0 & 0 \end{bmatrix}.$

   (b) $\begin{bmatrix} 1 & 0 & 0 & 0 & 0 \\ 0 & 1 & 0 & 0 & 0 \\ 0 & 0 & 1 & 0 & 0 \end{bmatrix}.$

Section 1.6, p. 56

3. (a) $\begin{bmatrix} 1 & -4 & 0 & 0 \\ 0 & 1 & 0 & 0 \\ 0 & 0 & 1 & 0 \\ 0 & 0 & 0 & 1 \end{bmatrix}.$ (b) $\begin{bmatrix} 1 & 0 & 0 & 0 \\ 0 & 0 & 1 & 0 \\ 0 & 1 & 0 & 0 \\ 0 & 0 & 0 & 1 \end{bmatrix}.$

   (c) $\begin{bmatrix} 1 & 0 & 0 & 0 \\ 0 & 1 & 0 & 0 \\ 0 & 0 & 4 & 0 \\ 0 & 0 & 0 & 1 \end{bmatrix}.$

5. (a) $\begin{bmatrix} 1 & 0 & -1 \\ 1 & -1 & 2 \\ -1 & 1 & -1 \end{bmatrix}.$ (b) $\begin{bmatrix} \frac{7}{3} & -\frac{1}{3} & -\frac{1}{3} & -\frac{2}{3} \\ \frac{4}{9} & -\frac{1}{9} & -\frac{4}{9} & \frac{1}{9} \\ -\frac{1}{9} & -\frac{2}{9} & \frac{1}{9} & \frac{2}{9} \\ -\frac{5}{3} & \frac{2}{3} & \frac{2}{3} & \frac{1}{3} \end{bmatrix}.$

   (c) No inverse.

(d) $\begin{bmatrix} \frac{3}{2} & -1 & \frac{1}{2} \\ \frac{1}{2} & 0 & -\frac{1}{2} \\ -\frac{3}{2} & 1 & \frac{1}{2} \end{bmatrix}$.

(e) No inverse.

7. (a) $\begin{bmatrix} 1 & 0 & 0 \\ 0 & 1 & 0 \\ 0 & 0 & -3 \end{bmatrix}$. (b) $\begin{bmatrix} 1 & 0 & 0 \\ 0 & 0 & 1 \\ 0 & 1 & 0 \end{bmatrix}$. (c) $\begin{bmatrix} 1 & 0 & -5 \\ 0 & 1 & 0 \\ 0 & 0 & 1 \end{bmatrix}$.

9. (a) and (b) have nontrivial solutions.

10. (a) and (d) are singular.

(b) $\begin{bmatrix} 1 & 3 \\ -2 & 6 \end{bmatrix}^{-1} = \begin{bmatrix} \frac{1}{2} & -\frac{1}{4} \\ \frac{1}{6} & \frac{1}{12} \end{bmatrix}$.

(c) $\begin{bmatrix} 1 & 2 & 3 \\ 1 & 1 & 2 \\ 0 & 1 & 2 \end{bmatrix}^{-1} = \begin{bmatrix} 0 & 1 & -1 \\ 2 & -2 & -1 \\ -1 & 1 & 1 \end{bmatrix}$.

12. $A^{-1} = \begin{bmatrix} 1 & -1 & 0 \\ \frac{3}{2} & \frac{1}{2} & -\frac{3}{2} \\ -1 & 0 & 1 \end{bmatrix}$.

15. $A^{-1} = \begin{bmatrix} 1 & -1 & 0 & -1 \\ 0 & -\frac{1}{2} & 0 & 0 \\ -\frac{1}{5} & 1 & \frac{1}{5} & \frac{3}{5} \\ \frac{2}{5} & -\frac{1}{2} & -\frac{2}{5} & -\frac{1}{5} \end{bmatrix}$.

Section 1.7, p. 62

2. (a) $\begin{bmatrix} I_4 \\ 0 \end{bmatrix}$. (b) I_3.

(c) $\begin{bmatrix} I_2 & 0 \\ 0 & 0 \end{bmatrix}$. (d) I_4.

8. Possible answers are

(a) $\begin{bmatrix} 1 & -2 & 3 & 0 \\ 0 & -1 & 4 & 3 \\ 0 & 2 & -5 & -2 \end{bmatrix}$. (b) $\begin{bmatrix} 1 & 0 \\ 0 & 0 \end{bmatrix}$.

(c) $\begin{bmatrix} 1 & 0 & 0 & 0 & 0 \\ 0 & 1 & -2 & 0 & 2 \\ 0 & 5 & 5 & 4 & 4 \end{bmatrix}$.

Chapter 2

Section 2.1, p. 75

3. Properties 3, 4, and (b).
5. Properties 5, 6, and 8.

7. Property 8.

15. (b), (c), (e), and (f).

27. (c).

29. (a) $x = 2 + 2t, y = -3 + 5t, z = 1 + 4t.$
    (b) $x = -3 + 8t, y = -2 + 7t, z = -2 + 6t.$

Section 2.2, p. 93

1. No.

5. (a) and (c) are linearly dependent; (b) is linearly independent.
   (a) $[3 \quad 6 \quad 6] = 2[1 \quad 1 \quad 0] + 1[0 \quad 2 \quad 3] + 1[1 \quad 2 \quad 3].$
   (c) $[0 \quad 0 \quad 0] = 0[1 \quad 1 \quad 0] + 0[0 \quad 2 \quad 3] + 0[1 \quad 2 \quad 3].$

7. (a) and (b) are linearly independent; (c) is linearly dependent: $t + 13 = 3(2t^2 + t + 1) - 2(3t^2 + t - 5).$

10. (a) is a basis for R^3 and
$$\begin{bmatrix} 2 \\ 1 \\ 3 \end{bmatrix} = \frac{3}{2}\begin{bmatrix} 1 \\ 1 \\ 1 \end{bmatrix} + \frac{1}{2}\begin{bmatrix} 1 \\ 2 \\ 3 \end{bmatrix} - \frac{3}{2}\begin{bmatrix} 0 \\ 1 \\ 0 \end{bmatrix}.$$

11. (a) A possible answer is $\left\{ \begin{bmatrix} 1 \\ 0 \\ 2 \end{bmatrix}, \begin{bmatrix} 1 \\ 0 \\ 0 \end{bmatrix}, \begin{bmatrix} 0 \\ 1 \\ 0 \end{bmatrix} \right\}.$

   (b) A possible answer is $\left\{ \begin{bmatrix} 1 \\ 0 \\ 2 \end{bmatrix}, \begin{bmatrix} 0 \\ 1 \\ 3 \end{bmatrix}, \begin{bmatrix} 1 \\ 0 \\ 0 \end{bmatrix} \right\}.$

12. A possible answer is that $\left\{ \begin{bmatrix} -3 \\ 0 \\ 1 \\ 1 \\ 0 \end{bmatrix}, \begin{bmatrix} -2 \\ 1 \\ 0 \\ 0 \\ 0 \end{bmatrix} \right\}$ is a basis; dim $V = 2.$

15. A possible answer is that $\left\{ \begin{bmatrix} 1 \\ 2 \\ 2 \end{bmatrix}, \begin{bmatrix} 3 \\ 2 \\ 1 \end{bmatrix} \right\}$ is a basis; dim $W = 2.$

19. 3.

22. Yes.

23. (a) and (c) form bases.
    $5t^2 - 3t + 8 = 5(t^2 + t) - 8(t - 1)$
    $5t^2 - 3t + 8 = -3(t^2 + t) + 8(t^2 + 1).$

26. Zero.

Section 2.3, p. 104

4. Let $L: R_n \to R^n$ be defined by $L([a_1 \ a_2 \ \dots \ a_n]) = \begin{bmatrix} a_1 \\ a_2 \\ \vdots \\ a_n \end{bmatrix}$. Verify that L is an isomorphism.

5. Let $L: P_2 \to R^3$ be defined by $L(at^2 + bt + c) = \begin{bmatrix} a \\ b \\ c \end{bmatrix}$. Verify that L is an isomorphism.

6. (a) Let $L: {}_2R_2 \to R^4$ be defined by $L\left(\begin{bmatrix} a & b \\ c & d \end{bmatrix}\right) = \begin{bmatrix} a \\ b \\ c \\ d \end{bmatrix}$. Verify that L is an isomorphism.

   (b) 4.

8. (a) $[\alpha]_S = \begin{bmatrix} 3 \\ 2 \\ -7 \end{bmatrix}$; $[\beta]_S = \begin{bmatrix} 2 \\ 3 \\ -3 \end{bmatrix}$.

   (b) $P = \begin{bmatrix} 2 & 1 & 0 \\ 1 & -\frac{2}{5} & \frac{3}{5} \\ 0 & \frac{2}{5} & \frac{2}{5} \end{bmatrix}$.

   (c) $[\alpha]_T = \begin{bmatrix} 8 \\ -2 \\ -2 \end{bmatrix}$; $[\beta]_T = \begin{bmatrix} 7 \\ -1 \\ 0 \end{bmatrix}$.

   (d) Same as (c).

   (e) $Q = \begin{bmatrix} \frac{1}{3} & \frac{1}{3} & -\frac{1}{2} \\ \frac{1}{3} & -\frac{2}{3} & 1 \\ -\frac{1}{3} & \frac{2}{3} & \frac{3}{2} \end{bmatrix}$.

   (f) $[\alpha]_S = Q[\alpha]_T = \begin{bmatrix} 3 \\ 2 \\ -7 \end{bmatrix}$; $[\beta]_S = Q[\beta]_T = \begin{bmatrix} 2 \\ 3 \\ -3 \end{bmatrix}$; same as (a).

10. (a) $[\alpha]_S = \begin{bmatrix} 1 \\ 1 \\ 1 \\ 0 \end{bmatrix}$; $[\beta]_S = \begin{bmatrix} 2 \\ -2 \\ 1 \\ -1 \end{bmatrix}$.

    (b) $P = \begin{bmatrix} 1 & 0 & 0 & 1 \\ \frac{1}{3} & \frac{2}{3} & -\frac{2}{3} & 0 \\ \frac{1}{3} & -\frac{1}{3} & \frac{1}{3} & 0 \\ -\frac{1}{3} & \frac{1}{3} & \frac{2}{3} & 0 \end{bmatrix}$.

(c) $[\alpha]_T = \begin{bmatrix} 1 \\ \frac{1}{3} \\ \frac{1}{3} \\ \frac{2}{3} \end{bmatrix}$; $[\beta]_T = \begin{bmatrix} 1 \\ -\frac{4}{3} \\ \frac{5}{3} \\ -\frac{2}{3} \end{bmatrix}$.

(d) Same as (c).

(e) $Q = \begin{bmatrix} 0 & 1 & 2 & 0 \\ 0 & 1 & 0 & 1 \\ 0 & 0 & 1 & 1 \\ 1 & -1 & -2 & 0 \end{bmatrix}$.

(f) Same as (a).

11. If α is any vector in V, then $\alpha = ae^t + be^{-t}$, where a and b are scalars. Let $L: V \to R^2$ be defined by $L(\alpha) = \begin{bmatrix} a \\ b \end{bmatrix}$. Verify that L is an isomorphism.

13. (a) $\begin{bmatrix} 5 \\ 1 \\ 1 \end{bmatrix}$. (b) $\begin{bmatrix} 1 \\ 1 \\ 0 \end{bmatrix}$. (c) $\begin{bmatrix} 1 \\ 1 \\ 2 \end{bmatrix}$.

15. $\begin{bmatrix} 1 \\ 1 \\ -1 \end{bmatrix}$.

Section 2.4, p. 113

1. A possible answer is $\left\{ \begin{bmatrix} 1 \\ 0 \\ 0 \end{bmatrix}, \begin{bmatrix} 0 \\ 1 \\ 0 \end{bmatrix}, \begin{bmatrix} 0 \\ 0 \\ 1 \end{bmatrix} \right\}$.

3. A possible answer is $\left\{ \begin{bmatrix} 1 & 0 \\ 0 & 0 \end{bmatrix}, \begin{bmatrix} 0 & 1 \\ 0 & 0 \end{bmatrix}, \begin{bmatrix} 0 & 0 \\ 1 & 0 \end{bmatrix}, \begin{bmatrix} 0 & 0 \\ 0 & 1 \end{bmatrix} \right\}$.

5. (a) 3. (b) 5. (c) 2.

7. (a) 3. (b) 2. (c) 2.

9. B and C are equivalent; A, D, and E are equivalent.

11. (b) and (c) are nonsingular.

13. (a) 3. (b) 3. (c) 2.

Chapter 3

Section 3.1, p. 125

1. (a) 1. (b) 0. (c) $\sqrt{14}$. (d) $\sqrt{26}$. (e) $\sqrt{21}$.

3. (a) $\sqrt{155}$. (b) $\sqrt{67}$. (c) $\sqrt{155}$.

4. (a) 0. (b) 0. (c) 0. (d) -12. (e) 19.

5. (a) $\dfrac{-32}{\sqrt{14}\sqrt{77}}$. (b) $\dfrac{12}{\sqrt{14}\sqrt{77}}$. (c) $\dfrac{-32}{\sqrt{14}\sqrt{77}}$.

9. $\sqrt{35}$.

17. (b).

19. (a) $-15\mathbf{i} - 2\mathbf{j} + 9\mathbf{k}$.

    (b) $-3\mathbf{i} + 3\mathbf{j} + 3\mathbf{k}$.

    (c) $7\mathbf{i} + 5\mathbf{j} - \mathbf{k}$.

    (d) $0\mathbf{i} + 0\mathbf{j} + 0\mathbf{k}$.

25. (a).

27. (a) $x - z + 2 = 0$.

    (b) $3x + y - 14z + 47 = 0$.

29. $4x - 4y + z + 16 = 0$.

31. $x = -2 + 2t, y = 5 - 3t, z = -3 + 4t$.

Section 3.2, p. 137

9. (a) $\frac{13}{6}$.   (b) 3.   (c) 4.   (d) $\frac{3}{2}$.   (e) 1.

10. (a) $\sqrt{22}$.   (b) $\sqrt{18}$.   (c) 1.   (d) $\sqrt{12}$.   (e) $\sqrt{68}$.

11. (a) $-\frac{1}{2}$.   (b) 1.   (c) $\frac{4}{7}\sqrt{3}$.   (d) 1.   (e) 0.

16. The vectors in (b) are orthogonal.

21. (a) $\dfrac{\sin 1}{\sqrt{2}}$.   (b) $\sqrt{1/30}$.   (c) $\sqrt{783/60}$.   (d) $\sqrt{3}$.

22. (a) is orthonormal, (c) is orthogonal.

23. (a) $a = 0$.   (b) $3a = -5b$.

Section 3.3, p. 144

1. $\left\{ \dfrac{1}{\sqrt{3}} [1 \;\; 1 \;\; -1 \;\; 0], \dfrac{1}{\sqrt{33}} [-2 \;\; 4 \;\; 2 \;\; 3] \right\}$.

2. $\{\sqrt{3}t, 2 - 3t\}$.

6. $\left\{ \dfrac{1}{\sqrt{3}} \begin{bmatrix} 1 \\ 1 \\ 1 \end{bmatrix}, \sqrt{3/2} \begin{bmatrix} -\frac{2}{3} \\ \frac{1}{3} \\ \frac{1}{3} \end{bmatrix}, \sqrt{2} \begin{bmatrix} 0 \\ -\frac{1}{2} \\ \frac{1}{2} \end{bmatrix} \right\}$.

7. Possible answer: $\left\{ \begin{bmatrix} 0 \\ 0 \\ 1 \end{bmatrix}, \dfrac{1}{\sqrt{2}} \begin{bmatrix} 1 \\ 1 \\ 0 \end{bmatrix}, \sqrt{2} \begin{bmatrix} -\frac{1}{2} \\ \frac{1}{2} \\ 0 \end{bmatrix} \right\}$.

9. Possible answer: $\{[1/\sqrt{3} \;\; -1/\sqrt{3} \;\; 1/\sqrt{3}], [1/\sqrt{6} \;\; 2/\sqrt{6} \;\; 1/\sqrt{6}]\}$.

11. (a) $\left\{ \begin{bmatrix} 1 \\ 0 \\ 1 \end{bmatrix}, \begin{bmatrix} -5 \\ 1 \\ 5 \end{bmatrix} \right\}$.

    (b) $\left\{ \begin{bmatrix} 1/\sqrt{2} \\ 0 \\ 1/\sqrt{2} \end{bmatrix}, \begin{bmatrix} -5/\sqrt{51} \\ 2/\sqrt{51} \\ 5/\sqrt{51} \end{bmatrix} \right\}$.

15. (a) $\sqrt{14}$.

(b) $\left\{ \dfrac{1}{\sqrt{5}} \begin{bmatrix} 1 \\ 0 \\ -2 \end{bmatrix}, \dfrac{1}{3} \begin{bmatrix} -2 \\ 2 \\ -1 \end{bmatrix} \right\}$.

(c) $[\alpha]_T = \begin{bmatrix} \sqrt{5} \\ 3 \end{bmatrix}$, so $|[\alpha]_T| = \sqrt{5 + 9} = \sqrt{14}$.

Chapter 4

Section 4.1, p. 153

3. (b) is a linear transformation.

5. (a) $\begin{bmatrix} 2 \\ 15 \end{bmatrix}$.   (b) $\begin{bmatrix} 2a_1 + 3a_2 + 2a_3 \\ -4a_1 - 5a_2 + 3a_3 \end{bmatrix}$.

7. (a) $\begin{bmatrix} 15 & 5 & 4 & 8 \\ -5 & -1 & 10 & 2 \end{bmatrix}$.

10. $a =$ any real number, $b = 0$.

12. (a) Reflection about the y-axis.

(b) Reflection about the origin.

(c) Counterclockwise rotation through $\pi/2$.

Section 4.2, p. 164

1. (a) [0   0].   (b) Yes.   (c) No.

2. (a) A possible answer is $\{[1 \quad -1 \quad -1 \quad 1]\}$.   (b) 1.

(c) A possible answer is $\{[1 \quad 0 \quad 1], [1 \quad 0 \quad 0], [0 \quad 1 \quad 1]\}$.   (d) 3.

3. (a) A possible basis for kernel L is $\{1\}$ and dim ker $L = 1$.

(b) A possible basis for range L is $\{2t^3, t^2\}$, and dim range $L = 2$.

9. (a) A possible basis for ker L is $\left\{ \begin{bmatrix} -2 \\ 0 \\ 1 \\ 1 \\ 0 \end{bmatrix}, \begin{bmatrix} 0 \\ 1 \\ 0 \\ 0 \\ 0 \end{bmatrix} \right\}$; dim ker $L = 2$.

(b) A possible basis for range L is $\left\{ \begin{bmatrix} 1 \\ 0 \\ 0 \\ 1 \end{bmatrix}, \begin{bmatrix} 0 \\ 1 \\ 0 \\ -1 \end{bmatrix}, \begin{bmatrix} 0 \\ 0 \\ 1 \\ 0 \end{bmatrix} \right\}$; dim range $L = 3$.

10. No.

12. (b) $L^{-1}\left(\begin{bmatrix} 2 \\ 3 \\ 4 \end{bmatrix} \right) = \begin{bmatrix} \frac{3}{2} \\ -\frac{1}{2} \\ \frac{1}{2} \end{bmatrix}$.

15. $L[a_1 \quad a_2] = [a_1 + 3a_2 \quad -a_1 + a_2 \quad 2a_1 - a_2]$.

18. (a) 2.   (b) 1.

Section 4.3, p. 171

1. (a) $\begin{bmatrix} 1 & 2 \\ 2 & -1 \end{bmatrix}$.   (b) $\begin{bmatrix} 1 & -\frac{1}{2} \\ 1 & \frac{3}{4} \end{bmatrix}$.   (c) $\begin{bmatrix} 3 & 2 \\ -4 & 4 \end{bmatrix}$.

(d) $\begin{bmatrix} -2 & 2 \\ \frac{1}{2} & 2 \end{bmatrix}$.   (e) $\begin{bmatrix} 5 \\ 0 \end{bmatrix}$.

2. (a) $\begin{bmatrix} 1 & 0 & 0 & 0 \\ 0 & 1 & 1 & 0 \\ 0 & 0 & 1 & 1 \end{bmatrix}$.   (b) $\begin{bmatrix} 0 & -1 & 1 & -1 \\ 0 & 1 & 0 & 3 \\ 1 & 1 & 0 & 1 \end{bmatrix}$.

(c) $[2 \quad 0 \quad 2]$.

5. (a) $\begin{bmatrix} 1 & 2 & 1 \\ 1 & 0 & 0 \\ 0 & 1 & 1 \end{bmatrix}$.   (b) $\begin{bmatrix} 8 \\ 1 \\ 5 \end{bmatrix}$.

7. (a) $\begin{bmatrix} 10 \\ 5 \\ 5 \end{bmatrix}$.   (b) $\begin{bmatrix} 4 \\ 2 \\ 2 \end{bmatrix}$.

9. $\begin{bmatrix} 0 & 1 & 0 & 0 \\ 0 & 0 & 0 & 0 \\ 0 & 0 & 1 & 1 \\ 0 & 0 & 0 & 1 \end{bmatrix}$.

11. (a) $\begin{bmatrix} 0 & -3 & 2 & 0 \\ -2 & -3 & 0 & 2 \\ 3 & 0 & 3 & -3 \\ 0 & 3 & -2 & 0 \end{bmatrix}$.   (b) $\begin{bmatrix} 0 & 3 & -2 & 3 \\ 0 & -9 & -2 & -6 \\ 0 & 3 & 6 & 0 \\ 0 & 4 & 0 & 3 \end{bmatrix}$.

(c) $\begin{bmatrix} 0 & 3 & -2 & 0 \\ -3 & 0 & 3 & 3 \\ 3 & -6 & 1 & -3 \\ 1 & 3 & -1 & -1 \end{bmatrix}$.   (d) $\begin{bmatrix} 0 & -3 & 2 & -3 \\ 0 & -5 & -2 & -3 \\ 0 & 3 & 6 & 0 \\ 0 & 3 & -2 & 3 \end{bmatrix}$.

13. (a) $\begin{bmatrix} 1 & 0 \\ 0 & -1 \end{bmatrix}$.   (b) $\begin{bmatrix} 0 & -1 \\ -1 & 0 \end{bmatrix}$.

(c) $\begin{bmatrix} \frac{1}{2} & -\frac{1}{2} \\ -\frac{1}{2} & -\frac{1}{2} \end{bmatrix}$.   (d) $\begin{bmatrix} 1 & -1 \\ -1 & -1 \end{bmatrix}$.

17. (a) $\begin{bmatrix} 1 & 0 \\ 0 & 1 \end{bmatrix}$.   (b) $\begin{bmatrix} 1 & 0 \\ 0 & 1 \end{bmatrix}$.

(c) $\begin{bmatrix} \frac{3}{5} & -\frac{2}{5} \\ \frac{1}{5} & \frac{1}{5} \end{bmatrix}$.   (d) $\begin{bmatrix} 1 & 2 \\ -1 & 3 \end{bmatrix}$.

19. $\begin{bmatrix} 0 & -1 \\ 1 & 0 \end{bmatrix}$.

Section 4.4, p. 180

4. (a) $[-a_1 + 4a_2 - a_3 \quad 3a_1 - a_2 + 3a_3 \quad 4a_1 + 3a_2 + 5a_3]$.
   (b) $[5 \quad -4 \quad -4]$.
   (c) $\begin{bmatrix} -1 & 4 & -1 \\ 3 & -1 & 3 \\ 4 & 3 & 5 \end{bmatrix}$.
   (d) $[2a_1 + 2a_3 \quad -4a_1 - 2a_2 - 2a_3 \quad -2a_1 - 4a_2 - 6a_3]$.
   (e) $[14 \quad -28 \quad 14]$.
   (f) $\begin{bmatrix} 2 & 16 & 2 \\ 0 & -10 & 6 \\ 10 & 0 & 2 \end{bmatrix}$.

7. (a) $[-3a_1 - 5a_2 - 2a_3 \quad 4a_1 + 7a_2 + 4a_3 \quad 11a_1 + 3a_2 + 10a_3]$.
   (b) $[8a_1 + 4a_2 + 4a_3 \quad -3a_1 + 2a_2 + 3a_3 \quad a_1 + 5a_2 + 4a_3]$.
   (c) $\begin{bmatrix} -3 & -5 & -2 \\ 4 & 7 & 4 \\ 11 & 3 & 10 \end{bmatrix}$. (d) $\begin{bmatrix} 8 & 4 & 4 \\ -3 & 2 & 3 \\ 1 & 5 & 4 \end{bmatrix}$.

8. (a) $\begin{bmatrix} 2 & 8 & -2 \\ 4 & 2 & 6 \\ 2 & -2 & 4 \end{bmatrix}$. (b) $\begin{bmatrix} 7 & 8 & -2 \\ 4 & 7 & 6 \\ 2 & -2 & 9 \end{bmatrix}$.

10. (a) 6. (b) 6. (c) 12. (d) 12.

13. (a) $L(\varepsilon_1) = \begin{bmatrix} 1 \\ 3 \end{bmatrix}, L(\varepsilon_2) = \begin{bmatrix} 2 \\ 4 \end{bmatrix}, L(\varepsilon_3) = \begin{bmatrix} -2 \\ -1 \end{bmatrix}$.

   (b) $\begin{bmatrix} a_1 + 2a_2 - 2a_3 \\ 3a_1 + 4a_2 - a_3 \end{bmatrix}$. (c) $\begin{bmatrix} -1 \\ 8 \end{bmatrix}$.

19. $\begin{bmatrix} \frac{1}{2} & -\frac{1}{2} & \frac{1}{2} \\ -\frac{3}{2} & \frac{3}{2} & -\frac{1}{2} \\ \frac{1}{2} & \frac{1}{2} & -\frac{1}{2} \end{bmatrix}$.

21. $\begin{bmatrix} -\frac{9}{2} & -6 & 2 \\ \frac{1}{2} & 1 & 0 \\ \frac{5}{2} & 3 & -1 \end{bmatrix}$.

Section 4.5, p. 188

2. (a) $\begin{bmatrix} 1 & 0 & 1 & 0 \\ 0 & 0 & 1 & 1 \\ 0 & 0 & 0 & 1 \\ 1 & 1 & 0 & 0 \end{bmatrix}$. (b) $\begin{bmatrix} 1 & -1 & 1 & 0 \\ -1 & 1 & -1 & 1 \\ 0 & 1 & -1 & 0 \\ 0 & 0 & 1 & 0 \end{bmatrix}$.
   (c) $\begin{bmatrix} 1 & 0 & 1 \\ 1 & 1 & 0 \\ 0 & 0 & 1 \end{bmatrix}$. (d) $\begin{bmatrix} 0 & -1 & 1 & -1 \\ 0 & 1 & 0 & 3 \\ 1 & 1 & 0 & 1 \end{bmatrix}$. (e) 3.

5. (a) $A = \begin{bmatrix} 1 & 0 \\ 0 & -1 \end{bmatrix}$.   (b) $B = \begin{bmatrix} \frac{1}{3} & \frac{4}{3} \\ \frac{2}{3} & -\frac{1}{3} \end{bmatrix}$.

    (c) We seek a nonsingular matrix P so that $B = P^{-1}AP$ or $PB = AP$. A

    possible matrix is $P = \begin{bmatrix} 1 & 1 \\ 1 & -2 \end{bmatrix}$.

9. $\begin{bmatrix} \frac{13}{2} & \frac{1}{2} & -\frac{3}{2} \\ -\frac{5}{2} & \frac{1}{2} & -\frac{1}{2} \end{bmatrix}$.

12. $\begin{bmatrix} 4 & 3 & 0 & 3 \\ -6 & -5 & -4 & -3 \\ 3 & 3 & 7 & 0 \\ 8 & 6 & 4 & 4 \end{bmatrix}$.

Chapter 5

Section 5.1, p. 194

1. (a) 5.   (b) 7.   (c) 4.   (d) 4.   (e) 7.   (f) 0.
3. (a) $-$.   (b) $+$.   (c) $-$.   (d) $-$.   (e) $+$.
5. (a) 7.   (b) 2.   (c) 0.
7. (a) 9.   (b) 0.   (c) 144.
9. (a) $t^2 - 3t - 4$.   (b) $t^2 - 8t - 20$.
11. (a) $t = 4, t = -1$.   (b) $t = 10, t = -2$.

Section 5.2, p. 203

1. (a) -24.   (b) 24.   (c) -120.   (d) 3.
7. (a) 4.   (b) 72.   (c) 0.
9. 3.
12. (a) 2.   (b) -120.   (c) $t^3 - 6t^2 + 11t - 6$.
    (d) $t^2 - 2t - 11$.
16. 1.
22. 32.
23. (a), (b), and (d) are nonsingular.
25. The system has a nontrivial solution.

Section 5.3, p. 208

1. (a) 1.   (b) 7.   (c) 2.   (d) 10.
2. (a) -3.   (b) 0.   (c) 3.   (d) 6.
6. (a) 2.   (b) 29.   (c) -30.   (d) 4.   (e) -30.
10. $t = 1, t = -1, t = -2$.

Section 5.4, p. 213

2. (a) $\begin{bmatrix} 2 & -7 & -6 \\ 1 & -7 & -3 \\ -4 & 7 & 5 \end{bmatrix}$.   (b) -7.

4. $\begin{bmatrix} -\frac{2}{7} & 1 & \frac{6}{7} \\ -\frac{1}{7} & 1 & \frac{3}{7} \\ \frac{4}{7} & -1 & -\frac{5}{7} \end{bmatrix}$.

7. (a) $-\dfrac{1}{28} \begin{bmatrix} -30 & -5 & 9 & 46 \\ -32 & 4 & 4 & 36 \\ -12 & -2 & -2 & 24 \\ 16 & -2 & -2 & -32 \end{bmatrix}$.

    (b) $\begin{bmatrix} \frac{3}{14} & -\frac{3}{7} & \frac{1}{7} \\ \frac{1}{7} & \frac{5}{7} & -\frac{4}{7} \\ -\frac{1}{14} & \frac{1}{7} & \frac{2}{7} \end{bmatrix}$.   (c) $\begin{bmatrix} \frac{2}{9} & -\frac{1}{9} \\ \frac{1}{6} & \frac{1}{6} \end{bmatrix}$.

9. $\begin{bmatrix} \dfrac{d}{ad-bc} & \dfrac{-b}{ad-bc} \\[2mm] \dfrac{-c}{ad-bc} & \dfrac{a}{ad-bc} \end{bmatrix}$.

11. $\begin{bmatrix} \frac{1}{4} & 0 & 0 \\ 0 & -\frac{1}{3} & 0 \\ 0 & 0 & \frac{1}{2} \end{bmatrix}$.

Section 5.5, p. 218

1. $x_1 = -2, x_2 = 0, x_3 = 1$.
3. Yes.
5. $x_1 = 1, x_2 = 2, x_3 = -2$.
6. No.
10. $|A| = 0$, cannot use Cramer's rule.
11. Yes.

Chapter 6

Section 6.1, p. 237

1. The only eigenvalue of L is $c = -1$. Every nonzero vector in R^2 is an eigenvector of L associated with c.

3. (a) $f(t) = t^3$. The eigenvalues of A are $c_1 = c_2 = c_3 = 0$. Associated eigenvectors are

$$\alpha_1 = \alpha_2 = \alpha_3 = r \begin{bmatrix} 1 \\ 0 \\ 0 \end{bmatrix},$$

where $r =$ any nonzero real number.

(b) $f(t) = (t - 1)(t - 3)(t + 2)$. The eigenvalues are $c_1 = 1$, $c_2 = 3$, and $c_3 = -2$. Associated eigenvectors are

$$\alpha_1 = \begin{bmatrix} 6 \\ 3 \\ 8 \end{bmatrix}, \quad \alpha_2 = \begin{bmatrix} 0 \\ 5 \\ 2 \end{bmatrix}, \quad \text{and} \quad \alpha_3 = \begin{bmatrix} 0 \\ 0 \\ 1 \end{bmatrix}.$$

(c) $f(t) = t^2 - 2t$. The eigenvalues are $c_1 = 0$ and $c_2 = 2$. Associated eigenvectors are

$$\alpha_1 = \begin{bmatrix} 1 \\ -1 \end{bmatrix} \quad \text{and} \quad \alpha_2 = \begin{bmatrix} 1 \\ 1 \end{bmatrix}.$$

5. (a) $f(t) = t^2 - 5t + 6$. The eigenvalues are $c_1 = 2$ and $c_2 = 3$. Associated eigenvectors are

$$\alpha_1 = \begin{bmatrix} 1 \\ -1 \end{bmatrix} \quad \text{and} \quad \alpha_2 = \begin{bmatrix} 1 \\ -2 \end{bmatrix}.$$

(b) $f(t) = t^3 - 7t^2 + 14t - 8$. The eigenvalues are $c_1 = 1$, $c_2 = 2$, and $c_3 = 4$. Associated eigenvectors are

$$\alpha_1 = \begin{bmatrix} -1 \\ 1 \\ 1 \end{bmatrix}, \quad \alpha_2 = \begin{bmatrix} 1 \\ 0 \\ 0 \end{bmatrix}, \quad \text{and} \quad \alpha_3 = \begin{bmatrix} 7 \\ -4 \\ 2 \end{bmatrix}.$$

(c) $f(t) = t^3 - 5t^2 + 2t + 8$. The eigenvalues are $c_1 = -1$, $c_2 = 2$, and $c_3 = 4$. Associated eigenvectors are

$$\alpha_1 = \begin{bmatrix} 1 \\ 0 \\ -1 \end{bmatrix}, \quad \alpha_2 = \begin{bmatrix} -2 \\ -3 \\ 2 \end{bmatrix}, \quad \text{and} \quad \alpha_3 = \begin{bmatrix} 8 \\ 5 \\ 2 \end{bmatrix}.$$

7. L is not diagonalizable. The eigenvalues of L are $c_1 = c_2 = c_3 = 0$. The set of associated eigenvectors do not form a basis for P_2.

8. (a) Diagonalizable. The eigenvalues are $c_1 = -3$ and $c_2 = 2$. The result follows by Theorem 6.4.

(b) Not diagonalizable. The eigenvalues are $c_1 = c_2 = 1$. Associated eigenvectors are

$$\alpha_1 = \alpha_2 = r \begin{bmatrix} 0 \\ 1 \end{bmatrix},$$

where $r =$ any nonzero real number.

(c) Diagonalizable. The eigenvalues are $c_1 = 0$, $c_2 = 2$, and $c_3 = 3$. The result follows by Theorem 6.4.

(d) Diagonalizable. The eigenvalues are $c_1 = 1$, $c_2 = -1$, and $c_3 = 2$. The result follows by Theorem 6.4.

(e) Not diagonalizable. The eigenvalues are $c_1 = c_2 = c_3 = 3$. Associated eigenvectors are

$$\alpha_1 = \alpha_2 = \alpha_3 = r \begin{bmatrix} 0 \\ 0 \\ 1 \end{bmatrix},$$

where $r =$ any nonzero real number.

11. The eigenvalues of A^2 are $c_1 = 9$ and $c_2 = 4$. Associated eigenvectors are

$$\alpha_1 = \begin{bmatrix} 1 \\ -1 \end{bmatrix} \quad \text{and} \quad \alpha_2 = \begin{bmatrix} 4 \\ 1 \end{bmatrix}.$$

14. (a) There is no such P. The eigenvalues of A are $c_1 = 1$, $c_2 = 1$, and $c_3 = 3$. Associated eigenvectors are

$$\alpha_1 = \alpha_2 = r \begin{bmatrix} 1 \\ 0 \\ -1 \end{bmatrix}, \text{ where } r = \text{any nonzero real number, and}$$

$$\alpha_3 = \begin{bmatrix} 5 \\ -2 \\ 3 \end{bmatrix}.$$

(b) $P = \begin{bmatrix} 1 & 0 & 1 \\ 0 & -2 & 0 \\ 0 & 1 & 1 \end{bmatrix}$. The eigenvalues of A are $c_1 = 1$, $c_2 = 1$, and $c_3 = 3$. Associated eigenvectors are the columns of P.

(c) $P = \begin{bmatrix} 1 & -3 & 1 \\ 0 & 0 & -6 \\ 1 & 2 & 4 \end{bmatrix}$. The eigenvalues of A are $c_1 = 4$, $c_2 = -1$, and $c_3 = 1$. Associated eigenvectors are the columns of P.

(d) $P = \begin{bmatrix} 1 & 1 \\ -1 & -2 \end{bmatrix}$. The eigenvalues of A are $c_1 = 1$ and $c_2 = 2$. Associated eigenvectors are the columns of P.

(e) $P = \begin{bmatrix} 1 & 2 & 1 \\ 0 & 1 & 0 \\ 0 & 0 & -3 \end{bmatrix}$. The eigenvalues of A are $c_1 = 3$, $c_2 = 2$, and

$c_3 = 0$. Associated eigenvectors are the columns of P.

16. The eigenvalues of L are $c_1 = 2$, $c_2 = -1$, and $c_3 = 3$. Associated eigenvectors are $\alpha_1 = [1 \quad 0 \quad 0]$, $\alpha_2 = [1 \quad -1 \quad 0]$, and $\alpha_3 = [3 \quad 1 \quad 1]$.

19. The eigenvalues of L are $c_1 = 2$, $c_2 = -3$, and $c_3 = 4$. Associated eigenvectors are $\alpha_1 = t^2$, $\alpha_2 = t^2 - 5t$, and $\alpha_3 = 9t^2 + 4t + 14$.

24. (a) A possible answer is $\left\{ \begin{bmatrix} 3 \\ -3 \\ 1 \\ 0 \end{bmatrix} \right\}$.

(b) A possible answer is $\left\{ \begin{bmatrix} 1 \\ 0 \\ 0 \\ 0 \end{bmatrix} \right\}$.

Section 6.2, p. 253

2. (a) $x_1 = 1$, $x_2 = 2 + 3i$. (b) $3 - 5i$.

4. (a) A'. (b) B'.

17. A is similar to $D = \begin{bmatrix} 0 & 0 \\ 0 & 4 \end{bmatrix}$ and $P = \begin{bmatrix} 1/\sqrt{2} & 1/\sqrt{2} \\ -1/\sqrt{2} & 1/\sqrt{2} \end{bmatrix}$.

19. A is similar to $D = \begin{bmatrix} 0 & 0 & 0 \\ 0 & 0 & 0 \\ 0 & 0 & 4 \end{bmatrix}$ and $P = \begin{bmatrix} 1 & 0 & 0 \\ 0 & -1/\sqrt{2} & 1/\sqrt{2} \\ 0 & 1/\sqrt{2} & 1/\sqrt{2} \end{bmatrix}$.

21. A is similar to

$D = \begin{bmatrix} -2 & 0 & 0 \\ 0 & 1 & 0 \\ 0 & 0 & 1 \end{bmatrix}$ and $P = \begin{bmatrix} 1/\sqrt{3} & -1/\sqrt{2} & -1/\sqrt{6} \\ 1/\sqrt{3} & 1/\sqrt{2} & -1/\sqrt{6} \\ 1/\sqrt{3} & 0 & 2/\sqrt{6} \end{bmatrix}$.

23. A is similar to $D = \begin{bmatrix} 3 & 0 \\ 0 & 1 \end{bmatrix}$.

25. A is similar to $D = \begin{bmatrix} 1 & 0 & 0 \\ 0 & 2 & 0 \\ 0 & 0 & 0 \end{bmatrix}$.

27. A is similar to $D = \begin{bmatrix} 1 & 0 & 0 \\ 0 & 0 & 0 \\ 0 & 0 & 2 \end{bmatrix}$.

29. A is similar to $D = \begin{bmatrix} 2 & 0 & 0 \\ 0 & -2 & 0 \\ 0 & 0 & 4 \end{bmatrix}$.

Section 6.3, p. 265

1. (a) $[x \quad y]\begin{bmatrix} -3 & \frac{5}{2} \\ \frac{5}{2} & -2 \end{bmatrix}\begin{bmatrix} x \\ y \end{bmatrix}$.

(b) $[x_1 \quad x_2 \quad x_3]\begin{bmatrix} 2 & \frac{3}{2} & -\frac{5}{2} \\ \frac{3}{2} & 0 & \frac{7}{2} \\ -\frac{5}{2} & \frac{7}{2} & 0 \end{bmatrix}\begin{bmatrix} x_1 \\ x_2 \\ x_3 \end{bmatrix}$.

(c) $[x_1 \quad x_2 \quad x_3]\begin{bmatrix} 3 & \frac{1}{2} & -1 \\ \frac{1}{2} & 1 & -2 \\ -1 & -2 & -2 \end{bmatrix}\begin{bmatrix} x_1 \\ x_2 \\ x_3 \end{bmatrix}$.

5. (a) $\begin{bmatrix} -1 & 0 & 0 \\ 0 & 2 & 0 \\ 0 & 0 & 0 \end{bmatrix}$. (b) $\begin{bmatrix} 3 & 0 & 0 \\ 0 & 0 & 0 \\ 0 & 0 & 0 \end{bmatrix}$.

7. $3x'^2 - 2y'^2$.

9. $x'^2 - z'^2$.

11. $-2y_1^2 + 5y_2^2 - 5y_3^2$.

13. $y_1^2 + y_2^2$.

15. $y_1^2 + y_2^2 + y_3^2$.

19. $9y_1^2 - y_2^2 = 1$, a hyperbola.

21. $y_1^2 - y_2^2 = 1$, rank = 2, signature = 0.

23. (a) and (b) are positive definite. The eigenvalues of the matrices are
(a) 1, 3. (b) 1, 3. (c) 1, -1. (d) 0, 2. (e) 0, 4.

Chapter 7, p. 279

1. (a) $X(t) = \begin{bmatrix} x_1(t) \\ x_2(t) \\ x_3(t) \end{bmatrix} = \begin{bmatrix} b_1 e^{-3t} \\ b_2 e^{4t} \\ b_3 e^{2t} \end{bmatrix} =$

$$b_1 \begin{bmatrix} 1 \\ 0 \\ 0 \end{bmatrix} e^{-3t} + b_2 \begin{bmatrix} 0 \\ 1 \\ 0 \end{bmatrix} e^{4t} + b_3 \begin{bmatrix} 0 \\ 0 \\ 1 \end{bmatrix} e^{2t}.$$

(b) $\begin{bmatrix} 3e^{-3t} \\ 4e^{4t} \\ 5e^{2t} \end{bmatrix} = 3\begin{bmatrix} 1 \\ 0 \\ 0 \end{bmatrix} e^{-3t} + 4\begin{bmatrix} 0 \\ 1 \\ 0 \end{bmatrix} e^{4t} + 5\begin{bmatrix} 0 \\ 0 \\ 1 \end{bmatrix} e^{2t}.$

3. $X(t) = b_1 \begin{bmatrix} 6 \\ 2 \\ 7 \end{bmatrix} e^{4t} + b_2 \begin{bmatrix} 0 \\ 7 \\ -1 \end{bmatrix} e^{-5t} + b_3 \begin{bmatrix} 0 \\ 0 \\ 1 \end{bmatrix} e^{2t}.$

5. $X(t) = b_1 \begin{bmatrix} 1 \\ 0 \\ 0 \end{bmatrix} e^{5t} + b_2 \begin{bmatrix} 0 \\ 3 \\ 1 \end{bmatrix} e^{5t} + b_3 \begin{bmatrix} 0 \\ -3 \\ 1 \end{bmatrix} e^{-5t}.$

7. $X(t) = b_1 \begin{bmatrix} 1 \\ 0 \\ 0 \end{bmatrix} e^{-3t} + b_2 \left(\begin{bmatrix} 1 \\ 0 \\ 0 \end{bmatrix} te^{-3t} + \begin{bmatrix} 0 \\ -\frac{1}{5} \\ \frac{1}{5} \end{bmatrix} e^{-3t} \right) + b_3 \begin{bmatrix} 1 \\ -2 \\ 1 \end{bmatrix} e^{-4t}.$

9. (a) $\begin{bmatrix} \dot{x}_1 \\ \dot{x}_2 \end{bmatrix} = \begin{bmatrix} 0 & 1 \\ -1 & 2 \end{bmatrix} \begin{bmatrix} x_1 \\ x_2 \end{bmatrix}.$

   (b) $\begin{bmatrix} \dot{x}_1 \\ \dot{x}_2 \\ \dot{x}_3 \end{bmatrix} = \begin{bmatrix} 0 & 1 & 0 \\ 0 & 0 & 1 \\ 1 & -1 & 2 \end{bmatrix} \begin{bmatrix} x_1 \\ x_2 \\ x_3 \end{bmatrix}.$

   (c) $\begin{bmatrix} \dot{x}_1 \\ \dot{x}_2 \\ \dot{x}_3 \end{bmatrix} = \begin{bmatrix} 0 & 1 & 0 \\ 0 & 0 & 1 \\ 1 & 0 & 0 \end{bmatrix} \begin{bmatrix} x_1 \\ x_2 \\ x_3 \end{bmatrix}.$

11. (a) $x(t) = b_1 e^{-t} + b_2 e^{3t}.$ **(b)** $x(t) = e^{-t} + e^{3t}.$

12. $x(t) = b_1 e^t + b_2 te^t + b_3 e^{3t}.$

15. $x(t) = b_1 e^{2t} + b_2 te^{2t} + b_3 e^{3t} + b_4 te^{3t}.$

17. $x(t) = b_1 e^{2t} + b_2 e^{-3t} + b_3 e^{4t}.$

List of Frequently Used Symbols

Greek Alphabet

α	alpha	η	eta	ν	nu	τ	tau
β	beta	θ	theta	ξ	xi	υ	upsilon
γ	gamma	ι	iota	o	omicron	ϕ	phi
δ	delta	κ	kappa	π	pi	χ	chi
ε	epsilon	λ	lambda	ρ	rho	ψ	psi
ζ	zeta	μ	mu	σ	sigma	ω	omega

$A = [a_{ij}]$	an $m \times n$ matrix, p. 10
$_mA_n$	an $m \times n$ matrix, p. 11
$A_n,$	an $n \times n$ matrix, p. 11
$[A \vdots B]$	an augmented matrix, p. 15
A'	the transpose of A, p. 16
$_mO_n$	the zero $m \times n$ matrix, p. 19
$O_n,$	the zero $n \times n$ matrix, p. 19
I_n	the $n \times n$ identity matrix, p. 25
A^{-1}	the inverse of the matrix A, p. 30
E, F	elementary matrices, p. 50

P, Q	nonsingular matrices, p. 58 (Exercise 17), 62
V, W	vector spaces, p. 68, 71
α, β, γ	vectors in a vector space, p. 68
θ, θ_V	the zero vector in the vector space V, p. 68
R^n	the vector space of all $n \times 1$ matrices, p. 68
$_mR_n$	the vector space of all $m \times n$ matrices, p. 69
R_n	the vector space of all $1 \times n$ matrices, p. 69
$p(t)$	a polynomial in t, p. 70
P	the vector space of all polynomials, p. 70
P_n	the vector space of all polynomials of degree $\leq n$ and the zero polynomial, p. 72
S, T, S', T'	bases for a vector space, p. 84
ε_i	an $n \times 1$ matrix with a 1 in the ith row, zeros in all other rows, p. 84
$\{\varepsilon_1, \varepsilon_2, \ldots, \varepsilon_n\}$,	the natural basis for R^n, p. 84
$\{\varepsilon'_1, \varepsilon'_2, \ldots, \varepsilon'_n\}$	the natural basis for R_n, p. 84, 91
$\{\mathbf{i}, \mathbf{j}, \mathbf{k}\}$,	the natural basis for R^3, p. 85
dim V	the dimension of the vector space V, p. 89
$[\alpha]_S$	the coordinate vector of the vector α with respect to the ordered basis S for the vector space V, p. 96
$|\alpha|$	the length of the vector α, p. 116, 132
$|\alpha - \beta|$	the distance between the vectors α and β, p. 117, 135
$\cos \phi$	the cosine of the angle between two nonzero vectors α and β, p. 118–119, 134
(α, β)	an inner product, p. 119, 128
$\times$	the cross product operation, p. 120
L, L_1, L_2	linear transformations, p. 147
$p'(t)$	the derivative of $p(t)$ with respect to t, p. 153
ker L	the kernel of the linear transformation L, p. 155
range L	the range of the linear transformation L, p. 157
$j_1 j_2 \cdots j_n$	a permutation of $S = \{1, 2, \ldots, n\}$, p. 190
$|A|$	the determinant of the matrix A, p. 191
$|M_{ij}|$	the minor of a_{ij}, p. 205
A_{ij}	the cofactor of a_{ij}, p. 205
adj A	the adjoint of the matrix A, p. 210
D	a diagonal matrix, p. 223
c_j	an eigenvalue of the linear transformation L or matrix A, p. 224, 226
α_j	an eigenvector of L or A associated with the eigenvalue c_j, p. 224, 226
$|tI_n - A| = f(t)$	the characteristic polynomial of A, p. 229–230
$|tI_n - A| = f(t) = 0$	the characteristic equation of A, p. 230
$c = a + bi$	a complex number, p. 240
$\bar{c} = a - bi$	the conjugate of $c = a + bi$, p. 241
$g(X) = X'AX$	a real quadratic form in n variables, p. 256

$$X(t) = \begin{bmatrix} x_1(t) \\ x_2(t) \\ \vdots \\ x_n(t) \end{bmatrix}$$ an $n \times 1$ matrix whose entries are functions of t, p. 268

$$\dot{X}(t) = \begin{bmatrix} \dot{x}_1(t) \\ \dot{x}_2(t) \\ \vdots \\ \dot{x}_n(t) \end{bmatrix}$$ p. 268

Page Index to Lemmas, Theorems, and Corollaries

Index